BREAST CANCER IN WOMEN OF AFRICAN DESCENT

Breast Cancer in Women of African Descent

Edited by

Senior Editor:

CHRISTOPHER KWESI O. WILLIAMS

British Columbia Cancer Agency and University of British Columbia, Nanaimo and Vancouver, Canada

Editors:

OLUFUNMILAYO I. OLOPADE

University of Chicago, Chicago, IL, U.S.A.

and

CARLA I. FALKSON

University of Alabama at Birmingham Comprehensive Cancer Center, Birmingham, AL, U.S.A.

Springer

A C.I.P. Catalogue record for this book is available from the Library of Congress.

ISBN-10 1-4020-3682-5 (HB)
ISBN-10 1-4020-3664-7 (e-book)
ISBN-13 978-1-4020-3682-8 (HB)
ISBN-13 978-1-4020-3664-4 (e-book)

Published by Springer,
P.O. Box 17, 3300 AA Dordrecht, The Netherlands.

www.springeronline.com

Printed on acid-free paper

Printed in the Netherlands.

TABLE OF CONTENTS

PREFACE

Cancer is a global problem and, increasingly, cancer research is a collaborative international enterprise. Breast Cancer in Women of African Descent is an important contribution to our understanding of the global dimensions of this problem in black women around the world. Not only does it summarize key information and place it in context, it helps define many of the relevant research questions and methodological issues that must be addressed if black women are to benefit fully from the many advances that are being made against breast cancer.

For some time, it has been recognized that although black women in the United States have a lower incidence of breast cancer than white women, they have a higher mortality rate. The importance and relevance of race for the genetics and biology of the disease are controversial and poorly understood and remain the subject of research. However, it appears certain that some of the disparity in outcome can be attributed to lack of access to high quality medical care, both for screening and early detection and for treatment. Less access to, or utilization of, screening services, such as mammograms, may explain in part the tendency toward more advanced disease at presentation that contributes to a higher mortality rate in black women.

This book highlights the even greater obstacles faced by black women in the developing world where modern technology and treatments may be largely unavailable. It also addresses the context of cultural attitudes and beliefs in which breast cancer in black women must be addressed.

Breast cancer research studying black women is fraught with many methodological issues and challenges which are discussed throughout this text. They include the relatively small numbers of black women participating in research studies and clinical trials which makes even most randomized Phase 3 trials grossly underpowered to detect significant differences in outcome, or other factors, attributable to race. Often potentially relevant factors such as socioeconomic status, age, and important tumor characteristics, which could confound our understanding of the outcomes, are not well studied or controlled.

Complexities in performing risk assessments and providing management advice for black women based on genetic differences of unknown significance, such as mis-sense variants in BRCA 1 and 2 are also highlighted. Furthermore, many of the models used to determine risk and develop molecular diagnostic markers have been developed in largely non-hispanic white populations and their direct applicability to women from other racial groups may be uncertain. Eventually, technology may enable a molecular classification of each individual's risk and each patient's tumor and render race a trivial and irrelevant biological issue. Until then, this book provides useful and provocative insights into possible biological variables, while at the same time reminding us of some of the critical social and cultural issues that contribute to a poorer outcome for black women.

Beyond the challenges of doing effective research regarding breast cancer in black women is the whole area of effective treatment. Because black women are distributed throughout the Diaspora, cultural issues may differ significantly from community to community. Several chapters describe the impact of cultural beliefs and traditions on the treatment and course of breast cancer. In addition, issues related to attitudes toward and access to end of life care are raised that clearly have implications beyond breast cancer.

For health care professionals whose experience is limited primarily to the United States, many of the attitudes that influence patient behavior in other cultures may be particularly unfamiliar. The ramifications of shame are discussed, not only for the cancer patient but potentially also for their family and may have a significant bearing on how patients relate to their physicians and health care providers and ultimately on the management of their cancer. The importance of increasing the cultural competence of medical professionals as a whole and of increasing the number of non-white physicians and other health care professionals who may understand relevant cultural issues and bring that insight to research and practice settings is also discussed in a number of chapters.

The answers to some of the important questions regarding this common cancer in black women lie in rigorous research to help us better understand the underlying complex biology and will be limited by our general understanding of cancer biology. However, this book also draws attention to the critical role that appropriate social policies might play to address the barriers that limit the access black women have to high quality medical care, both in the United States and in many other parts of the world. By placing breast cancer in black women in context and highlighting many of the most challenging issues, new research approaches will be apparent and it is likely that this new text will stimulate high quality research that will address some of the many remaining questions utilizing rigorous methodologies. Issues of importance in practice management and the care of breast cancer patients are also covered and will make this book of interest to practitioners. Hopefully, it will also be useful to policy-makers who are working to solve some of the access problems that contribute to disparate outcomes for black women with breast cancer and other life-threatening diseases.

Michaele C. Christian, M.D.
Associate Director
Division of Cancer Treatment and Diagnosis
Cancer Therapy Evaluation Program
National Cancer Institute, National Institutes of Health

LIST OF CONTRIBUTORS

Olusegun Agboola MB, BS (IB); DABR (T); FRCPC
Ottawa Regional Cancer Centre and University of Ottawa, Ottawa, Ontario, Canada

Kathy S. Albain, MD
Loyola University Chicago Medical Center, Cardinal Bernardin Cancer Center, Maywood, IL, USA

Steven K. H. Aung, MD, FAAFP
Departments of Medicine and Family Medicine, University of Alberta, Edmonton, Canada

David F. Chhieng, MD
Department of Pathology, University of Alabama at Birmingham, Birmingham, Alabama, USA

Andra R. Frost, MD
Department of Pathology, University of Alabama at Birmingham, Birmingham, Alabama, USA

Paul E. Goss, MD, FRCPC, FRCP (UK), PhD
Department of Medical Oncology and Hematology, Princess Margaret Hospital, University Health Network, Toronto, ON, Canada, and Harvard Medical School, Boston, MA, USA.

Stephanie Green, PhD
Southwest Oncology Group Statistical Center, Seattle, WA, USA

William E. Grizzle, MD, PhD
Department of Pathology, University of Alabama at Birmingham, Birmingham, Alabama, USA

Offiong Francis Ikpatt, MBBS, FMCPath (Nigeria), PhD
Section of Hematology/Oncology, Department of Medicine, University of Chicago, Chicago, Illinois, USA.

Hussein M. Khaled, MD
Medical Oncology Department, National Cancer Institute, Cairo University, Cairo, Egypt

Paul H. Levine, MD
The George Washington University School of Public Health and Health Services, Washington, DC, USA

Rhonda J. Moore, Ph.D.
Social Science Researcher, NHTSA/DOT, NTI-131, Traffic Injury control, Program Delivery and Development, Office of Research and Technology Behavioural Research Division, Washington, DC, USA

Joseph O'Donnell, MD
Dartmouth Medical School and the Norris Cotton Cancer Center, Hanover, NH, USA.

Olufunmilayo I. Olopade, MB, BS (Ibadan), FACP
Section of Hematology/Oncology, Department of Medicine, University of Chicago, Chicago, Illinois, USA.

D. Max Parkin, BSc, MD, MRCP (UK), DipSocMed (Edin), FFCM
Unit of Descriptive Epidemiology, International Agency for Research on Cancer, Lyon, France

B. L. Rapoport, MD, MMed. (Int Med)
The Medical Oncology Centre of Rosebank, Johannesburg, South Africa

Kathrin Strasser-Weippl, MD
First Medical Department with Medical Oncology, Wilhelminen Hospital, Montleartstr. 37, A-1160 Vienna, Austria.

Lynya I. Talley, PhD
Biostatistics Unit of the Comprehensive Cancer Center, University of Alabama at Birmingham, Birmingham, Alabama, USA
Joseph M, Unger, MS
Southwest Oncology Group Statistical Center, Seattle, WA, USA

Carmela Veneroso, MPH
The George Washington University School of Public Health and Health Services, Washington, DC, USA

Christopher K. O. Williams, MD, FRCPC, DABIM
B.C. Cancer Agency, Nanaiamo, British Columbia, and University of British Columbia, Vancouver, Canada

Carla Falkson, MD
University of Alabama at Birmingham Comprehensive Cancer Center, University of Alabama at Birmingham, Birmingham, Alabama, USA

ACKNOWLEDGEMENTS

We would like to acknowledge the help of Carrie P. Hunter, MD. MPH, formerly of the Lombardi Cancer Center, Georgetown University Medical Center, Washington, DC, USA. Dr. Hunter made significant suggestions on the design of the book structure. We are also grateful to Mrs. Anne McIntyre for secretarial assistance.

INTRODUCTION

Christopher K.O. Williams
British Columbia Cancer Agency, Vancouver Island Centre, Nanaimo Cancer Clinic and University of British Columbia, Vancouver, BC, Canada

Breast cancer is primarily a disease of women, regardless of their racial background, ethnicity, religion and background. Less than 10% of cases of the disease occur in men. The world literature on breast cancer is disproportionately skewed in the coverage of the nature and management of the disease in Caucasian women, to the extent that members of other ethnic groups tend to view the disease as an exclusive Caucasian health problem. This is in spite of the multicultural and multiracial diversity of the composition of the population of Western societies, among whom the disease is so prevalent. Available literature on breast cancer has little relevance, not only to the women of African descent living in these societies, but also to those of them living in the emerging economies of Africa, the Caribbean islands, Mexico, Central America and South America.

Breast cancer is probably a disease of lifestyle although the aspects of lifestyle that predisposes to it are poorly understood. The incidence of the disease has continued to increase at a steady rate in the Western societies since the end of the Second World War (*1-4*). Changes in lifestyle consonant with the economic gains of the period have been cited as being contributory. Given the current global trend of "westernization" of lifestyles in non-western societies and the societal advances made by women the world over, the world may well be at the threshold of augmentation of the international breast cancer burden. Already, breast cancer has become the commonest cancer among Nigerian women (*5*), and it is likely to become a much more common disease in developing countries within the foreseeable future (see the contribution of Max Parkin in this book). Breast cancer is a disease that causes a great deal of suffering especially because a substantial proportion of its victims succumb to the disease at the prime of their lives. It is only a matter of time that "non-western" women worldwide, and the underserved women of the Western societies would join the current activism of the breast cancer lobby in demanding culturally sensitive approach to their mutual plight. Their demand would include provision of culturally relevant medical literature for this category of breast cancer patients and survivors.

Historical events of the last four to five centuries that have resulted in the dispersion of people of African descent from their ancestral continental geographical homeland to other parts of the world are well documented. The United Nations Educational and Scientific Organisation (UNESCO) has documented humanity's dark history of man's inhumanity to man at its heritage sites in the slave castles along the coast of West Africa. It was from such sites that about 12 million to 25 million of "unwilling African immigrants" were transferred in shackles and transported primarily to the New World of the western hemisphere. "Brazil received fully one-third of all Africans; the Carribean islands together received the second third; and the final third was spread throughout the rest of the Americas. Scholars agree that the smallest number of Africans came to what is now the United States and Canada. Estimates range between 1.5 to 2 million (that) arrived there between 1620 and 1860...During the 20th century, there has been much voluntary immigration: from the Carribean to England and the US, and from Africa to Europe, the US and the Caribbean" (culled from the Museum of Cape Coast, Ghana, an UNESCO World Heritage site).

In spite of considerable admixture with other racial entities, the mere presence of recognisable physical features of African descent tends to translate invariably to the global effect of social deprivation and low socio-economic status. Thus, it is not always clear whether the adverse clinical features and outcomes of breast cancer that is so common to women of African descent wherever they may be in the Diaspora is the result of a shared genetic susceptibility or their shared socio-cultural milieu.

The advent of genomics in medicine is promising to resolve several long-held assumptions about the role of genetics in predisposition and outcomes of management of diseases (*6,7*). In the first place, we know that "race" has no biological meaning (*8,9*). Furthermore, most population geneticists concur that the bulk of genetic variations (90% - 95%) occur within, not among continental populations (*6,10,11*). Genomics has also shown that race, at the continental level, has not been shown to provide a useful categorisation of genetic information about the response to drugs, diagnosis or causes of disease (*6*).

This book addresses the various aspects of breast cancer as it is documented in most standard literature on the disease. However, special effort is made to emphasise the nature of the disease and its management in women of Africa and their descendants in Europe and the Americas. The book explores the differences in the manifestations of the disease, not only in women of African descent, but also those who, though not racially related to them, share their lifestyle and the impact of their social marginalisation. These are women in much of the developing world as well as new immigrants in their new chosen homelands in the developed world. This is important, because breast cancer, like many other cancers, and as already mentioned above, is a disease of lifestyle. Hence, the book highlights the differences of the epidemiology of the disease in relation to the various lifestyles in the African Diaspora. It examines the role of genetic susceptibility in the different populations for the variation of the behaviour of the cancer in different groups of women. It explores the differences in prognostic and predictive factors of the disease. The book examines the challenges of management of breast cancer in the underserved populations of Africa, and of the Western societies. It explores the differences in the experience of breast cancer and other chronic illnesses by women outside the mainstream of the Western societies. It also discusses the role of traditional health care methods in the management of breast cancer.

REFERENCES

(1) Yeole BB, Kurkure AP, Lacey JV, Jr., Devesa SS, Brinton LA, Ziegler RG, et al: An epidemiological assessment of increasing incidence and trends in breast cancer in Mumbai and other sites in India, during the last two decades. Recent trends in breast cancer incidence and mortality. Relative weight, weight change, height, and breast cancer risk in Asian-American women. Asian Pac J Cancer Prev 2003; 4:51-6.
(2) Lacey JV, Jr., Devesa SS, Brinton LA, Ziegler RG, Hoover RN, Nomura AM, et al: Recent trends in breast cancer incidence and mortality: Relative weight, weight change, height, and breast cancer risk in Asian-American women. Environ Mol Mutagen 2002; 39:82-8.
(3) Seow A, Duffy SW, McGee MA, Lee J, Lee HP: Breast cancer in Singapore: trends in incidence 1968-1992. Int J Epidemiol 1996; 25:40-5.
(4) Ziegler RG, Hoover RN, Nomura AM, West DW, Wu AH, Pike MC, et al: Relative weight, weight change, height, and breast cancer risk in Asian-American women. J Natl Cancer Inst 1996; 88:650-60.
(5) Adebamowo CA, Ajayi OO: Breast cancer in Nigeria. West Afr J Med 2000; 19:179-91.
(6) Cooper RS, Kaufman JS, and Ward R: Race and genomics. N Engl J Med 2003; 348:1166-70.
(7) Burchard EG, Ziv E, Coyle N, Gomez SL, Tang H, Karter AJ, et al: The importance of race and ethnic background in biomedical research and clinical practice. N Engl J Med 2003; 348:1170-5.
(8) Freeman HP: The meaning of race in cancer of the breast. Cancer J Sci Am 1997; 3:76-7.
(9) Freeman HP: The meaning of race in science--considerations for cancer research: concerns of special populations in the National Cancer Program. Cancer 1998; 82:219-25.

(10) Rosenberg NA, Pritchard JK, Weber JL, Cann HM, Kidd KK, Zhivotovsky LA, et al: Genetic structure of human populations. Science 2002; 298:2381-5.
(11) Owens K, King MC: Genomic views of human history. Science 1999; 286:451-3.

Chapter 1

BURDEN OF BREAST CANCER IN DEVELOPING AND DEVELOPED COUNTRIES

D. Max Parkin BSc, MD, MRCP (UK), Dip.Soc.Med. (Edin), FFCM
Unit of Descriptive Epidemiology, International Agency for Research on Cancer, Lyon, France

1. INTRODUCTION

This chapter will describe the current burden of breast cancer in year 2000 and the projected burden of breast cancer up to year 2050 for women of African descent. The overall trends and distributions of cancers in developing and developed countries and changes in cancer trends will be described. The impact of an increasing incidence of breast cancer in African populations on the health, economic and social well being of affected populations will be discussed

1.1 Indices of "burden"

Various statistics are available for assessing the importance (burden) of cancer, and of different types of cancer, in the population, either through quantifying the disease itself (the 'need' for services) or the demand that it places upon them (1).

Incidence is the number of new cases occurring. It can be expressed as the annual <u>number</u> of cases (the volume of new patients presenting for treatment) or as a rate per 100,000 persons per year. Rates are necessary if we wish to compare the <u>risk</u> of disease between populations (countries, ethnic groups, or different time periods within a country). Changes in incidence of cancer are the appropriate indicator of the impact of primary prevention strategies. Incidence data are produced by population-based

C. Williams et al. (eds.), Breast Cancer in Women of African Descent, 1-22.

cancer registries (2). Registries may cover national populations or, more often, certain regions. In developing countries in particular, coverage is often confined to the capital city and its environs. It was estimated that, in 1995, about 16.3% of the world population were covered by registries, 52.4% of developed countries and 7.5% of developing countries. The latest volume of "Cancer Incidence in Five Continents" (8th) contains comparable incidence information from 186 registries in 57 countries, mainly over the period 1993-1997 (3).

Mortality is the number of deaths occurring, and the mortality rate the number of deaths per 100,000 persons per year. The number of deaths provides an unambiguous measure of the outcome, or impact of cancer. It is the product of incidence and fatality (the inverse of survival) of a given cancer. Mortality rates measure the average risk to the population of dying from a specific cancer, while fatality (1-survival) represents the probability that an individual with cancer will die from it. Mortality data derive from vital registration systems, where the fact and "underlying" cause of death are certified, usually by a medical practitioner. Their great advantage is comprehensive coverage, and availability. By 1995, about 29% of the world population was covered by vital registration systems producing mortality statistics on cancer. This includes all of the developed countries, and many of the developing countries. National level statistics are collated and made available by the WHO (http://www-depdb.iarc.fr/who/menu.htm).

Survival statistics are produced by cancer registries; they require follow-up of registered cancer cases, either actively or by matching death certificates against cancer notifications and assuming that unmatched cases are still alive. Population-based figures are published by registries in many developed countries: for example, the SEER program covering 10% of the US population (4), and the EUROCARE II project, including 17 countries of Europe (5). Survival data from populations of China, the Philippines, Thailand, India and Cuba have been published by Sankaranarayanan *et al*, (6).

Prevalence is the proportion of a population that has the disease at a given point in time (7). For many diseases for which the time of onset is imprecise, (e.g., hypertension, diabetes), prevalence may be used as a substitute for incidence in comparative studies between populations. This is not necessary for cancer, and prevalence is not a useful indicator if the focus of interest is disease risk, and its possible causes. Even as an indicator of burden (and need for services) total prevalence ("ever had a cancer") is not useful; the figure will include many persons diagnosed in the past, some of whom have been "cured" and no longer have an excess risk of death. A pragmatic alternative is *"partial prevalence"* (8), which refers to cases diagnosed within a defined period following diagnosis (one, three and five years, say). The calculation can be made if data on incidence (or mortality) and survival are made.

Compound indicators involving duration or severity of disease have also been used, especially for priority setting.

Almost fifty years ago, the concept of person-years of life lost (PYLL) (9) was introduced, in order to refine the traditional mortality rates, by providing a weighting for deaths at different ages. These methods started to become more widely used from the late 1970's in health services planning (10). There are many variations in the calculations used, depending upon the weights used (the value of years of life at different ages), the 'normal' lifespan against which to compare premature death (fixed upper limit, or life table expectations of life), and the 'discount rate' to apply to life years (taking into consideration the life years that would have been lived in the future). Discounting gives decreasing weights to the life years saved over time, admitting that life-years in the future are valued less highly than at present. The weighting of life at different ages has appealed particularly to economists, who are, implicitly at least, interested in economic productivity of individuals, which varies of course by age. Nevertheless, it might also be argued that social roles, including ages of raising children, should be taken into account when age-weighting (11).

This approach has been taken a step further, with the development of indices such as "Quality Adjusted Life Years (lost)"(QALYs) or "Disability Adjusted life years (lost)"(DALYs) (12-13). Essentially, these admit that, between onset of a disease, and death or recovery, there is a spectrum of morbidity, which can be quantified in terms of its duration and severity. Three elements are needed in calculating these indices, therefore- the incidence of the disease, its mean duration (or, equivalently, survival probability), and a measure of life "quality" in between onset and end of disease. Since the latter is highly subjective (and in any case, likely to vary from time of diagnosis, or in different cultural settings), the values obtained for DALYs and QUALYs are likely to be more dependant upon the investigator and his assumptions, than the actual data (on incidence of, and survival from cancer).

2. ESTIMATION

Over the last 12 years, a series of estimates of the global burden of cancer have been published by the International Agency for Research on Cancer. The methods have evolved and been refined, but basically they rely upon the best available data on cancer incidence and/or mortality at country level, to build up the global picture. The results are more or less accurate, for different countries, depending on the extent and accuracy of locally available data. The most recent estimates are of incidence, mortality, and prevalence data, for five broad age groups (0-14, 15-44, 45-54, 55-64 and 65 and over) and sex for all countries of the world, for 24 different types of cancer (14). These form the basis of much of the data presented in this chapter.

3. BURDEN 2000

Breast cancer is the second most frequent cancer in the world, and by far the most common malignancy of women, with an estimated 1.05 million cases in the year 2000 (22% of all new cancer cases in women). Incidence rates are high in all of the 'developed' areas (except for Japan), with the highest age-standardised incidence in the Netherlands (91.6 per 100,000) and the United States (91.4) (Figure 1). High rates are also observed in the south of South America, especially Uruguay and Argentina; in contrast, low rates are found in most African and Asian populations, although they are increasing, and in some Asian populations they are already the same as in Southern Europe, and in some cases (e.g. the Philippines) even higher.

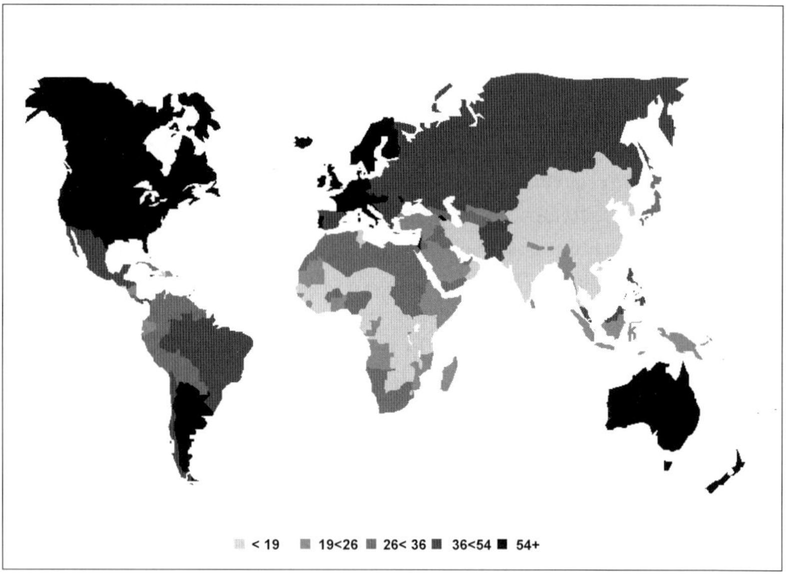

Figure 1. Incidence of Breast cancer (age std. rate per 10^5).

Worldwide, the ratio of mortality to incidence is about 36%, which implies that, broadly speaking, almost 2/3 of diagnosed breast cancer cases will survive. Survival in developed countries is, of course, much better than this. In Europe, survival from breast cancer is 91% at one year and 65% at five years (5). Stage of disease at diagnosis is the most important prognostic variable. For the SEER registries in the U.S.A., five-year survival for localized cases is 96.8%, while for cases with metastases it is only 20.6% (4). Even in developing countries, the differences by stage at diagnosis are very marked (6).

Because of this relatively favourable prognosis, breast cancer ranks much lower as a cause of death from cancer than in terms of numbers of new cases. Overall, it is numerically fifth in importance considering both sexes together, but is the leading cause of cancer mortality in women, in whom the 370,000 annual deaths represent 13.9% of all cancer deaths. Another consequence of the relatively good prognosis of breast cancer is that breast cancer is the most prevalent cancer in the world today; there are an estimated 3.6 million women alive who have had breast cancer diagnosed within the last five years (compared with just 1.4 million survivors – male or female – from lung cancer).

A comparison between breast cancer, as a cause of mortality in women aged 25-64, as compared with other cancers, and other causes of death, has been taken (15). Loss of life in women in this age range is particularly significant, as such women bear major family and economic responsibilities (12); there are also practical considerations of deficiencies in the availability and validity of mortality data in the elderly. For comparative purposes, breast cancer mortality is compared with that from three other cancers: cervix cancer (the major cancer of women in sub Saharan Africa, as in many developing countries), stomach cancer (third in importance in women, in terms of both incidence and mortality), and lung cancer (the most common cancer worldwide, in both sexes combined, and the second most common cause of death from cancer in women (16). Three other important causes of mortality in women are also used for comparison: tuberculosis, which is estimated to be the second most important cause of death worldwide in women aged 15-59 and a major priority for preventive intervention by the World Health Organization, maternal conditions, estimated to be the leading cause of death in women aged 15-44 worldwide (17), and AIDS, which is currently responsible for some 3 million deaths worldwide, with almost 80% of the total estimated to occur in sub Saharan Africa (18). Numbers of deaths were obtained from the WHO Mortality Database (19), or from the Global Burden of Disease (GBD) study (17). For Africa and the Caribbean, cancer deaths were estimated from the data in GLOBOCAN 2000 (16). For the USA, the data were obtained from CDC/NCHS (20).

Table 1 shows the estimated number of deaths by cause in women aged 25-64, and the percentage that each disease contributes to total mortality, worldwide, and in 3 areas of interest to the study of women of African descent: sub-Saharan Africa, the Caribbean, and the USA (including the US black population). AIDS was the major cause of death in women in this age range worldwide (12.4% deaths). Breast cancer is the most important cause of death from cancer worldwide, (3.0% deaths), but cancer of the cervix is quantitatively more important in sub-Saharan Africa and in the Caribbean.

Table 1. Number of deaths by cause (thousands) and percentage of total deaths among women aged 25-64 years, in the year 2000.

Region Country	Breast cancer	Cervical cancer	Lung cancer	Stomach cancer	Tuber-culosis	Maternal conditions	AIDS	All causes
Sub-Saharan Africa	12.1 (0.7%)	19.0 (1.1%)	1.3 (0.1%)	5.2 (0.3%)	62.0 (3.7%)	128.4 (7.8%)	688.0 (41.6%)	1653.6
Zimbabwe	0.2 (0.3%)	0.5 (0.9%)	0.0 (0.1%)	0.1 (0.2%)	1.2 (2.2%)	2.7 (5.0%)	14.2 (26.6%)	53.5
Caribbean	1.3 (3.1%)	1.6 (3.8%)	0.6 (1.4%)	0.4 (1.0%)			5.5 (12.9%)	42.7
United States	17.1 (8.5%)	2.6 (1.3%)	17.9 (8.8%)	1.2 (0.6%)	0.1 (0.05%)	0.3 (0.15%)	3.4 (1.7%)	202.5
Black women	0.3 (7.2%)	0.5 (1.3%)	2.3 (5.5%)	0.3 (0.8%)	0.04 (0.1%)	0.1 (0.3%)	2.3 (5.5%)	41.8
All areas	211.1 (3.0%)	151.6 (2.1%)	94.9 (1.3%)	778.7 (1.1%)	322.8 (4.5%)	345.0 (4.8%)	879.9 (12.4%)	7124.0

Years of life lost were calculated, based on the expected life span at the age of death given by a standard life table (the Coale and Demeny West Level 26 (21) life table, with an expectation of life at birth of 82.5 years). An age weighting was applied, to give greater weight to life-years lost in younger women (the weight at 25 years of age was 1.88, relative to age 65 (12), and the YLL were discounted at 3% per year. Table 2 shows the results, expressed as absolute numbers of (weighted) YLL by cause and region/country and the percentage that each individual cause contributes to the total YLL in the population. Breast cancer contributes about 3.8 million YLLs among women between the ages of 25 and 64 worldwide, representing around 2.7% of the YLL from all causes of disease.

Table 2. Weighted years of life lost (thousands) and percentage of all causes among women aged 25-64 in 2000.

Region country	Breast cancer	Cervical cancer	Lung cancer	Stomach cancer	Tuber-culosis	Maternal conditions	AIDS	All Causes
Sub-Saharan Africa	222.2 (0.6%)	337.8 (0.9%)	22.3 (0.1%)	89.4 (0.2%)	1407.7 (3.7%)	3275.7 (8.5%)	17900.0 (46.5%)	38500.0
Zimbabwe	3.2 (0.2%)	8.5 (0.6%)	0.8 (0.1%)	1.4 (0.1%)	28.3 (2.1%)	68.6 (5.2%)	374.7 (28.1%)	1332.3
Caribbean	23.1 (3.0%)	29.5 (3.8%)	9.8 (1.3%)	6.7 (0.9%)			139.7 (18.1%)	773.3
United States	281.7 (8.4%)	48.5 (1.4%)	255.1 (7.6%)	20.1 (0.6%)	1.9 (0.0%)	8.6 (0.3%)	77.8 (2.3%)	3360.1
Black women	52.5 (7.1%)	10.1 (1.4%)	34.2 (4.7%)	5.0 (0.7%)	0.8 (0.1%)	2.9 (0.4%)	53.0 (7.2%)	734.8
All areas	3752.5 (2.7%)	2708.6 (2.0%)	1498.3 (1.1%)	1290.7 (0.9%)	6674.3 (4.9%)	9411.0 (6.9%)	22800.0 (16.6%)	137000.0

The results are summarized in Figure 2, which contrasts the cause-specific total years of life lost in developing countries and developed countries. Breast cancer is the most important cause of YLL in this age range in developed countries, while in developing countries, the YLL is similar to that from cervical cancer.

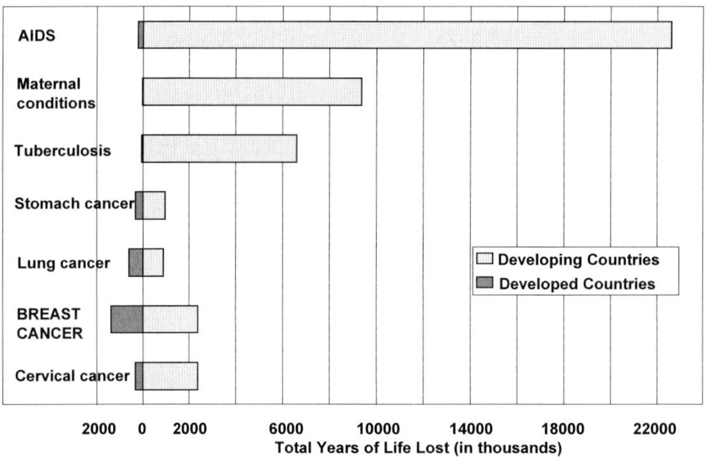

Figure 2. Cause-specific total years of life lost among women aged 25-64 years for developing and developed countries, in 2000.

4. BREAST CANCER IN AFRICA

Table 3 shows age-standardized incidence rates from cancer registries in Africa, with comparison data from Europe and North America. Table 4 shows incidence data from time periods in the 1960s and 1970s. The estimated national incidence rates shown in Figure 1 are based on these data. In total, there were an estimated 59 000 new cases in the year 2000, comprising 18% of cancers in women, (compared with 67000 cases of cervix cancer). This corresponds to an estimated 26,600 deaths (13.6% of cancer deaths in African women) and 179000 survivors of breast cancer alive within 5 years of diagnosis.

There is a fair amount of regional variation in incidence, however, and in North Africa, breast cancer is considerably more common than cervical cancer, accounting for 27% of cancers in women, compared with 15.7% in sub-Saharan Africa. In sub-Saharan Africa, breast cancer is more common in urban rather than rural settings (22). Current incidence rates are highest in cities such as Abidjan (Côte d'Ivoire) and Harare (Zimbabwe). It also seems that incidence rates in the more recent registry series are higher than those reported in the past.

Table 3. Incidence of breast cancer in Africa (Parkin et al, 2003).

AFRICANS			**Age std. Rate**
NORTH			**(per 100,000)**
Algeria:			
	Algiers	(1993-96)	21.2
	Setif	(1990-95)	17.0
	Oran	(1996-98)	34.5
Tunisia, Sousse		(1987-95)	22.7
WEST			
Gambia		(1997-98)	6.9
Guinea, Conarkry		(1996-99)	15.5
Mali, Bamako		(1988-97)	14.9
Niger, Niamey		(1993-99)	24.6
Nigeria, Ibadan		(1998-99)	25.2
CENTRAL			
Congo, Brazzaville		(1996-99)	22.3
EAST			
Kenya, Eldoret		(1998-2000)	13.7
Malawi, Blantyre		(2000-01)	11.4
Uganda, Kyadondo		(1993-97)	21.0
SOUTH			
South Africa (White)		(1989-92)*	62.9
South Africa (Black)		(1988-95)*	13.5
South Africa, Transekei		(1996-98)	6.5

Swaziland	(1996-99)	12.0
Zimbabwe, Harare, Black	(1994-97)	20.3
Zimbabwe, Harare, European	(1990-97)	121.2
EUROPE/US		
SEER (white)	(1993-97)	92.4
SEER (black)	(1993-97)	83.4
England and Wales	(1993-97)	75.6
French cancer regs. (8)	(1993-97)	81.1

The incidence in white women living in Africa is much higher than in black Africans. This is clear in the old data in Table 4. Recent data (1993–95) from the National Cancer Registry of South Africa (23) indicate age-standardized rates of 70.2 per 100 000 in white females, and 11.3 per 10^5 in blacks (although some of this variation may represent differences in rates of treatment or biopsy, since the data are based on histologically diagnosed cases only). In Harare, Zimbabwe, in 1990–92, age-standardized incidence was 127.7 per 100 000 in white females and 20.4 per 100 000 in blacks.

Table 4. Incidence of breast cancer in Africa, 1950's-1970's.

REGISTRY	PERIOD	ASR (Per 100,000)	SOURCE
Senegal, Dakar	1969-74		4
Mozambique, Lour. Mar.	1956-1960	3.2	1
Nigeria, Ibadan	1960-1969	15.3	2
SA, Cape Prov. : Bantu	1956-1959	13.6	1
SA, Cape prov. : Coloured	1956-1959	25.9	1
SA, CAPE Prov. : White	1956-1959	57.4	1
SA, Johannesburg, Bantu	1953-1955	15.3	1
SA, Natal Prov. : African	1964-1966	11.9	2
SA, Natal Prov. : Indian	1964-1966	19.9	2
Uganda, Kyandondo	1954-1960	9.7	1
Uganda, Kyandondo	1960-1971	10.8	5
Zimbabwe, Bulawayo	1963-1972	13.4	6

1: CI5 vol I
2: CI5 vol II
3: CI5 vol III
4: CI5 vol IV
5: Wabinga et al, 2000
6: Skinner et al,1993

The risk of breast cancer increases with age, but the rate of increase slows down after the menopause. In low-incidence countries, the slope of the age–incidence curve after the menopause may be very flat, or even negative. Almost certainly, this reflects increasing risks in successive generations of women, rather than a true decline in risk with age (24). The young age structure of African populations, coupled with this rather flat age–incidence curve, means that the average age at diagnosis of cases in Africa is lower than in European and American populations. This is often remarked upon in clinical series from Africa (25-27), but it has no etiological or prognostic significance.

Figure 3 shows age-specific incidence rates from four African registries, two in North Africa (Oran, Algeria; North Tunisia), and two in sub Saharan Africa (Kampala, Uganda; Harare, Zimbabwe), in comparison with the black population covered by the SEER program of the United States.

Migrant studies suggest that migrants from north and east Africa to France retain their relatively low breast cancer mortality rates on migration to France (28-29). However, African migrants to England and Wales do not have a significantly low mortality (30).

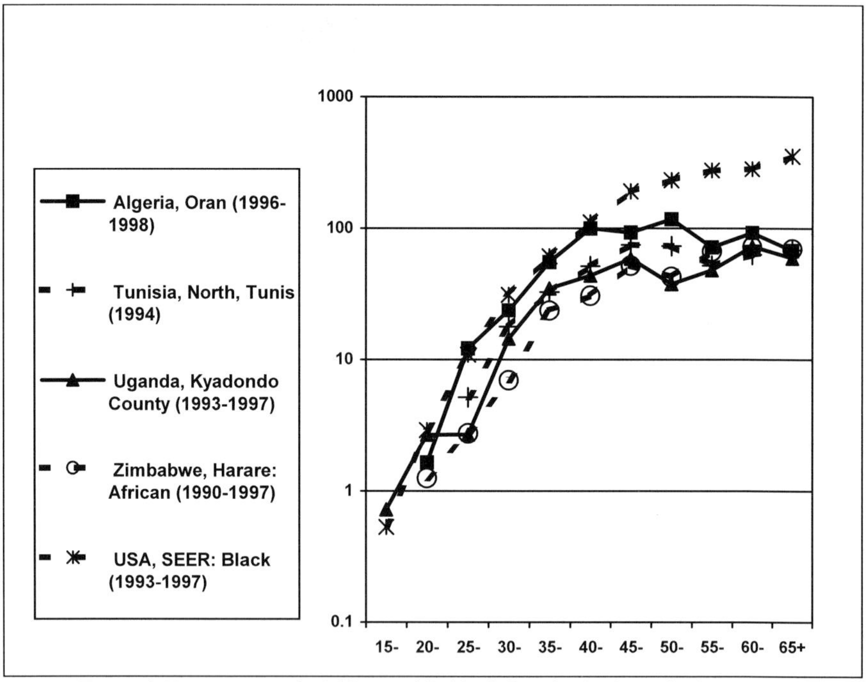

Figure 3. Breast cancer in Africa, and in black females in the United States.

Stage at presentation of tumours in African women is generally very advanced (31-32). Until recently, there have been no-population-based data on cancer survival in Africa, the only descriptions coming from case series amassed by different clinicians. However, follow up of unselected patients diagnosed in 1993-1997 among the population of Harare by the Zimbabwe Cancer registry has now been completed. The study shows that relative survival of breast cancer among black (African) women was 32.6% at 5 years, compared with 58.2% among white women in the same city (33).

5. UNITED STATES

Breast cancer was the most common cancer of women in the USA in 2001, responsible for 192,200 new cases, 31% of cancers in women. It was the second leading cause of death from cancer (after lung cancer), with 40,200 deaths (15% of all cancer deaths in women) (34). The black (African American) population comprises about 13% of the total in the US. The estimated number of new breast cancer cases in 2001 (19,300) is a little over 10% of all breast cancer cases, and comprises 31% of new cancers among black females. The number of deaths was estimated to be 5,800, some 19% of all cancer deaths (35).

There are extensive data comparing breast cancer incidence, mortality and survival by race. These consistently show the black women have lower overall incidence rates compared with white women, but are diagnosed with later stage disease, have shorter survival, and have the highest rate of breast cancer mortality of all racial-ethnic groups in the USA (36). In 1993-1997, the age standardized incidence (world standard) of breast cancer in women in the populations registered in the eleven cancer registries of the US SEER program were 92.1 per 10^5 for whites and 83.1 per 10^5 for blacks; mortality rates were 19.5 per 10^5 for whites and 27.2 per 10^5 for blacks (37). Figure 4 shows the age specific incidence rates. Incidence in black females is slightly higher (about %) than in whites before age 50, after which the rates in white women are substantially higher.

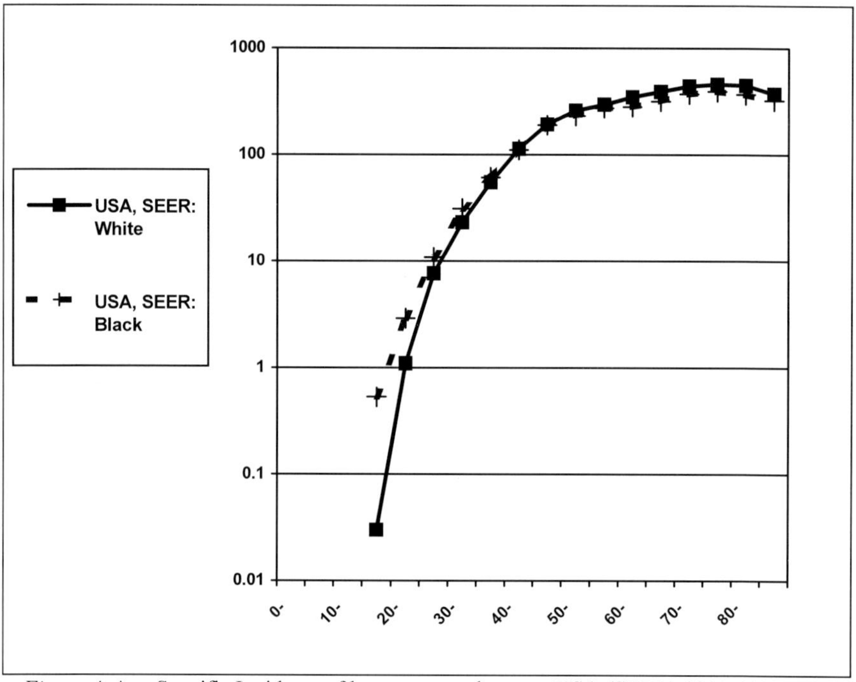

Figure 4. Age Specific Incidents of breast cancer by race, USA (SEER) 1993-1997 (37).

Table 5 shows stage distribution and survival of breast cancer cases in the SEER registries in 1992-1997. They show that survival among black females is poorer than that in whites, even within the same categories of stage. Differences in the biology of breast carcinomas between black and white women have been sought to explain these observations. Few comparative studies have adjusted for important confounding variables such as age, stage, treatment, and social status, and when these are taken into account, there is little or no evidence for differences in histopathological type (38). In a careful review of 963 tumours, adjusting for such factors (39) found that cancers from black patients had a higher grade of nuclear atypia and a higher mitotic activity.

Table 5. Distribution and 5-year relative survival with invasive breast cancer by race/ethnicity and stage at diagnosis: SEER Program, 1992-1997 (40).

Stage at	Black		White	
diagnosis	**% cases**	**Survival**	**% cases**	**survival**
Stage I	29	96	43	100
Stage II	36	76	32	85
Stage III	10	37	6	58
Stage IV	7	13	4	17

6. CARIBBEAN

The population of the Caribbean region was estimated to be about 38 million in 2000 (UN, 2002), of which 71% lived in the two large islands of Cuba (11 million) and Hispaniola (Haiti and Dominican Republic, 8 million each). The peoples of the region have a large component of African ancestry. While the populations of the smaller islands, and of Haiti are predominantly African-derived, the populations of Cuba and Dominican Republic are mainly mixed race (mulatto) with 11% described as "black".

There were an estimated 6200 new cases of breast cancer in the region in 2000 (18.7% of all cancers in women) and 2300 deaths (13% cancer deaths). This is slightly fewer than for cancer of the cervix (6700 new cases, 3100 deaths). Breast cancer is, however, the most prevalent form of cancer in the region, with 18200 survivors within 5 years of diagnosis.

Figure 5 shows rates of incidence and mortality in some Caribbean populations, in comparison with the USA (white and black populations). Mortality rates in the island countries (Barbados, Bahamas, Trinidad & Tobago) are not much inferior to those in the black population of the USA. However, incidence rates are rather lower (some 60-70% of the US rates). The higher ratios of mortality to incidence imply poorer survival in the Caribbean countries, and cases series suggest that stage at diagnosis is relatively late in many cases (41, 42). Incidence in Cuba is rather lower than in the other countries (Figure 5); (43) observed that mortality rates from breast cancer among Cuban-born women in the USA were intermediate between the lower rates in Cuba and the higher rates among U.S. whites.

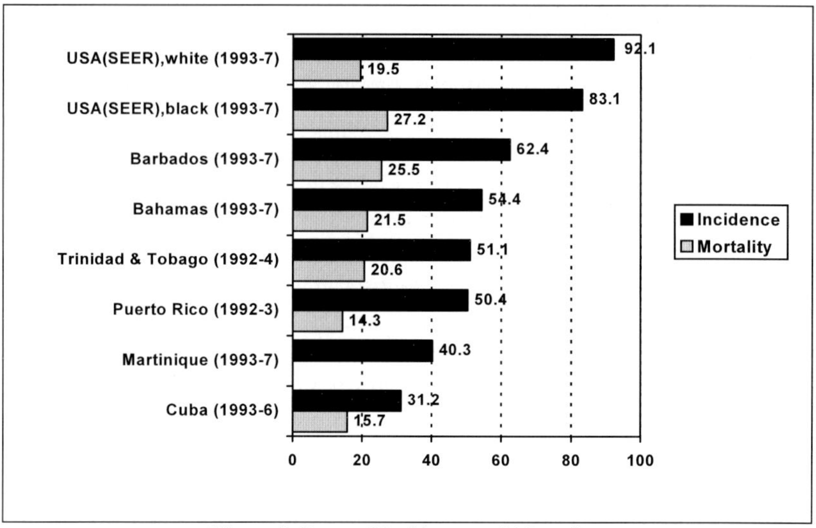

Figure 5. Rates of Incidence and Mortality.

7. BRAZIL

Bouchardy et al (44) presented data from the Sao Paolo cancer registry from the years 1969-1974. Because accurate population denominators were not available by ethnic group, comparison of risk was done by calculating odds ratios for specific cancers, compared to all others. For breast cancer, the risk in blacks was 0.8 (95% c.i. 0-7-0.9), and in mulattos 0.6 (95% c.i. 0.5-0.7), relative to whites.

8. THE FUTURE

Planning and evaluation of cancer control activities (prevention, early detection, treatment, rehabilitation and palliative care) requires knowledge of the current pattern of occurrence, and an estimate of the likely evolution of the cancer burden in the future. Projections of future cancer burden generally rely upon assumptions concerning the continuation of trends of incidence and mortality observed in the past. Recent methodological advances in forecasting allowed the use of Bayesian approaches to specify smooth variation in cancer rates and to give more weight to recent, rather than distant, changes (45), to select appropriate models on the basis of recent trends (as well as through the usual goodness-of-fit tests) and provide sensible confidence intervals for the prediction (46-47). However, sophisticated projection methods cannot be used for most of the populations considered in this chapter, since detailed information on the evolution of past trends in age-specific incidence and mortality is not available. In any case, projections based on historic patterns are not always a sound basis for future predictions. There can be quite abrupt changes in trends in incidence and/or mortality, for example, following the introduction of screening, or improved forms of treatment. Both of these have profoundly affected time trends for cancers of the breast in many western countries, and it is not easy to foresee what further changes of this type will occur in the next decade, let alone in the next 50 years. Therefore, in this section, the emphasis is upon the impact that estimated population growth and ageing will have on the burden of breast cancer in the next few decades. Then, after a brief review of time trends in breast cancer incidence and mortality, the effect that these may have on the demographically driven changes. The population projections have been taken from the United Nations publication, "World Population Prospects" (48), using figures based on the "medium-fertility variant". "Older" and "elderly" persons are used as synonyms for those persons aged 65 or over.

The major characteristics of the evolution of the world population in the next 50 years are the consequences of a projected gradual decline in fertility, and increase in life expectancy. These imply that, although the world population will continue to grow (from 6 billion in 2000 to 8.9 billion in 2050), the rate of increase will slow, and there will be a progressive ageing, with the numbers of elderly people increasing from 7% of the world population in 2000 to over 16% by 2050. This slowing of population growth will occur in both developing and developed areas, though amongst the least developed areas, the demographic transition is some 20 to 25 years behind other developing regions. By 2050, 56 countries (including all European countries, Japan and China) are projected to have a negative growth rate. The total population of the developed countries will peak around 2020 and then decline – by 2050 the overall population should be some 2% less than the 2000 estimate. This contrasts with less developed countries – a 63% increase in the overall population is expected between 2000 and 2050. The expansion is particularly evident in Africa – the population is forecast to double by 2030 – its current 13% share of the world population is set to rise to one fifth of the global population by 2050. Conversely, the world population share of Northern America is projected to decline from 5.2% to 4.7% during this period.

Figure 6 shows the number of new breast cancer cases and Figure 7 the number of deaths from breast cancer if four regions, that are projected to occur at three future periods in the next half century. The projected numbers of cases and deaths, by age group, for the World, Developing and Developed areas are shown in Table 6.

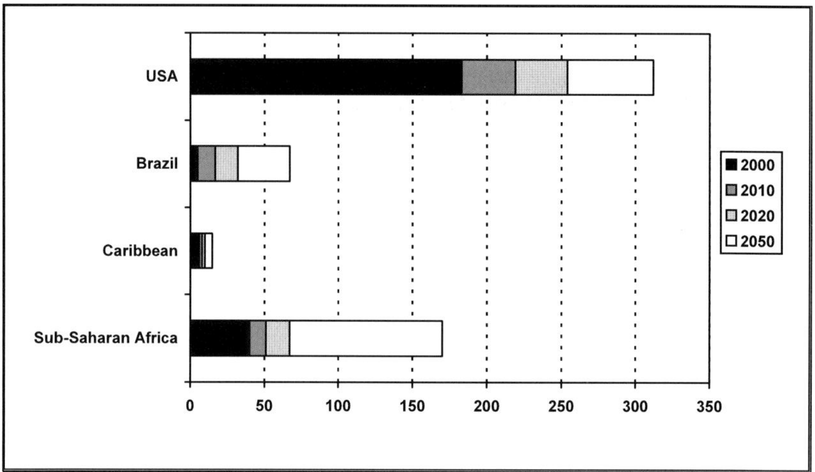

The number of new cases (in thousands) of female breast cancer				
REGION	**2000**	**2010**	**2020**	**2050**
World	10100	12100	14300	19700
Sub-Saharan Africa	40	51	67	170
Caribbean	6	8	10	15
Brazil	35	47	62	97
U.S.A.	183	219	254	312

Figure 6. Projections of breast cancer incidence (thousands) in four regions, 2000-2050.

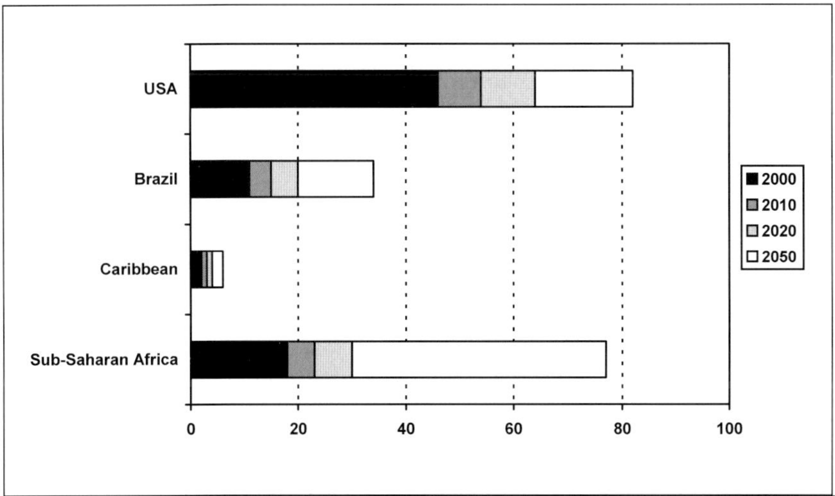

The number of deaths (in thousands) from female breast cancer				
Region	**2000**	**2010**	**2020**	**2050**
World	380	460	550	810
Sub-Saharan Africa	18	23	30	77
Caribbean	2	3	4	6
Brazil	11	15	20	23
U.S.A.	46	54	64	82

Figure 7. Projections of breast cancer deaths (thousands) in four regions, 2000-2050.

Table 6. Age-specific number of breast cancer cases and deaths (x 100,000) in the World, developed countries and developing countries.

		World				More Developed Regions				Less Developed Regions			
	Year	**0-44**	**45-64**	**65+**	**all ages**	**0-44**	**45-64**	**65+**	**all ages**	**0-44**	**45-64**	**65+**	**all ages**
Cases	**2000**	2	4.7	3.4	10.1	0.8	2.6	2.4	5.8	1.3	2.1	0.9	4.3
	2010	2.2	6	3.9	12.1	0.7	3	2.7	6.4	1.5	2.9	1.2	5.6
	2020	2.3	7	5	14.3	0.7	3.1	3.2	7	1.6	3.8	1.8	7.2
	2050	2.3	8.6	8.8	19.7	0.6	2.7	4.3	7.6	1.7	5.9	4.5	12.1
Deaths	**2000**	0.6	1.6	1.5	3.7	0.2	0.7	1.1	2	0.5	0.9	0.4	1.8
	2010	0.7	2.1	1.7	4.5	0.1	0.9	1.2	2.2	0.6	1.3	0.6	2.5
	2020	0.8	2.6	2.2	5.6	0.1	0.9	1.4	2.4	0.6	1.7	0.8	3.1
	2050	0.8	3.4	3.9	8.1	0.1	0.8	1.8	2.7	0.7	2.7	2.1	5.5

Breast cancer is projected to remain the most common cancer in women in the next half century. Changes in global population structure suggest that the one million cases estimated worldwide in 2000 will rise to over 1.4 million cases in 2020, and by 2050, the number of new cases is projected to reach nearly two million. Mortality, though considerably lower than incidence, will reach 0.8 million by 2050. In 2000 there were more breast cancer cases (57%) and deaths (54%) occurring in developed than in less-developed countries. As noted above, the biggest changes in the demography of the world in the next 50 years will take place in developing areas. As a result, more and more of the relative future cancer burden will be in these regions – the number of cases in developing countries will almost triple between 2000 and 2050, compared with an estimated 30% increase in developed regions. There is a similar dramatic projected increase in the numbers of deaths from breast cancer. As seen in Figure 7, the numbers of deaths from breast cancer would be almost the same in sub-Saharan Africa (77,000) as in the USA (82,000).

In most populations, incidence of breast cancer is increasing, and, if this trend continues, the burden of this disease will be very much greater than projections based on demographic change alone imply.

In developed countries, incidence of breast cancer is, in general, still increasing at all ages, although there are some exceptions. In the USA, incidence increased by 29% in whites and 39% in blacks in the 25-year period 1973-1997, with rather faster rates of increase in pre-menopausal women. However, trends in mortality from breast cancer are less straightforward, and in many countries there is evidence for a decline in death rates in recent years. This was first remarked upon in the U.S.A. (49), where national mortality data show a decline in overall rates in white women, of 15% between 1969 and 1997, while the rates for black females increased by 22% (40).

In almost all developing countries with relatively low rates of incidence of breast cancer, the risk appears to be increasing (50). There are very few data from Africa. As can been seen from Tables 3 and 4, rates seem to be rather higher in more recent data than in older series. In Ibadan, Nigeria, incidence in 1998–99 was 24.7 per 100 000, compared with 13.7 per 100 000 in 1960–69. In Kampala, Uganda, there has been a significant increase in incidence since the 1960s (Figure 8), (51). Mortality rates in Mauritius have also been increasing since the 1960s.

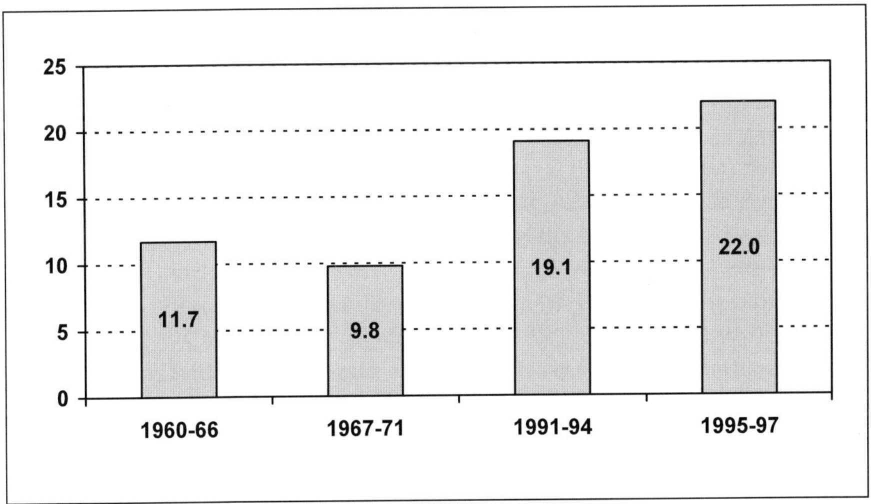

Figure 8. UGANDA, Kampala. Trends in breast cancer incidence (ASR per 105) (51).

Some of these changes may be related to declining fertility, as well as to better nutrition and greater body size; quite possibly both will continue into the future, though the potential for further change is clearly greatest in the developing world. As far as mortality rates are concerned, it is likely that any increases will be less than for incidence, as the improvements in survival (due to earlier diagnosis and better treatment), observed in the USA and U.K., become more widespread.

REFERENCES

(1) Hakama M, Hakulinen T, Teppo L, Saxen E. Incidence, mortality or prevalence as indicators of the cancer problem. Cancer 1975; 36:2227-31.

(2) Jensen OM, Parkin DM, Maclennan R, Muir CS, Skeet RG, editors. Cancer Registration Principles and Methods. IARC Sci Publ 1991; 95.

(3) Parkin DM, Whelan SL, Ferlay J, Teppo L, Thomas DB, editors. Cancer Incidence in Five Continents,Vol. VIII. IARC Sci Publ 2002; 155.

(4) Wingo PA, Ries LA, Rosenberg HM, Miller DS, Edwards BK. Cancer incidence and mortality, 1973-1995: a report card for the U.S. Cancer 1999; 82:1197-1207.

(5) Berrino F, Capocaccia R, Estève J, Gatta G, Micheli A, Hakulinen T, et al. Survival of Cancer Patients in Europe: The EUROCARE-2 Study. IARC Sci Publ 1991; 151.

(6) Sankaranarayanan R, Black RJ, Parkin DM, editors. Cancer Survival in Developing Countries, IARC Sci Publ 1998; 145.

(7) Rothman, KJ, Greenland S. Modern Epidemiology, 2nd edition. Philadelphia, (PA): Lippincot-Raven; 1998.

(8) Pisani P, Parkin DM, Bray F, Ferlay J. Estimates of the worldwide mortality from 25 cancers in 1990. Int J Cancer 1999; 83:18-29.

(9) Dempsey, M., Decline In Tuberculosis. The Death Rate Fails To Tell The Entire Story. Amer Rev Tubercul 1947; 56:157-164.

(10) Murray JL, Axtell LM. Impact of cancer years of life lost due to cancer mortality. J Natl Cancer Inst 1974; 52:3-7.

(11) Daniels N. Just Health Care, New York: Cambridge University Press; 1985.

(12) Murray CJ. Quantifying The Burden Of Disease: The Technical Basis For Disability-Adjusted Life Years. Bull.World Health Organ 1994; 72:429-445.

(13) Morrow RH, Bryant JH. Health policy approaches to measuring and valuing human life: conceptual and ethical issues. Am J Public Health 1995; 85:1356-60.

(14) Parkin DM, Bray F, Ferlay J, Pisani P. Estimating the world cancer burden: Globocan 2000. Int J Cancer 2001; 94:153-6.

(15) Yang BH, Bray FI, Parkin DM, Sellors JW, Zhang ZF. Cervix cancer as a priority for prevention in different world regions. An evaluation using Years of Life Lost. Int J Cancer (in press); 2003.

(16) Ferlay J, Bray F, Pisani P, Parkin DM. GLOBOCAN 2000: Cancer Incidence, Mortality And Prevalence Worldwide, Version 1.0. IARCancerbase No. 5; 2001.

(17) Murray CJ, Lopez AD. Mortality by cause for eight regions of the world: Global burden of Disease Study. Lancet 1997; 349:1269-1276.

(18) UNAIDS/WHO (2002) AIDS Epidemic Update: December 2002. Joint United Nations Programme on HIV/AIDS (UNAIDS); 2002.

(19) World Health Organization. The WHO Statistical Information System (WHOSIS). http://www3.who.int/whosis/menu.cfm; 2002.

(20) CDC. National Center for Health Statistics. Mortality Data from the National Vital Statistics System. http://www.cdc.gov/nchs/about/major/dvs/mortdata.htm; 2003.

(21) Coale A, Guo G. Revised regional model life tables at very low levels of mortality. Population Index 1989; 55:613-643.

(22) Isaacson C, Selzer G, Kaye V, Greenberg M, Woodruff JD, Davies J, et al. Cancer in the urban blacks of South Africa. S Afr Cancer Bull 1978; 22:49–84.

(23) Sitas F, Madhoo J, Wessie J. Incidence of histologically diagnosed cancer in South Africa, 1993–1995. Johannesburg, National Cancer Registry, South African Institute for Medical Research; 1998.

(24) Moolgavkar SH, Stevens RG, Lee JA. Effect of age on incidence of breast cancer in females. J Natl Cancer Inst 1979; 62:493–501.

(25) Ihekwaba FN. Breast cancer in Nigerian women. Br J Surg 1992; 79:771–775.

(26) Muguti GI. Experience with breast cancer in Zimbabwe. J R Coll Surg Edin 1993; 37, 159–161.

(27) Maalej M, Frikha H, Ben Salem S, Daoud J, Bouaouina N, Ben Abdallah M, et al. Le cancer du sein en Tunisie: étude clinique et épidémiologie [Breast cancer in Tunisia: clinical and epidemiological study]. Bull Cancer 1999; 86:302–306.

(28) Bouchardy C, Wanner P, Parkin DM. Cancer mortality among sub-Saharan African migrants in France. Cancer Causes Control 1995; 6:539–544.

(29) Bouchardy C, Parkin DM, Wanner P, Khlat M. Cancer mortality among North African migrants in France. Int J Epidemiol 1996; 25:5–13.

(30) Grulich AE, Swerdlow AJ, Head J, Marmot MG. Cancer mortality in African and Caribbean migrants to England and Wales. Br J Cancer 1992; 66:905–911.

(31) Walker AR, Walker BF, Tshabalala EN, Isaacson C, Segal I. Low survival of South African urban black women with breast cancer. Br J Cancer 1984; 49:241–244.

(32) Dansey RD, Hessel PA, Browde S, Lange M, Derman D, Nissenbaum M, et al. Lack of a significant independent effect of race on survival in breast cancer. Cancer 1988; 61:1908–1912.

(33) Gondos A, Chokunonga E, Brenner H, Parkin DM, Sankila R, Borok MZ, et al. Cancer survival in a southern African urban population [in preparation].

(34) American Cancer Society. Cancer Facts and Figures, 2001. Atlanta: ACS; 2001.

(35) American Cancer Society. Cancer Facts and Figures for African Americans 2000-2001. Atlanta: ACS; 2001.

(36) Ries LA, Eisner MP, Kosary CL, Hankey BF, Miller BA, Clegg L, et al., editors. SEER cancer statistics review, 1975-2000. Bethesda (MD) : National Cancer Institute; 2003.

(37) Parkin DM, Ferlay J, Hamdi-Chérif M, Sitas F, Thomas J, Wabinga H, Whelan S (2003). Cancer in Africa: Epidemiology and Prevention IARC Sci Publ 2003; 153.

(38) Middleton LP, Chen V, Perkins GH, Pinn V, Page D. Histopathology of breast cancer among African-Amrican women. Cancer 2003; 97 (1 Suppl.): 253-257.

(39) Chen VW, Correa P, Kurman RJ, Wu XC, Eley JW, Austin D, et al. Histological characteristics of breast carcinoma in blacks and whites. Cancer Epidemiol Biomarkers Prevention 1994; 3:127-135.

(40) Clarke CA, West DW, Edwards BK, Figgs LW, Kerner J, Schwartz AG. Existing data on breast cancer in African-American women: what we know and what we need to know. Cancer 2003; 97(1 Suppl):211-21.

(41) Raju GC, Naraynsingh V. Breast cancer in West Indian women in Trinidad. Trop Geogr Med 1989; 41:257-60.

(42) Fernandez L, Graupera M, Galan Y, Lezcano M, Garcia M, Camacho R. Cancer Survival in Cuba. In: Sankaranarayanan R, Black RJ, Parkin DM, editors. Cancer Survival in Developing Countries, IARC Sci Publ 1998; 145:51-60.

(43) Shai D. Cancer mortality in Cuba and among the Cuban-born in the United States: 1979-81. Public Health Rep 1991; 106:68-73.

(44) Bouchardy C, Mirra AP, Khlat M, Parkin DM, de Souza JM, Gotlieb SL Ethnicity and cancer risk in Sao Paulo, Brazil. Cancer Epidemiol Biomarkers Prev 1991; 1:21-7.

(45) Berzuini C, Clayton D. Bayesian analysis of survival on multiple time scales. Stat Med 1994; 13:823-38.

(46) Hakulinen T, Dyba T. Precision of incidence predictions based on Poisson distributed observations. Stat Med 1994; 13:1513-23.

(47) Dyba T, Hakulinen T, Paivarinta L. A simple non-linear model in incidence prediction. Stat Med 1997; 16:2297-309.

(48) United Nations. World Population Prospects. The 2000 Revision. Population Division, Department of Economic and Social Affairs. New York: United Nations; 2001.

(49) Blot WJ, Fraumeni JF, Jr. Time-Related Factors in Cancer Epidemiology. J Chronic Dis 1987; 40:1S-8S.

(50) Parkin DM. Cancer in developing countries. Cancer Surv 1994; 19/20:519-561.

(51) Wabinga HR, Parkin DM, Wabwire-Mangen F, Nambooze S. Trends in cancer incidence in Kyadondo county, Uganda, 1960-1997. Br J Cancer 2000; 82:1585–1592.

Chapter 2

GENETICS OF BREAST CANCER IN WOMEN OF AFRICAN DESCENT: AN OVERVIEW

Offiong Francis Ikpatt, MBBS, FMCPath (Nigeria), PhD, Olufunmilayo I. Olopade, MBBS, FACP
Section of Hematology/Oncology , Department of Medicine, Universithy of Chicago, Chicago, Il, USA

1. EPIDEMIOLOGY

In 2003, an estimated 20,000 new cases of breast cancer were diagnosed among black women in the United States, making it the most common malignancy in this population (*1*). The rates of breast cancer incidence in Caucasian women and African American women are 113.2 and 99.3 cases per 100,000, respectively (*1*). In West Africa, the founder population of most African Americans, breast cancer has been considered to be a rare virulent disease of young women. According to the International Agency of Research on Cancer, breast cancer incidence in seven African countries has doubled, going from an average of 15.3 cases per 100,000 in 1976 to 33.6 per 100,000 in 1998 (*2*) (also see the contribution of Max Parkin in this book). This could be due to an actual increase in incidence or may be the result of improved reporting of cases. Unfortunately, the large numbers of resource poor nations in Africa make it extremely difficult to have accurate estimates of the number of breast cancer cases diagnosed in those countries. Nevertheless, the incidence of cancer, in general, appears to have increased in Africa and may likely be related to the changes in social conditions, life-style, and emergence of the AIDS epidemic (*3*). Although the AIDS epidemic has resulted in a striking increase in the incidence of Kaposi's sarcoma (now the most common cancer in Africa), there has not been an increase in incidence of AIDS-associated breast cancer (*4,5*).

C. Williams et al. (eds.), Breast Cancer in Women of African Descent, 23-37.

Cancer of the cervix remains the most common female cancer in sub-Saharan Africa (see the contribution of Max Parkin in this book), where the most common cancers are those thought to have infectious etiologies (*6*). Nonetheless, breast cancer is emerging as the most common cancer among women in some regions of Africa, such as Nigeria and the Ivory Coast (*7,8*). In North Africa, nasopharyngeal and bladder cancers appear to be more prominent but breast cancer still accounts for about 33% of cancers in Egyptian women according to some reports (*9,10*).

Cancer patterns in women of European descent living in Africa resemble those of the Caucasian populations of Europe and North America (*4*). The African Diaspora consists of populations that share a common genetic background, yet reside in widely varying social settings. Heritable aspects of breast cancer should be detectable in cross-cultural comparisons of Blacks in Africa, the Caribbean and North America. Potential protective factors in the environment could be identified because of the sharp contrasts in environmental exposure. However, such comparisons are extremely complex, and there have been limited resources to develop an infrastructure that can support unified assessment of the contribution of environmental and genetic factors to the incidence and clinical outcomes of breast cancer among blacks of African descent.

2. GENETICS

There is paucity of data on genetics of breast cancer in Africa but the fact that black women of African descent develop breast cancer at a younger age than white women suggest that risk factors for early onset breast cancer may be more common in black women as a whole. A number of breast cancer predisposition genes such as *TP53*, *PTEN*, have been identified but the two most clinically relevant ones appear to be *BRCA1* and *BRCA2* genes. According to the Breast Cancer Linkage Consortium (1997) (*11*), the phenotype of breast cancers in women carrying BRCA1 mutations differs from that of women carrying BRCA2 mutations and sporadic cases. Breast cancer that arises in the setting of a *BRCA1* mutation tends to be of higher grade, have a higher proportion classified as having atypical medullary features and a lower proportion with tubular differentiation, all of which are poor prognostic features. Multiple regression analyses of the pathological features of BRCA1-associated breast cancers further indicate that the main features which reflect poor outcome are higher mitotic rate, continuous pushing edges and lymphocytic infiltration. (Breast Cancer Linkage Consortium, 1997) (*11*). *BRCA2* associated tumors are more similar to sporadic breast tumors in that they are usually of intermediate grade, are frequently hormone receptor positive, and occur at later ages than *BRCA1*-associated tumors.

Hedenfalk and colleagues used DNA micro-array technology to determine gene-expression profiles for *BRCA1* and *BRCA2* associated breast cancers (*12*). They found differential clustering of genes expressed by breast cancers that occur in the setting of germline *BRCA1* and *BRCA2* mutations. These observations suggest that breast cancers due to *BRCA1* mutations have a different natural history than those that occur sporadically or those that are due to *BRCA2* mutations.

3. PARALLELS BETWEEN BRCA-ASSOCIATED BREAST CANCERS (BABC) AND BREAST CANCERS IN BLACK WOMEN

In contrast to sporadic tumors, breast cancers associated with germ line mutations in *BRCA1 or BRCA2* have features that are similar to those observed in a significant proportion of early onset breast cancer that develops among black women. These features include:

1. An earlier mean age at onset (30-49 years)
2. Aggressive features such as poor differentiation, high tumor grade, aneuploidy, high S-phase fraction, hormone-receptor negativity, and high frequency of p53 mutations
3. Higher than expected rates of medullary or atypical medullary carcinoma

Because of the strong association between age of onset and genetic susceptibility, a significant proportion of breast cancer cases in women of African descent might be suspected to be attributable to inherited mutations in breast cancer susceptibility genes such as *BRCA1*or *BRCA2* genes but there are very few studies that have evaluated this possibility.

3.1 BRCA1 Mutations in African-Americans

To date, 1947 distinct variants of *BRCA1* and 2253 variants of *BRCA2* have been found across all population groups, of which 710 (36.5%) of the *BRCA1* and 872 (38.7%) of the *BRCA2* variants have been identified as deleterious (Breast Cancer Information Core Database) (*13*). Of these pathogenic variants, only 24 *BRCA1* and 18 *BRCA2* mutations are reported to be among populations of African descent.

BRCA1 and *BRCA2* mutations in high-risk African-American families appear at a lower frequency than in their Caucasians counterparts but African Americans have a higher frequency of Unclassified Variants. It is also clear that Africans and African Americans exhibit a unique spectrum of deleterious *BRCA1* and *BRCA2* mutations and variations in comparison to other ethnic groups (*14*) as shown in Tables 1a and 1b. Of the 24 distinct *BRCA1* pathogenic mutations detected in African Americans or Africans, six (25%) have been detected in more than one family of African ancestry. However, only two mutations (M1775R and 943ins10) have been observed in more than two unrelated families. These mutations are classified as founder mutations among African-Americans on the basis of a shared haplotype in the genomic region containing the gene (*15*).

Pathogenic BRCA1 mutations (missense mutations M1775R, C64G), which segregate with breast cancer in large African-American families, were first described in 1994 (*16,17*). Subsequently, multiple groups reported BRCA1 frame shift mutations. In Florida, Arena and colleagues (*18*) identified 3 frame-shift mutations in African-American families with a strong family history of breast cancers. These included a 10 base pair (bp) duplication (943ins10) which was later also detected by Stoppa-Lyonnet et al. (*19*) in one family who immigrated from the Ivory Coast.

A number of studies have evaluated population based African American breast cancer cases as well as families. The findings are summarized in Table 1 and 2.

Table 1a. BRCA1 Mutations in African-American Breast Cancer Patients.

***BRCA1* mutations**	***No. Af ances fam**	**No. non-Af ances fam**
155del4	1	0
943ins10	7	2
1625del5	1	0
1832del5	2	0
E673X	1	0
2418delA	2	0
3331insG	1	0
3450del4	1	9
3875delGTCT	1	35
3883insA	1	0
3888delGA	1	0
K1290X	1	1
4160delAG	1	0
Y1463X	1	2
4730insG	1	0
4794insA	1	0
5296del4	2	9
Disease associated missense mutations		
C61G	2	65
C64G	1	2
W1718C	1	1
M1775R	5	7

***BRCA1* mutations**	***No. Af ances fam**	**No. non-Af ances fam**
Splicing mutations		
IVS4 > IG/T	1	9
IVS13 > IG/A	1	0
IVS22 > 5G/T	1	0
Unclassified variants and polymorphisms		
K38K (G233A), S186Y, M2971, R315G, K355R, Q356R (A1186G), 1379M, A521T, S616delS, N723D, L771L (T2430C), K820E, P871L (C2731T), P938P (A2933G), E1038G, S1040N, S1140G (A3537G), K1183R (A3667G), Q1200H, L1260L (T4932C), S1297F (C4009T), L1564P, L1605L (T4932C), Q1785H, E1794D, IVS1 > 10T/C, IVS8 > 58delT, IVS12 > 12delGT, IVS16 > ? 68G/A, IVS18 > ? A/G P 34, IVS19 > 85delT, IVS20 > 59ins12, IVS22 > 7 T/C, IVS22 > 8 T/A, IVS22 > 8 T/C, IVS22 > 67 T/C, IVS22 > 78 C/A, IVS23 > 10C/A, C5817G (3>UTR).		

*Af African
Anc Ancestry
Fam Family

Table 1b. BRCA1 Mutations in African Breast Cancer Patients.

***BRCA1* mutations**	**No. Af ances fam**	**No. non-Af ances fam**
Protein-truncating mutations		
943ins10	1	2
1742insG	1	0
Q1090X	1	0
Unclassified variants and polymorphisms		
1379M	1	4
K820E	1	13
K1183R (A3667G)	1	186

*Af African
Anc Ancestry
Fam Family

Table 2a. BRCA2 Mutations in African-American Breast Cancer Patients.

BRCA2	**No. Af ances fam**	**No. non-Af ances fam**
1536del4	1	0
1882 del T	1	2
1991delATAA	1	0
1993delAA	1	0
2001delTTAT	1	0
2816insA	2	5
4075delGT	1	16
4088delA	1	1
6696delTC	1	2
Q2342X N	1	0
7436del4	1	0
7795delCT	2	0
7907delTT	1	0
8643 delAT	1	0
9481insA	2	2

BRCA2	No. Af ances fam	No. non-Af ances fam
R3128X	1	7
Splicing mutations		
IVS13 > 2A/G	1	0
Unclassified variants and polymorphisms		
P46S, P59A, N108H, Q147H, A248T, N289H, Q713L, L929S, S976I, N987I, N991D, S1172L, C1290Y, Q1396R, T1414M, D1420Y, D1781G, N1880K, N1880R, T1980I, H2074N, H2116R, K2339N, Q2384K, H2440R, A2466V, S2835P , I2944F, S3020C, M3118T, G3212R, V3244I, T3357I, I3412V, IVS6 > 19 C > T, IVS26 > 10 T > G, IVS26 > 106 delT.		

*Af African
Anc Ancestry
Fam Family

Table 2b. ***BRCA2*** Mutations in African Breast Cancer Patients.

BRCA2	No. Af ances fam	No. non-Af ances fam
Protein-truncating mutations		
3034del4 (3036del4)	1	35
Unclassified variants and polymorphisms		
A248T, N289H, L929S, N9871, N991D, T1414M, N1880K, N1880R, Q2348K, A2466V, I2944F, G3212R, IVS6 > 19 C > T, IVS11 > 73 T >A, IVS18 > 109 G >A, IVS20 > 36 C > G, IVS24 >T, IVS26 > 106 delT		

Table 3. Examples of BRCA1 and BRCA2 founder mutations.

Population	Mutation	Reference
Ashkenazi Jews	BRCA1 185delAG	Neuhausen et al 1996 (*25*),
	BRCA1 5382insC	Struewing et al. 1995 (*37*)
	BRCA2 6174delT	Tonin et al. 1995 (*38*)
Belgians	BRCA1 IVS5 +3A>G	Claes et al. 1999 (*39*)
Dutch	BRCA1 2804delAA	Petrij-Bosch et al.1997 (*40*)
	BRCA1 IVS 21-36del510	
	BRCA1 IVS 12-1643 del3835	
	BRCA2 5573insA	
Finnish	BRCA1 3745delT	Huusko et al. 1998 (*41*)
	BRCA1 IVS 11-2 A>G	
	BRCA2 999del5	
	BRCA2 IVS23-2A>G	
Swedes	BRCA1 Q563X	Johannsson et al. 1996 (*42*)
	BRCA1 3166ins5	
	BRCA1 1201del11	
	BRCA1 2594delC	
	BRCA2 4486delG	
Norwegians	BRCA1 1675delA	Dorum et al. 1999 (*43*)
	BRCA1 1135insA	
Icelanders	BRCA2 999del5	Thorlacius et al. 1996 (*44*)
Germans	BRCA1 5382insC	Backe et al. 1999 (*45*)

Population	Mutation	Reference
	BRCA1 C61G	
Russians	BRCA1 5382insC	Gayther et al. 1997 (*46*)
	BRCA1 4153delA	
Latvians	BRCA1 C61G	Csokay et al. 1999 (*47*)
	BRCA1 5382insC	
	BRCA1 4153delA	
French-Canadians	BRCA1 R1443X	Tonin et al. 1999 (*48*)

3.2 BRCA2 mutations in African-American

Eighteen distinct pathogenic *BRCA2* mutations have been detected in families of African ancestry, of which 56% (10 of 18) are novel and probably unique to this group (*13,14*). Seventeen percent (3 of 18) of the pathogenic mutations have been detected in more than one African-American or African family. No *BRCA2* mutations were carried by more than two families of African ancestry (*14*). Tables 2a and 2b show the *BRCA1* in breast cancers of African-American and African women, respectively.

A recurrent *BRCA2* frame shift mutation, 2816insA, has been identified in African American families with breast and ovarian cancers and male breast cancer (*20,21*). Ganguly et al. (*22*) later reported three disease-related *BRCA2* mutations (7436del4, 7907delTT, IVS13 - 2A/G) among 10 African-American patients with a family history of breast cancer (*22*). Additional pathogenic *BRCA2* mutations were described by Kanaan et al. and Whitfield-Broome et al. (1882 del T, 1991delATAA, 1993delAA, 2001delTTAT, 2816insA, 4075delGT, 4088delA, 8643 delAT) (*20,23*). Distinct disease-related *BRCA2* mutations in African Americans have also been reported by Myriad (*13*). These include 9481insA, 9481insA, and Q2342X. (Table 2a).

3.3 Polymorphisms/Sequence Variations

In addition to the deleterious mutations observed in *BRCA1* and *BRCA2*, there are several common polymorphisms in the *BRCA1/2* genes that generate amino acid substitutions (Tables 1a, 1b, 2a and 2b). A higher frequency of these variations/polymorphisms occurs in African Americans as compared to Caucasians (*14*). Dunning and colleagues (*24*) first raised the question of whether these common polymorphisms could be related to increased risk of developing breast or ovarian cancer. They concluded that the most common polymorphisms of the *BRCA1* gene do not make a significant contribution to breast or ovarian cancer risk and one such polymorphism, Q356R, may actually be protective as it is more frequently found in unaffected controls (p=0.01). The fact that sequence variants of unknown significance are more likely to be found among African Americans underscores the need for more studies clarifying the functional significance of such variants.

3.4 BRCA1 and BRCA2 Mutations among Africans

The entire coding regions and the intron/exon boundaries of *BRCA1* and *BRCA2* have been evaluated in 70 African breast cancer patients younger than 40 years of age and diagnosed at the University of Ibadan College of Medicine, Nigeria, with no selection for a family history of breast cancer (*14*). Two *BRCA1* truncating mutations and one *BRCA2* truncating mutation have been identified (Tables 1b, 2b). The protein truncating *BRCA1* mutations, Q1090X and 1742insG, are unique to this cohort. The *BRCA1* amino acid substitutions, however, have all been described in other populations. For example, E1038G and K1183R have both previously been described as benign polymorphisms and I379M and K820E have both been described as unclassified variants (BIC). The *BRCA2* truncating mutation, 3034del4, has been described as a mutational hotspot and the *BRCA2* missense variation, N1880R, appears to be unique to the African cohort. Others like the *BRCA2* alleles, G3212R, A248T, N987I, and L929S, have all been previously reported and are unclassified variants (*13*).

4. FOUNDER MUTATIONS AMONG FAMILIES OF AFRICAN ANCESTRY

When the same mutation is found in multiple unrelated families, this may be due to common ancestry from a small isolated group of founders or to an independent mutational event. A common haplotype among unrelated families around the gene of interest is evidence of a founder effect. The length of the haplotype is inversely related to the age of the mutation. The *BRCA1* 943ins10 mutation was associated with a single haplotype in five families from the Ivory Coast, Washington DC, Florida, South Carolina, and the Bahamas (*15,19*). The length of the common haplotype is about the same as the length of the Ashkenazi Jewish founder mutation, 185delAG, which has been estimated to be 760 years old (*25*). Therefore, the *BRCA1* 943ins10 mutation represents an ancient founder mutation of West African origin. A common haplotype was reported for two African-American families with the *BRCA1* 5296del4 mutation and for two African-American families with the *BRCA1* 1832del5 mutation, but more families with the same haplotypes are needed to determine if they represent founder mutations unique to African Americans (*14*).

Overall, the frequency of *BRCA1* and *BRCA2* germ-line deleterious mutations in selected at-risk families is quite low, suggesting the presence of yet unidentified genes or unidentified mechanisms for inactivating *BRCA1* and *BRCA2* that contribute to breast and/or ovarian cancer risk. Other molecular markers that have been evaluated in association with breast cancers in black women include: p53 tumor suppressor gene mutations, Her-2/*neu* gene amplifications, estrogen and progesterone receptor up-regulation, and *UGT1A1* gene polymorphisms.

4.1 TP53

The tumor suppressor gene *TP53* is a transcription factor that is implicated in the regulation of several biological pathways, such as cell growth, gene transcription, apoptosis, senescence and genomic stability (*26*). *TP53* mutations have been associated with tumorigenesis in many different types of human cancers. Of the observed *TP53* mutations, 80-90% involve exons 5 to 8, spanning the conserved region of the protein (domains II to V). In breast cancer, the reported frequency of *TP53* mutations varies considerably, ranging from 12 to 60% (*27*). In a comparative assessment of *TP53* mutations in breast cancers of the White- and African Brazilians, a relatively high frequency of *TP53* mutations was found in blacks (32.7%) as compared to whites (13.6%, p=0.01) (*28*). African-Brazilian women have a large proportion of mutations in exon 5 (29.4%) and exon 7 (41.2%), whereas white women have more mutations in exon 7 (27.3%) and exon 8 (39.4%). A:T→G:C transitions and G:C→C:G transversions were more common in African-Brazilians as compared to white women (27.3% vs. 9.5% and 27.3% vs. 4.8%, respectively). Whites were found to have a higher frequency of G:C→T:A transversion than their black counterparts (23.8% vs. 9%). There was a prevalence of G:C→A:T nucleotide transitions in both groups. Seventy-two percent of these transitions occurred at CpG dinucleotide sites. Transversions A:T→C:G and A:T→T:A were found only in the white population. The frequency of the neutral polymorphism at codon 213 of the *TP53* gene was 10.5% among whites and 3.8% among blacks. Similar findings were observed in studies involving African-Americans in North America (*29,30*). It is generally believed that *TP53* mutations occur more frequently in tumors with more aggressive features, so this difference may in part explain the difference in the prevalence of *TP53* mutations in both groups. The excess of A:T→G:C in black populations might result from an increased exposure to exogenous or endogenous nitric oxide and/or defective repair of deaminated adducts induced by nitric oxide.

Differences in inherited characteristics of the population, such as polymorphisms for enzymes involved in the activation and detoxification of carcinogens, might also contribute to differences in the *TP53* mutation spectrum between black and white populations.

4.2 ATM

The *ATM* gene, located on chromosome 11q22-23, is 157 kb in length and contains 66 coding exons. The ATM protein is approximately 370 kDa in size and is involved in cell cycle regulation and genomic stability. ATM germ line mutations result in the ataxia telangiectasia (AT) syndrome, an autosomal recessive disorder characterized by progressive cerebellar degeneration, immuno-deficiency, oculo-cutaneous telangiectasia, and increased sensitivity to ionizing radiation (*31*). Of the 149 DNA sequence alterations of the *ATM* gene identified in breast cancers of African Americans, four represent non-conservative missense mutations and are predicted to cause the substitution of one amino acid by another with a dissimilar side chain that could affect the function of the ATM protein (*31*). That is, the 2119C→T would substitute a hydrophobic amino acid for a polar amino acid, the 4400A→G would result in the substitution of a neutral for an acidic amino acid; the 4138C→T results in a neutral amino acid being substituted for a basic one, and the 2362A→C would switch a basic for a neutral amino acid.

4.3 UGT1A1

UGT1A1 is one of the major enzymes involved in estradiol glucuronization and also constitutes a major detoxification pathway for toxic or carcinogenic compounds. Unlike mutation in high penetrance genes such as *BRCA1* and *BRCA2*, genetic variations in *UGT1A1* are quite common among the general population. Because *UGT1A1* variations are much more common than *BRCA1/2* mutations, they could potentially contribute to a larger number of breast cancer cases, given the frequency at which they occur in the general population. Populations of African origin harbor four different alleles of *UGT1A1,* characterized by a variation in the number of TA repeats in the TATA box region. Non-African populations appear to have only two different alleles.

Increasing the number of repeats in the promoter region leads to a decrease in the rate of transcription of the *UGT1A1* gene. Lower expression of *UGT1A1* might lead to an increase in the level of estradiol and expose cells to a higher local concentration of active hormone and therefore have considerable impact on tumor initiation and growth.

Studies to assess the association between single nucleotide polymorphisms and a number of the genes in the *UGT1A* cluster and *UGTIA* polymorphisms as a potential modifier of *BRCA1* or *BRCA2* cancer risk are ongoing (*32,33*).

5. GENETIC COUNSELING

There is a great need to intensify efforts to integrate genetic counseling and testing into the clinical care of young African women already affected by cancer. Such an approach will serve not only to identify deleterious mutations in index breast and ovarian cancer cases, but to also provide the link to other at-risk family members and help develop better strategies to reduce their risk of dying of breast and/or ovarian cancer. An intensive outreach program focused on cancer control through genetic testing could improve awareness in the African-American community, and further research is needed to explore the mechanisms needed to increase the cultural sensitivity of education and counseling protocols (*34*).

Compared to Caucasian women, African-American women appear to have less knowledge but more positive attitudes about the benefits of genetic testing. Ethnic differences in knowledge and attitudes regarding genetic testing for breast and ovarian cancer risk may be attributable to differences in exposure to genetic information and in referrals by health care providers. Furthermore, baseline levels of spiritual faith, cancer-specific distress, perceived risk, and demographic factors like income may play into the decision-making process concerning genetic testing. Consideration of these factors is important in effectively designing risk assessment and education programs for minority women (*35*).

6. CONCLUSION

There are inherent limitations in the comparisons of cancer statistics generated by the cancer registries of different African countries, particularly with respect to methods, quality, and reporting of cancer incidence data. Nevertheless, the reported trends from multiple cancer registries suggest that breast cancer in Africa is characterized by its low incidence rate and its early age-of-onset, as has been observed in Nigeria. It is not yet clear how much of the variation between African countries is attributable to genetic factors, or what proportion of all cases of breast cancer or ovarian cancer are "familial."

Because of the relatively young ages of diagnosis of breast cancer in Africa, and because hereditary cancers occur disproportionately in young women, it is expected that alterations in *BRCA*1/2 mutations will account for a greater proportion of all breast cancers in these African countries.

Differences in frequency between African-American controls and African controls may be attributable to the small African sample size studied so far, African-American admixture with Europeans and Native Americans, and to differences in the composition of the African population. The African American biology is significantly shaped by periods of intermixture involving the incorporation of non-African genes into the gene pool. Subsequently a high heterogeneity is created. Furthermore, the selective pressures emanating from the unique environmental events in the United States of America may equally contribute to this evolutionary hypothesis (*36*).

REFERENCES

(1) Anonymous: American Cancer Society (2003): Breast Cancer Facts and Figures., 2003.

(2) Ijaduola TG, Smith EB: Pattern of breast cancer among white-American, African-American, and nonimmigrant west-African women. Journal of the National Medical Association. 1998;90:547-51.

(3) Thomas JO: Acquired immunodeficiency syndrome-associated cancers in Sub-Saharan Africa. Semin Oncol 2001;28:198-206.

(4) Bassett MT, Levy L, Chokunonga E, Mauchaza B, Ferlay J, Parkin DM: Cancer in the European population of Harare, Zimbabwe, 1990-1992. Int J Cancer 1995;63:24-8.

(5) Sitas F, Pacella-Norman R, Carrara H, Patel M, Ruff P, Sur R, et al: The spectrum of HIV-1 related cancers in South Africa. Int J Cancer 2000;88:489-92.

(6) Bah E, Parkin DM, Hall AJ, Jack AD, Whittle H: Cancer in the Gambia: 1988-97. Br J Cancer 2001;84:1207-14.

(7) Solanke TF, Adebamowo CA: Reports of the workshop of the states of art in oncology in Ibadan and Ife. Ibadan1: National Headquarters of Cancer Registry in Nigeria, University College Hospital., 1996.

(8) Echimane AK, Ahnoux AA, Adoubi I, Hien S, M'Bra K, D'Horpock A, et al: Cancer incidence in Abidjan, Ivory Coast: first results from the cancer registry, 1995-1997. Cancer 2000;89:653-63.

(9) Bouchardy C, Parkin DM, Wanner P, Khlat M: Cancer mortality among north African migrants in France. Int J Epidemiol 1996;25:5-13.

(10) Bedwani R, Abdel-Fattah M, El-Shazly M, Bassili A, Zaki A, Seif HA, et al: Profile of familial breast cancer in Alexandria, Egypt. Anticancer Res 2001;21:3011-4.

(11) Anonymous: Breast Cancer Linkage Consortium. Pathology of familial breast cancer: differences between breast cancers in carriers of BRCA1 and BRCA2 mutations and sporadic cases. Lancet 1997;349:1505-10.

(12) Hedenfalk I, Duggan D, Chen Y, Radmacher M, Bittner M, Simon R, et al: Gene-expression profiles in hereditary breast cancer. N Engl J Med 2001;344:539-48.

(13) Anonymous: Breast Cancer Information Core Database: www.nhgri.nih.gov/ Intramural_research/Lab_transfer/Bic/resources. Breast Cancer Linkage Consortium. Pathology of familial breast cancer: differences between breast cancers in carriers of BRCA1 and BRCA2 mutations and sporadic cases. Lancet 1997;349:1505-10.
(14) Olopade OI, Fackenthal JD, Dunston G, Tainsky MA, Collins F, Whitfield-Broome C: Breast cancer genetics in African Americans. Cancer 2003;97:236-45.
(15) Mefford HC, Baumbach L, Panguluri RC, Whitfield-Broome C, Szabo C, Smith S, et al: Evidence for a BRCA1 founder mutation in families of West African ancestry. Am J Hum Genet 1999;65:575-8.
(16) Miki Y, Swensen J, Shattuck-Eidens D, Futreal PA, Harshman K, Tavtigian S, et al: A strong candidate for the breast and ovarian cancer susceptibility gene BRCA1. Science 1994;266:66-71.
(17) Futreal PA, Liu Q, Shattuck-Eidens D, Cochran C, Harshman K, Tavtigian S, et al: BRCA1 mutations in primary breast and ovarian carcinomas. Science 1994;266:120-2.
(18) Arena J, Smith S, Plewinska M, al. e: BRCA1 mutations in African American women. American Journal of Human Genetics 1996;59 (Suppl.).
(19) Stoppa-Lyonnet D, Laurent-Puig P, Essioux L, Pages S, Ithier G, Ligot L, et al: BRCA1 sequence variations in 160 individuals referred to a breast/ovarian family cancer clinic. Institut Curie Breast Cancer Group (see comments). American Journal of Human Genetics 1997;60:1021-30.
(20) Whitfield-Broome C, Duston GM, Brody LC: BRCA2 mutations in African Americans (abstract). Annual Meeting of the American Association for Cancer Research, 1999, p. 269.
(21) Gao Q, Tomlinson G, Das S, Cummings S, Sveen L, Fackenthal J, et al: Prevalence of BRCA1 and BRCA2 mutations among clinic-based African American families with breast cancer. Hum Genet 2000;107:186-91.
(22) Ganguly T, Citron M, Stott J, al. e: Novel BRCA mutations in African American individuals with breast and ovarian cancer. American Journal of Human Genetics 1998;63 (Suppl.):A69.
(23) Kanaan Y, Kpenu E, Utley K, Brody L, Duston G, C W-B: BRCA2 mutations in African Americans. American Journal of Human Genetics 2000;67 (Suppl2):A455.
(24) Dunning AM, Chiano M, Smith NR, Dearden J, Gore M, Oakes S, et al: Common BRCA1 variants and susceptibility to breast and ovarian cancer in the general population. Hum Mol Genet 1997;6:285-9.
(25) Neuhausen S, Gilewski T, Norton L, Tran T, McGuire P, Swensen J, et al. Recurrent BRCA2 6174delT mutations in Ashkenazi Jewish women affected by breast cancer. Nat Genet 1996;13:126-8.
(26) Saifudeen Z, Dipp S, El-Dahr SS: A role for p53 in terminal epithelial cell differentiation. J Clin Invest 2002;109:1021-30.
(27) May P, May E, May PR, Tuma AH, Dixon WJ, Yale C, et al: Twenty years of p53 research: structural and functional aspects of the p53 proteinSchizophrenia. A follow-up study of the results of five forms of treatment. Oncogene 1999;18:7621-36.
(28) Nagai MA, Schaer Barbosa H, Zago MA, Araujo Silva W, Jr., Nishimoto IN, Salaorni S, et al: TP53 mutations in primary breast carcinomas from white and African-Brazilian patients. Int J Oncol 2003;23:189-96.
(29) Blaszyk H, Vaughn CB, Hartmann A, McGovern RM, Schroeder JJ, Cunningham J, et al: Novel pattern of p53 gene mutations in an American black cohort with high mortality from breast cancer. Lancet 1994;343:1195-7.

(30) Shiao YH, Chen VW, Lehmann HP, Wu XC, Correa P: Patterns of DNA ploidy and S-phase fraction associated with breast cancer survival in blacks and whites. Clin Cancer Res 1997;3:587-92.

(31) Telatar M, Teraoka S, Wang Z, Chun HH, Liang T, Castellvi-Bel S, et al: Ataxia-telangiectasia: identification and detection of founder-effect mutations in the ATM gene in ethnic populations. Am J Hum Genet 1998;62:86-97.

(32) Guillemette C, Millikan RC, Newman B, Housman DE: Genetic polymorphisms in uridine diphospho-glucuronosyltransferase 1A1 and association with breast cancer among African Americans. Cancer Res 2000;60:950-6.

(33) Beutler E, Gelbart T, Demina A: Racial variability in the UDP-glucuronosyltransferase 1 (UGT1A1) promoter: a balanced polymorphism for regulation of bilirubin metabolism? Proc Natl Acad Sci U S A 1998;95:8170-4.

(34) Cummings S, Sellergren S, Olopade O, Matthews A, Schum P: Cancer risk perception and genetic counseling among African American women. Annual Meeting of the American Society of Clinical Oncology, 2000, p. 617A.

(35) Schwartz MD, Hughes C, Roth J, Main D, Peshkin BN, Isaacs C, et al: Spiritual faith and genetic testing decisions among high-risk breast cancer probands. Cancer Epidemiol Biomarkers Prev 2000;9:381-5.

(36) Begg CB: On the use of familial aggregation in population-based casse probands for calculating penetrance. J Natl Cancer Inst 2002;94:1221-26.

(37) Struewing JP, Abeliovich D, Peretz T, Avishai N, Kaback MM, Collins FS, et al: The carrier frequency of the BRCA1 185delAG mutation is approximately 1 percent in Ashkenazi Jewish individuals. Nat Genet 1995;11:198-200.

(38) Tonin P, Serova O, Lenoir G, al. e: BRCA1 mutations in Ashkenazi Jewish women (letter). American Journal of Human Genetics 1995;57:189.

(39) Claes K, Machackova E, De Vos M, Poppe B, De Paepe A, Messiaen L: Mutation analysis of the BRCA1 and BRCA2 genes in the Belgian patient population and identification of a Belgian founder mutation BRCA1 IVS5 + 3A > G. Dis Markers 1999;15:69-73.

(40) Petrij-Bosch A, Peelen T, van Vliet M, van Eijk R, Olmer R, Drusedau M, et al: BRCA1 genomic deletions are major founder mutations in Dutch breast cancer patients. Nat Genet 1997;17:341-5.

(41) Huusko P, Paakkonen K, Launonen V, Poyhonen M, Blanco G, Kauppila A, et al: Evidence of founder mutations in Finnish BRCA1 and BRCA2 families. Am J Hum Genet 1998;62:1544-8.

(42) Johannsson O, Ostermeyer EA, Hakansson S, Friedman LS, Johansson U, Sellberg G, et al: Founding BRCA1 mutations in hereditary breast and ovarian cancer in southern Sweden. Am J Hum Genet 1996;58:441-50.

(43) Dorum A, Hovig E, Trope C, Inganas M, Moller P: Three per cent of Norwegian ovarian cancers are caused by BRCA1 1675delA or 1135insA. Eur J Cancer 1999;35:779-81.

(44) Thorlacius S, Olafsdottir G, Tryggvadottir L, Neuhausen S, Jonasson JG, Tavtigian SV, et al: A single BRCA2 mutation in male and female breast cancer families from Iceland with varied cancer phenotypes. Nat Genet 1996;13:117-9.

(45) Backe J, Hofferbert S, Skawran B, Dork T, Stuhrmann M, Karstens JH, et al: Frequency of BRCA1 mutation 5382insC in German breast cancer patients. Gynecol Oncol 1999;72:402-6.

(46) Gayther SA, Harrington P, Russell P, Kharkevich G, Garkavtseva RF, Ponder BA: Frequently occurring germ-line mutations of the BRCA1 gene in ovarian cancer families from Russia. Am J Hum Genet 1997;60:1239-42.

(47) Csokay B, Tihomirova L, Stengrevics A, Sinicka O, Olah E: Strong founder effects in BRCA1 mutation carrier breast cancer patients from Latvia. Mutation in brief no. 258. Online. Hum Mutat 1999;14:92.

(48) Tonin PM, Mes-Masson AM, Narod SA, Ghadirian P, Provencher D: Founder BRCA1 and BRCA2 mutations in French Canadian ovarian cancer cases unselected for family history. Clin Genet 1999;55:318-24.

Chapter 3

BIOLOGY OF BREAST CANCER

Molecular and Pathologic Features of Ductal Neoplasia of the Breast: Racial Considerations

[1]David F. Chhieng, MD, [1]Andra R. Frost, MD, [2]Lynya I. Talley, PhD, [1]William E. Grizzle, MD, PhD
[1]Department of Pathology; [2]Biostatistics Unit of the Comprehensive Cancer Center, University of Alabama at Birmingham, Birmingham, Alabama, USA

1. BIOLOGY OF BREAST CANCER - AN OVERVIEW

Breast carcinogenesis is a multi-step process that initially is recognized histopathologically as a series of preinvasive stages (intermediate stages) leading to invasive carcinoma, and ultimately to metastatic adenocarcinoma. Histologically the preinvasive stages of breast neoplasia consist of a sequence of epithelial changes which include hyperplasia, atypical hyperplasia, and carcinoma in-situ (1-2). All aspects of breast carcinogenesis are better characterized in ductal adenocarcinomas (the majority of breast cancers) than in lobular adenocarcinomas of the breast, which tend to have different molecular characteristics than ductal neoplasia. The subsequent discussion focuses only on ductal neoplasia of the breast.

Patients with hyperplastic ductal lesions of the breast (DH) have a moderately increased risk of subsequent invasive breast carcinoma and this risk doubles in patients with atypical ductal hyperplastic lesions (ADH) (3-4). There has been a marked increase in the numbers of both Caucasian and African-American women in whom ductal carcinoma in situ (DCIS) is detected because of the increased use of screening mammography. The proportion of women in whom DCIS is detected mammographically and who would have developed invasive tumors if the DCIS were untreated is unknown. Thus, there is controversy as to how aggressively to treat DCIS.

C. Williams et al. (eds.), Breast Cancer in Women of African Descent, 39-70.

Accompanying each of these histologic stages leading to ductal neoplasia are various characteristic molecular alterations (2), (5-6) and there is a general trend toward accumulation of molecular alterations as the process of carcinogenesis progresses (7-8). For example, there is evidence that an increase in the expression of cyclin D1 is associated with progression from ADH to DCIS and there is amplification of c-myc as a lesion progresses from DCIS to primary invasive ductal adenocarcinoma (IDC). However, some changes are maintained throughout different neoplastic stages. For example, when overexpression of $p185^{erbB-2}$ and cyclin D1 as well as mutations of p53 are detected in DCIS, these changes also are identified in the adjoining infiltrating carcinoma, and in metastases (5), (9-12). In addition, while certain p53 mutations are detected throughout the process of carcinogenesis, the frequency of such mutations increases with aggressiveness of tumors and correlates inversely with the phenotypic expression of Bcl-2 (Figure 1) (13-14).

Molecular alterations in breast cancers impact both intra- and intercellular functions. For example, neoplastic cells are stimulated by growth factors to proliferate. These growth factors may be present systemically in the circulation or may be secreted locally by the neoplastic cells and/or neighboring cells (e.g., stromal cells); the growth factors interact with their corresponding receptors expressed by the neoplastic cells resulting in their growth and proliferation. Additionally, molecular signaling between the neoplastic cells and the surrounding stroma provides an environment to potentiate carcinogenesis and to facilitate the progression of cancers. The importance of the interactions between epithelial cells and adjacent stromal cells on the development and progression of breast cancer has been recognized since the 1950s (15-19). The morphologic manifestations of stromal alterations that accompany most breast carcinomas are characterized by enhanced accumulation of fibroblasts and a modified, collagenized extracellular matrix (ECM). The increased ECM produced by stromal fibroblasts in response to the cells of carcinoma is referred to as desmoplasia, and bears similarity to the fibroblastic response in wound healing and scarring (20).

Breast cancers also induce an angiogenic response within the cancer and in surrounding tissues. Neo-angiogenesis (i.e., the formation of new blood vessels within and surrounding a cancer) is necessary for tumor growth beyond a few millimeters and the quantity of new vessels associated with a cancer has been reported to have a prognostic significance and is related to the metastatic potential of the cancer (21-23). Angiogenesis also provides a route for metastasis because of the lack of orderly structure of the newly formed vessels. Last but not least, molecular alterations may allow the tumor cells to escape the host immune system and/or develop resistance to therapeutic agents. Some of these molecular alterations have been found to correlate with clinicopathologic features and some possess prognostic and/or relative predictive value as to the benefits of treatment.

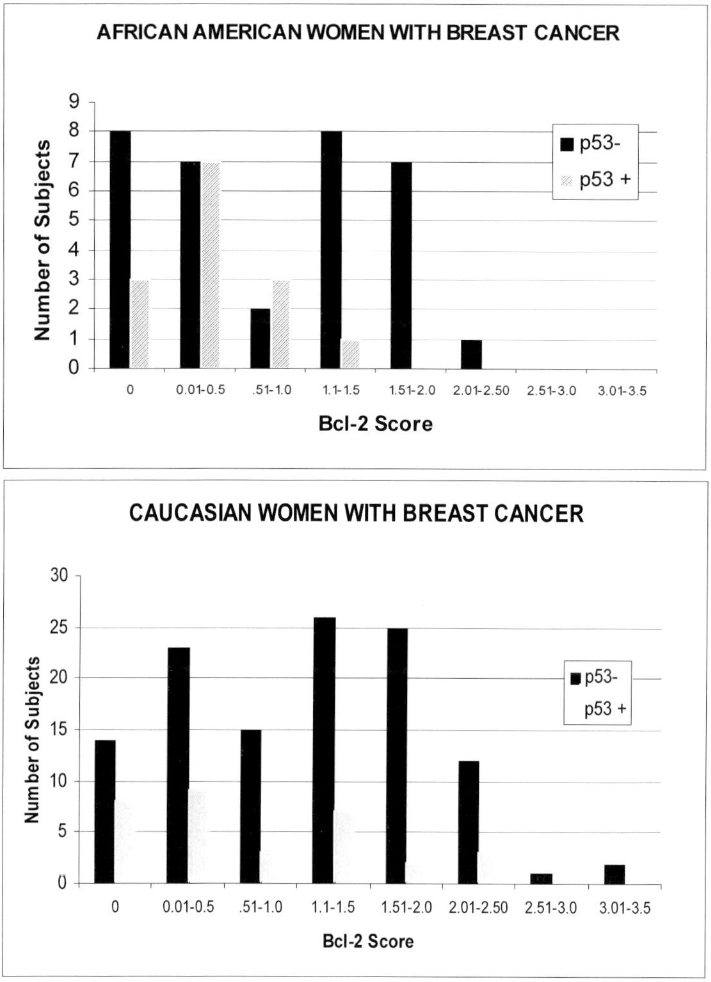

Figure 1. The top panel is a plot of African-American cases with p53 nuclear accumulation ($p53^{+}$) or without p53 nuclear accumulation ($p53^{-}$). It demonstrates that the cells of $p53^{-}$ tumors tend to express higher phenotypic levels of Bcl-2. The bottom panel also demonstrates that $p53^{-}$ tumors tend to express higher phenotypic levels of Bcl-2 and that overall Bcl-2 levels in Caucasian Americans are higher than Bcl-2 levels in African-Americans.

2. BREAST CANCER SIZE AND GRADE IN WOMEN OF AFRICAN DESCENT

Most studies have reported that African American women present with larger breast cancers and with tumors of a more advanced stage than Caucasian women (24-33). Additionally, the breast cancers in African American women are of higher nuclear or histologic grade (30), (34-39). However, many studies have failed to consider the effect of age on the pathology of breast cancers. Most studies have indicated that breast cancers in women under 40 exhibit more adverse characteristics, including a higher clinical stage at diagnosis, higher histologic grade, the absence of receptors to steroid hormones, and an increased rate of proliferation (40-42). Even after adjusting for age, stage and demographic features, such as a metropolitan domicile, Chen et al. found that cancers in African American women were more likely than those in Caucasian women to have a higher grade of nuclei (38). After categorization of women by age into three age ranges (i.e., 20 – 39, 40 – 59, and 60 –74 years), only African American women in the oldest age group had larger carcinomas and a more advance stage at presentation than Caucasian women. Additionally, only African American women in the youngest and middle-age groups were more likely to have higher grade carcinomas (43). Similarly, Shavers et al. reported that, in women under 35 years of age, both African American and Hispanic women had larger and higher grade breast cancers. These results suggest the potential importance of considering both age and race when comparing the biologic features of breast cancers. These results are confounded by differences in sociodemographic characteristics among different ethnicities. After statistical adjustment for sociodemographic factors, many differences in the biological features of breast cancers in both African Americans and Hispanics compared with those in non-Hispanic Caucasians are decreased, but not eliminated, suggesting that there are true biological differences associated with race and ethnicity (28-29), (44-46).

Comparisons of the local host response to breast cancer between women of African descent and other ethnicities are almost non-existent. However, breast cancers in African Americans have been reported to be less likely to exhibit marked desmoplasia than those in Caucasian Americans (38).

3. GENETICS OF FAMILIAL AND SPORADIC DUCTAL ADENOCARCINOMAS

The genetics of breast cancer have been clarified over the last two decades based on familial patterns of inheritance (Table 1). One of the first genes identified to be important in the development and progression of some forms of heritable breast cancer was p53. This gene also is very important in the progression of sporadic ductal adenocarcinoma of the breast. Subsequently PTEN (MMAC1), BRCA-1, and BRCA-2 have been accepted as important in the development of inheritable forms of breast cancer (reviewed in (47-48).

Table 1. Genes Involved in heriditary Cancers of the Breast.

Established Genes	Familial Syndrome
P53	Li-Fraumeni Syndrome
PTEN	Cowden's Syndrome
BRCA-1	Familial female breast and/or ovarian cancers
BRCA-2	Familial female and male breast and/or ovarian cancers
Likely Genes	**Familial Syndrome**
STK11 / LKB-1	Peutz-Jeghers
MLH-1 and MLH-2	Muir-Torre Syndrome

With all familial genes, racial differences are expected, at least in the types of mutations that occur. Differences among mutations in BRCA-1 and BRCA-2 in African-Americans versus Caucasians have been reviewed recently by Olopade et al (49) (also see Chapter 2 in this book). They noted that of mutations in BRCA-1 or BRCA-2 in families with breast and/or ovarian cancer, over 50% of the mutations in each gene were unique for African-Americans and thus these specific mutations had not been reported in Caucasian families. Also, Caucasians with breast cancers had a higher rate of both BRCA-1 or BRCA-2 mutations than African-Americans with breast cancer. The median age at diagnosis of African-Americans with breast cancer positive for BRCA-1 was 33.7 years and for BRCA-2 was 44.4 years. These ages at diagnosis of breast cancer were similar to those of Caucasians; however, for both races, the age at diagnosis of those women with BRCA-1 mutations was significantly different from non-carriers of BRCA-1, but the age at diagnosis of BRCA-2 carriers was not significantly different from that of non-carriers of BRCA-2.

Many cancers that occur in families may not be due to mutations in one gene but are secondary to changes in several genes. Familial cancers characteristically are inherited as a single mutated gene (germ line mutation) such as mutated BRCA-1 gene. This mutated gene alone will not cause cancers to develop but acts by uncertain mechanisms to increase the mutational rate of proliferating cells, which carry the one mutated copy of the gene. Because of the increased rate of mutational events and continued cellular proliferation, the likelihood of a second mutation in the other copy of BRCA-1 is increased greatly. When a mutation occurs in the second gene copy of BRCA-1, then a neoplastic transformation occurs in the proliferating cell causing a clonal expansion and a rapid progression of intraductal lesions through DH, ADH, DCIS, and to IDC (Figure 2). A more rapid rate of metastases may not be a characteristic of BRCA-1 related familial breast cancers.

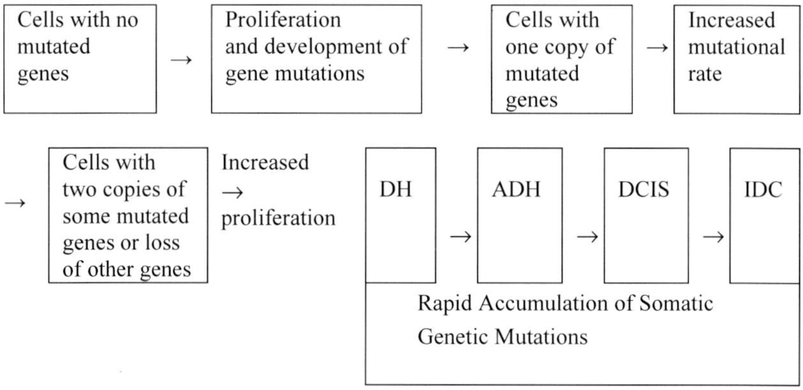

Figure 2. The cartoon demonstrates a potential genetic pathway by which sporadic breast neoplasia develops and accumulates sporadic mutations.

In contrast to familial cancers of the breast, most cases of sporadic ductal adenocarcinomas of the breast do not involve inheriting a mutated copy of any specific gene. As tumors proliferate, changes occur in the phenotype of the tumor cells (reviewed in (47) and (48). Some of the phenotypic changes develop because of somatic mutations in specific genes such as p53 (Figure 3). Other changes develop due to uncontrolled replication of copies of specific genes (gene amplification) such as myc and erbB-2 and are manifest as increased numbers of copies of genes in the nuclei of tumor cells and in increased expression of the coded protein of these genes (e.g., strong expression on cellular membranes of $p185^{erbB\text{-}2}$). Other changes in the cells of breast neoplasia present only as an increased expression of specific proteins without concomitant mutations of genes or gene amplification. Of great importance is that the function of promoters of some genes may be modulated and the gene silenced by methylation of C_pG islands in the promoter. Other epigenetic changes are induced by, for example, the microenvironment in which tumor cells are proliferating.

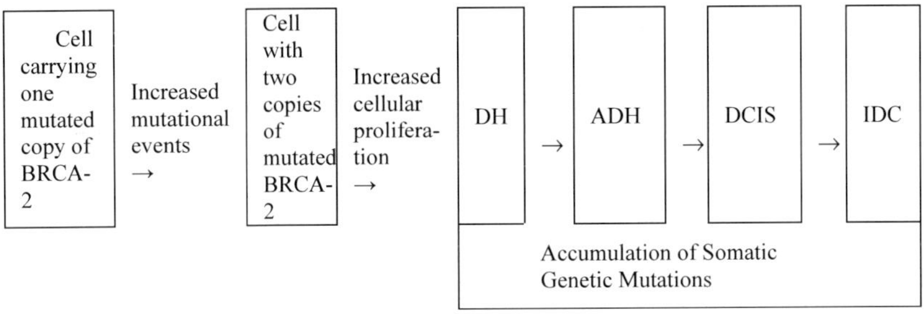

Figure 3. The cartoon demonstrates how breast neoplasia arises from inherited germ line mutations such as BRCA-1 and BRCA-2.

As mutational events accumulate in genes, proliferation increases as manifested by a high S-phase fraction and the genome may become unstable so that mitotic events do not progress normally resulting in different degrees of ploidy in cellular progeny. Cells may vary from diploid (2n) by becoming hypodiploid (<2n), hypotetraploid (>2n and <4n), tetraploid (4n) and hypertetraploid (>4n). In Caucasians specifically, and in breast cancer patients overall, the extent of cells in S-phase (flow cytometry) and variations of cells of breast cancers in DNA ploidy from diploid to aneuploid have been correlated with a poor prognosis (50-53). The S-phase fraction of breast cancers vary with race; the median S-phase fraction has been reported to be 6.9% for Caucasians, 8.3% for Hispanics and 8.6% for African-Americans (29). In this same study there was no statistically significant difference in ploidy among these three groups. When the prognostic usefulness of S-phase and ploidy were compared in breast cancers of Caucasian versus African-Americans, both S-phases and ploidy were prognostically important in Caucasians consistent with prior studies but neither S-phase nor ploidy of breast cancers was important prognostically in African-Americans (54) (The subject is also discussed in Chapter 8 in this book).

4. MOLECULAR MARKERS IN BREAST CARCINOMA

Some of the molecular alterations that occur in breast carcinomas have been explored as potential prognostic markers (i.e., indicators of overall or disease-free survival) or as predictive markers (i.e., indicators of potential responsiveness of a cancer to a particular therapy). Some of the more frequently investigated molecular markers and their known expression and significance in breast cancer in women of African descent are discussed subsequently.

4.1 Estrogen and Progesterone Receptors

Breast cancers can be divided broadly into two categories, those that are hormonally responsive and those that are not. Most hormonally responsive breast carcinomas express receptors to estrogen (ER) and approximately half of ER positive carcinomas also express receptors to progesterone (PR) (55-56). Expression of PR is induced by estrogens through their binding with ER; consequently, the presence of both receptors has been correlated with a functioning ER (57).

Many groups have demonstrated the prognostic usefulness of expression of ER and PR in breast cancers (3), (58-60). However, this prognostic usefulness of ER in IDC in women without positive lymph nodes is relatively modest in that the differences in 5 year relapse free survival and overall survival between the ER positive and ER negative groups is only 8-10% (61-62). Additionally, studies with longer follow-up suggest that the prognostic value of ER diminishes after follow-up of longer than 5 years (63-65). Expression of ER and PR correlates with other biological indicators of prognosis, such as histologic nuclear grade and rate of proliferation of cancer cells (i.e., ER and PR-positive carcinomas have a lower histologic grade as well as a lower rate of proliferation) (66-71).

ER expression is a stronger predictor of responsiveness to hormonal therapy than as an indictor of prognosis. In advanced disease, 50-60% of all ER-positive patients benefit from hormonal therapy, whereas, only 5 – 10% of ER-negative patients have a similar benefit (55-56), (72). Results of many studies of localized breast cancer, demonstrate a significant benefit of hormonal therapy, but only in ER-positive cancers (55-56), (72). As a marker of an intact, functional ER signaling pathway, PR expression also has been reported to predict responsiveness to hormonal therapy (71), (73).

Correlating with the reports that breast cancers in African American women are of higher grade is a higher incidence of breast cancers that are negative for receptors to ER and/or PR in African American women (29-30), (38), (43), (74-81). In a preliminary analysis of IDC in 46 African American and 151 Caucasian pre- and post-menopausal women, we also found a higher number of ER or PR negative cancers in African American women (ER negative = 63% of African American and 33% of Caucasian women ($p=0.0003$); PR negative = 76% of African American women and 50% of Caucasian women ($p=0.002$)) (Table 2).

A higher rate of ER or PR negativity also has been reported for breast cancers in American Indians and Hispanics (82-83); however, this finding is controversial for Hispanics (78). Many of the studies that also have controlled for socioeconomic status, stage and/or age have found these difference in the status of ER and PR to persist (75-78), (80), although others have found a higher incidence of ER and/or PR negative carcinomas only in younger age groups (33), (43). However, very few studies have found no differences in the expression of ER and/or PR in breast cancers between African American versus Caucasian women (24), (84-85).

4.2 Proliferation

The rate of proliferation of breast cancer cells is an accepted prognostic factor with a higher rate of proliferation correlating with a faster rate of tumor growth and a poorer disease free and overall survival (68), (86). Furthermore, the rate of proliferation can be indicative of responsiveness to endocrine therapy (better responsiveness in slowly proliferating cancers) or chemotherapy (better responsiveness in rapidly proliferating cancers) (87-88). There are a variety of methods for assessing the rate of proliferation, including assessing the number of mitotic figures present within a cancer, the proportion of cells in the S-phase of the cell cycle (i.e., the S-phase fraction), or the proportion of cells that express proteins associated with proliferating cells, including Ki-67/MIB-1 or proliferating cell nuclear antigen (PCNA). Some studies also have used the accumulation of bromodeoxyuridine (BrdU) in the DNA of neoplastic cells cultured immediately after their surgical removal.

Differences in the rate of proliferation of cells of breast cancers in women of African American descent versus Caucasians or other racial/ethnic groups have been assessed only occasionally. In most of these studies, S-phase fraction has been utilized as a measure of rate of proliferation and has been found to be significantly higher in breast cancers from African American than from Caucasian women (29), (54), (89). This also was reported for breast cancers from Hispanic women (29). Other investigators have found no significant difference in S-phase fraction in breast cancers from African American women (31) compared to Caucasians. In a preliminary analysis of IDC in 46 African American and 151 Caucasian pre- and post-menopausal women, we failed to detect a significant difference in rate of proliferation as assessed by the percentage of cancers expressing a high level of Ki-67 antigen (Table 2). However, a high level of Ki-67 expression was predictive of a poorer overall survival in African American, but not Caucasian women (Table 2).

Table 2. Comparison of Biomarker Expression in Carcinomas in African American and Caucasian Women.

	African American (n=46)	**Caucasian (n=151)**	**p-Value**
ER Positive*	37%	67%	**<0.001**
PR Positive*	24%	50%	**0.002**
High Ki-67**	62%	46%	0.067
High p27^{Kip1**}	47%	51%	0.618
p53 Positive*	30%	21%	0.196
High ErbB-2 (Membranous) **	30%	51%	**0.015**
High ErbB-2 (Cytoplasmic)**	37%	51%	0.096
High EGFR (Membranous)**	43%	55%	0.173
High EGFR (Cytoplasmic)**	39%	54%	0.072

*Positive is defined as ≥ 10% of cells staining.
** High is defined as an immunoscore ≥ the median immunoscore for that marker.

Most of the studies of proliferation and prognosis have not controlled for the age of the women. A higher rate of proliferation in invasive breast carcinomas in younger women has been documented in multiple studies using thymidine labeling indices (67), S-phase fraction (42), (68), (90-91) or the number of Ki-67 positive cells (92-93). In a study of women with breast carcinoma limited to individuals under 35 years of age, African American and Hispanic women were more likely to have a high S-phase fraction (>10%) compared with non-Hispanic white women (RR, 1.5;95% CI, 1.03-2.3 and RR2.4; 95%CI, 1.5-3.8, respectively) (33). However, after controlling for age at diagnosis, menopausal status and sociodemographic factors, Krieger et al. failed to detect a significant difference between these populations in rate of cellular proliferation as assessed by labeling for Ki-67 (85).

4.3 p27^{Kip-1}

The proliferative activity of cancers also can be assessed by evaluation of molecules involved in the regulation of the cell cycle. These include a variety of cyclins and cyclin dependent kinases and other molecules, such as p27^{Kip1}, that control progression of the cell through the cell cycle. p27^{Kip1} is a member of the Cip/Kip family of cyclin dependent kinase inhibitors which function to inhibit the transition from G1 to S phase of the cell cycle (94). The overexpression of $p27^{Kip1}$ blocks progression of cells through the G1 phase of the cell cycle (94). High levels of $p27^{Kip1}$ protein are expressed in normal human mammary epithelium. A low or absent nuclear expression of $p27^{Kip1}$ has been shown to be a clinical marker of disease that correlates directly with progression of several types of tumors, including breast cancer. Low $p27^{Kip1}$ has been associated with a worse relapse-free and overall survival in several univariate and multivariate analyses, most of which included proliferation rate and/or histologic grade in the multivariate model (95-98). In several other analyses, however, $p27^{Kip1}$ expression has not proven to be an independent predictor of survival (58), (90), (99)

Prior analyses of expression of $p27^{Kip1}$ has not included, to our knowledge, comparisons among races or ethnicities. In our preliminary analysis of infiltrating ductal breast carcinomas in 46 African American and 151 Caucasian pre- and post-menopausal women, we failed to detect a significant difference in $p27^{Kip1}$ expression in breast cancers between African American and Caucasian women (Tables 2 and 3) (100).

Table 3. Multivariate Analysis of Significant Predictors of Poor Overall Survival.

Marker	Adjusted Hazard Ratio	95% C.I.	P-Value
	Among Caucasian Women		
PR Negative	2.0	(1.09-4.17)	0.027
LOW $p27^{Kip1}$	2.5	(1.23-5.0)	0.001
Positive Lymph Nodes	1.9	(1.0-3.65)	0.045
	Among African American Women		
Positive Lymph Notes	2.9	(1.14-7.11)	0.024
High Rate of Proliferation	2.7	(1.02-7.37)	0.046

However, we did find that by multivariate analysis, a high level of $p27^{Kip1}$ expression was predictive of an improved over-all survival in Caucasian women, but not African American women. These differences in $p27^{Kip1}$ deserve further exploration and may explain in part the controversies as to the prognostic importance of $p27^{Kip-1}$.

4.4 Cyclin D1

Cyclin D1 is a 36 kD cell cycle associated nuclear protein and is encoded by the *CCND1* gene on chromosome 11q13 (101). It forms a complex with cyclin dependent kinases (cdk) 4 and 6. Following activation by cdk-activating kinases, the complex can then phosphorylate retinoblastoma protein (pRb), thus removing the block by Rb that prevents cells from entering into S-phase of the cell cycle (102-103). By in situ hybridization, an elevated level of mRNA for cyclin D1 was noted in only 18% of cases of atypical ductal hyperplasia but in 80% of cases of DCIS (11). This finding suggests that alterations in cyclin D1 may be a relatively early event in breast carcinogenesis. *CCND1* gene amplification was noted in 10-16% of carcinomas of the breast (104-106) and overexpression of cyclin D1 protein was found in 59% to 81% (11), (101), (107-109). It appears that more breast carcinomas demonstrate protein overexpression than gene amplification (102), (110-111). Possible explanations include post transcriptional and/or translational mechanisms as well as a reduction in the metabolic removal of the protein.

There have been multiple reports of a positive association between overexpression of cyclin D1 and ER status (99), (112,-113). However, reports of the prognostic usefulness of cyclin D1 for breast carcinomas are conflicting. Most studies have not found any significant relationship between cyclin D1 expression and patient outcome (107-108), (111), (113-116). In contrast, several authors have reported that women whose cancers demonstrated increased expression of cyclin D1 experienced a longer disease-free interval as well as increased overall survival when compared to those women with tumors with low or absent expression of cyclin D1 (117-118). Further, women with tumors demonstrating high expression of both cyclin D1 and ER were more likely to respond to tamoxifen than those lacking both proteins (117- 118).

Joe et al. showed that cyclin D1 overexpression tended to be more frequent in tumors from non-Caucasian patients including those of African-American background than their Caucasian counterparts (99). These authors suggested that differences in grade may explain this observation because there was a trend in the association between increased expression of cyclin D1 and tumor grade; in addition, the proportion of high grade tumors was significantly higher in non-Caucasian than in Caucasian patients.

4.5 Angiogenesis and VEGF

Angiogenesis is critical for all stages of development of breast cancers (119-120). Angiogenesis is often assessed in terms of microvessel density

(MVD) using an immunohistochemical stain against a specific endothelial marker such as antibodies against factor VIII related antigen, CD31, or CD34 (21-23), (110), (121-131). Most clinical studies, including those using multivariate analysis, have demonstrated that increased angiogenesis is an independent predictor of nodal and distant metastases as well as poor disease-free and overall survival (22-23), (110), (122-124). In addition, angiogenesis is useful in further stratifying other prognostic markers such as nodal metastases and ER status (123).

Angiogenesis results from complex interactions between many angiogenic peptides such as TGFα. Perhaps the best studied regulator of angiogenesis is vascular endothelial growth factor (VEGF). VEGF is a 34 to 42 kDa glycosylated heparin-binding glycoprotein that is produced by macrophages and endothelial, smooth muscle and tumor cells (132). VEGF acts selectively on endothelial cells by binding to specific class-III-receptor tyrosine kinases and by increasing intracellular calcium through calcium channel activation (133). This results in endothelial cell proliferation and an increase in vascular permeability (133). Using either immunohistochemistry or an immunoenzymatic assay, VEGF expression has been reported to be directly correlated with MVD in patients with primary breast carcinomas (126), (129), (134). There also is evidence that VEGF is an independent prognostic factor for both relapse-free and overall survival in multivariate analysis in Caucasian as well as African-American women (124), (128), (135-137). In a series of 125 women, predominantly of African-American as well as non-white Hispanic origin, with primary breast cancers, high VEGF plasma levels were associated with large tumor size and high tumor stage (136). These authors also noted that tamoxifen, along with chemotherapy, decreased levels of VEGF in patients with ER-positive cancers, but not in ER-negative cancers.

4.6 Growth Factors and Related Receptors: The Epidermal Growth Factor Receptor Family and Related Proto-oncogenes in Breast Cancer

4.6.1 Epidermal Growth Factor Receptor and its Ligands

Epidermal growth factor receptor (EGFr) is a 170 kDa molecule that spans the cytoplasmic membrane and that binds EGF, transforming growth factor alpha (TGFα), and other ligands such as amphiregulin. When ligands bind to the external domain of EGFr, the binding complex is autophosphorylated and the complex phosphorylates surrounding substrates sending signals via several signal transduction pathways; subsequently the ligand-receptor complex is internalized and in some cases EGFr is down-regulated (138-140). When internalized, the complex or components of the complex may move into the nucleus to provide additional regulatory signals. The internalization of EGFr following the binding of EGF is demonstrated in Figure 4.

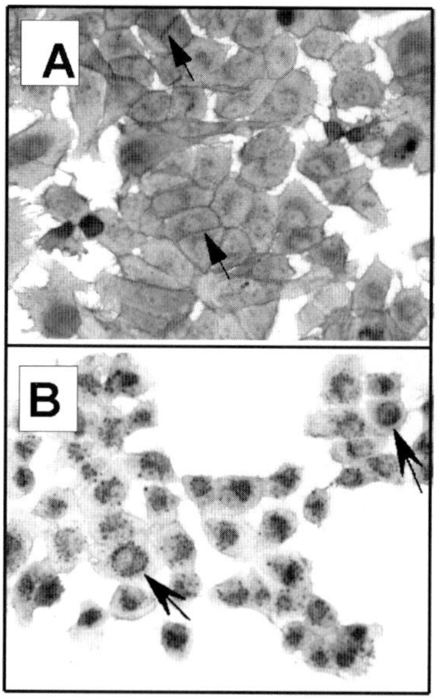

Figure 4. Panel A (X400 original magnification) demonstrates the phenotypic expression of EGF receptor in a lung cancer cell line (H228) grown on glass coverslips. Panel B (X400 original magnification) demonstrates the internalization of the EGF receptor in the same cells, 2 hours after the addition of TGFα at 10 ng/ml to the media.

Based upon these results, the pattern of EGFr expression depends upon the phenotypic expression of EGFr, of the ligands to EGFr and how these are able to interact via autocrine (cells and tissues) or paracrine (tissues) stimulation. EGFr also may be expressed in molecular forms that are consitutively activated.

Typically, the extent of EGFr phenotypic expression correlates inversely with the expression of ER in breast cancers and breast cancer cell lines and the strong phenotypic expression of EGFr usually occurs in breast cancers that do not express ER. In addition, both estrogens and androgens increase the expression of EGFr and TGFα (141-144).

In general, strong phenotypic expression of EGFr in IDC is correlated with a poor clinical outcome (123), (145-147). Our preliminary study indicates that the phenotypic expression of both cytoplasmic and membranous EGFr is slightly higher in Caucasian patients but this was not significant (Table 2). Other receptor-like molecules that are related to EGFr include $p185^{erbB-2}$ (HER2), $p180^{erbB-3}$ (HER3), and $p180^{erbB-4}$ (HER4).

4.6.2 $P185^{erbB-2}$ and Breast Cancer

No ligand that has been identified binds and activates the $p185^{erbB-2}$ protooncogene; yet amplification of the genes that code for this membrane bound protein in breast cancers is associated with a poor clinical outcome in patients with such breast cancers. Instead of being activated by a ligand, $p185^{erbB-2}$ likely is activated only after forming a heterodimer with another receptor molecule of the EGFr family including EGFr, $p180^{erbB-3}$, or $p180^{erbB-4}$. The dimer with EGFr is formed following the activation of EGFr by the binding of one of the EGFr ligands; the dimers with $p180^{erbB-3}$ or $p180^{erbB-4}$ are formed following their activation by the binding of one of their ligands, such as heregulin. In fact, because the internal component of $p180^{erbB-3}$ cannot be activated by phosphorylation and hence cannot produce a signal following the binding of heregulin to $p180^{erbB-3}$, the binding of heregulin with $p180^{erbB-3}$ is likely to induce a transduction signal only via heterodimerization of $p180^{erbB-3}$ with $p185^{erbB-2}$ and subsequent phosphorylated activation of the internal component of the $p185^{erbB-2}$ molecule.

The amplification of erbB-2 genes (i.e., increased copy numbers of erbB-2) in ductal adenocarcinomas of the breast, especially in node-positive cases, has been associated with a poor clinical outcome of patients with such tumors (148-151). Similarly, increased expression of $p185^{erbB-2}$ on cellular membranes is associated with constitutive phosphorylation of its tyrosine kinases and hence activation of $p185^{erbB-2}$. Also, stable transfection of neu, the activated rodent analogue of erbB-2 is transforming in cells and initiates tumors in transgenic mice.

When the phenotypic overexpression of $p185^{erbB-2}$ is evaluated using immunohistochemistry, several patterns of staining may be observed – 1) primarily on membranes, 2) primarily in the cytoplasm, and 3) a mixture of cytoplasmic and membranous immunostaining. These patterns are demonstrated in Figure 5. In general, only strong phenotypic expression of $p185^{erbB-2}$ at cytoplasmic membranes correlates with amplification of erbB-2 genes in that cell.

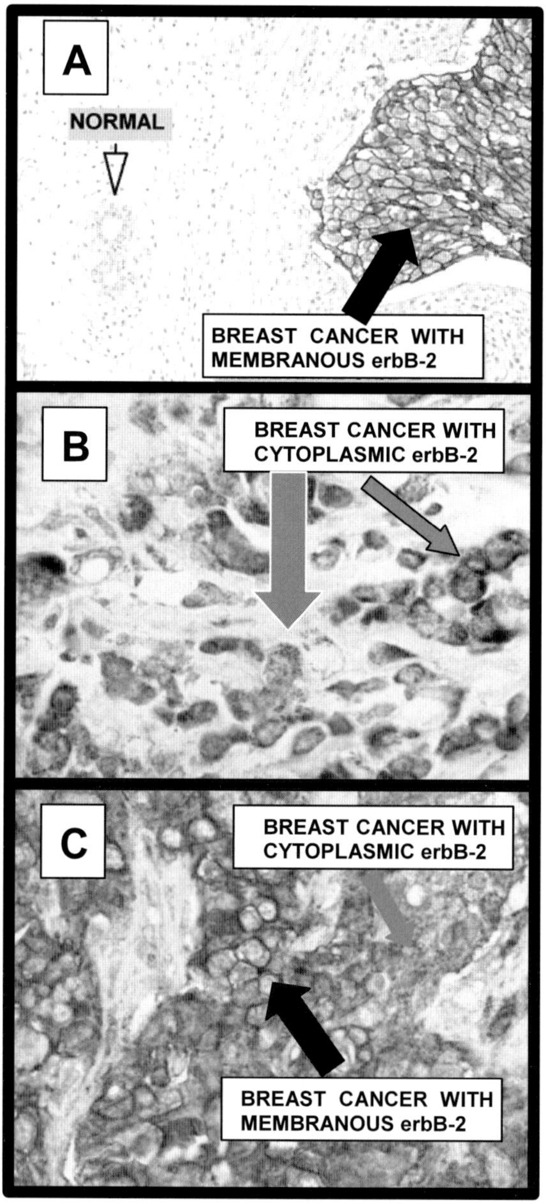

Figure 5. Panel A (original magnification X100) demonstrates a ductal adenocarcinoma of the breast with strong expression of p185^{erbB-2} on the cellular membranes (filled arrow) and very little expression on a normal duct of the breast (open arrow). Panel B (original magnification X400) demonstrates a ductal adenocarcinoma of the breast with primarily cytoplasmic expression of p185^{erbB-2}. Panel C (original magnification 400) demonstrates a ductal adenocarcinoma of the breast with focal areas of p185^{erbB-2} expression on membranes (black arrow) and other areas of primarily cytoplasmic expression of p185^{erbB-2} (gray arrow).

Because the overexpression of $p185^{erbB-2}$ in breast cancers has been associated with a poor clinical outcome, this molecule was an early target of immunotherapy. Antibodies to $p185^{erbB-2}$ have been demonstrated to inhibit the growth of cells expressing increased levels of $p185^{erbB-2}$; a recombinant humanized antibody to $p185^{erbB-2}$ has proved to be useful in therapy of selected breast cancers in which erbB-2 is amplified. This molecule, Herceptin, has been approved for therapy of breast cancer.

There have been several studies of racial differences in the expression of $p185^{erbB-2}$ in breast cancers. Elledge et al. (41) have reported no differences in phenotypic expression in breast cancers of $p185^{erbB-2}$ by Western blotting among Caucasians, African-Americans, and Hispanics (29). A review by Trock (28) concluded that the proportion of African-Americans over-expressing $p185^{erbB-2}$ was similar to the proportion of Caucasians over-expressing this molecular marker; however, our preliminary study found a significantly higher level of phenotypic expression of membranous $p185^{erbB-2}$ in Caucasians than in African-American women (Table 2). There also was a trend toward higher expression of cytoplasmic $p185^{erbB-2}$.

The extracellular component of the $p185^{erbB-2}$ receptor can be cleaved from the cell and can be detected in blood/plasma/serum along with any intact molecules of $p185^{erbB-2}$ that are released from damaged/dying cells. The cleaved fragment has been designated as $p105^{erbB-2}$. We have reported that $p105^{erbB-2}$ levels in serum correlates with the stage of prostatic adenocarcinomas (143) and in males without identified prostatic adenocarcinoma, that the concentration of $p105^{erbB-2}$ in serum is relatively constant over several months (152). In a study by Wu et al., plasma levels of erbB-2 protein were elevated when breast cancers were both PR positive plus greater than 5 cm in greatest dimension. In both Hispanics and African-Americans, plasma levels of erbB-2 were reported to be a strong predictor of overall survival, visceral metastases, and local recurrence (153).

4.6.3 $P180^{erbB-3}$ and $p180^{erbB-4}$ in Breast Cancers

One of the ligands to $p180^{erbB-3}$ as well as $p180^{erbB-4}$ is heregulin, the human homologue to the neu differentiation factor (NDF), a 44kD molecule found in rats that can stimulate the proliferation, differentiation and migration of breast cells. In contrast, the roles for erbB-3 and erbB-4 in breast carcinogenesis are still under study.

4.7 Apoptosis and Tumor Suppressor Genes in Breast Cancer

Apoptosis or programmed cell death is hypothesized to be an important process in tumor progression because reduced apoptosis permits mutational events to accumulate in proliferating cells and neoplastic cells to accumulate in a tumor without a balance to their proliferation. Several pathways have been identified which induce apoptosis including the binding of interleukin 3 to its receptor, the interaction of the receptor, Fas, with the Fas ligand, the binding of tumor necrosis factor to the tumor necrosis factor-related apoptosis-inducing ligand (TRAIL) and the toxic action of Granzyme B released by toxic lymphocytes. TRAIL includes the "death" receptors, DR5 and DR4, as well as decoy receptors. DR5 and DR4 are expressed on the surface of most breast cancer cells.

In the process of initiating apoptosis, a complex signal transduction pathway becomes activated and ultimately the output of this pathway initiates the endonuclease mediated cleavage of DNA at multiple sites, ultimately producing fragments of DNA consisting of 180 base pairs. This uniform pattern of DNA cleavage can be detected as a ladder effect on DNA separations using gels.

Apoptosis results in characteristic nuclear patterns in which the cell shrinks, chromatin condenses and the nuclear structure becomes very irregular and ultimately breaks up. The early aspects of apoptosis can be detected by several assays including the TUNEL assay.

The TUNEL technique is used to demonstrate the early breakdown of DNA during apoptosis; this breakdown results in the formation of numerous free 3′-OH ends in each DNA molecule. The TUNEL assay is based upon the detection of the free 3′-OH ends of double or single stranded DNA. This is achieved by the addition of digoxigenin labeled nucleotides to the 3′-OH ends of DNA using the enzyme terminal deoxyribonucleotide transferase. The incorporated digoxygenin is recognized by an antibody coupled to horseradish peroxidase. This technique works well in fresh or frozen tissue and on formalin fixed paraffin embedded tissue.

4.7.1 p53

The p53 gene is a suppressor gene which is involved primarily in the control of apoptosis, in regulating checkpoints in the cell cycle, and in modulating DNA repair; it is one of the most frequently somatically mutated genes in human cancers (154) and in some cancers, mutations in p53 have been demonstrated to be useful in identifying aggressive subsets of tumors. We have reported that mutations in p53 in proximal colorectal adenocarcinomas (CRC) results in a poor clinical outcome of Caucasian patients with such tumors, but not African-Americans with p53 mutations in either stage and grade matched proximal or distal CRCs (155-156) (Manne et al. unpublished). Similarly, in larger studies of ductal adenocarcinomas using immunohistochemistry of mostly Caucasian patients, p53 as determined by immunohistochemistry, has been shown to be useful prognostically (157-161).

Shiao et al. (162) reported that the mutational pattern of p53 was different in breast adenocarcinomas between African-Americans (only 3 (6.7%) non-silent missense point mutations) and Caucasians (7 (15.5%) non-silent missense point mutations) and that African-Americans but not Caucasians had a poor prognosis associated with specific mutations of p53 in their breast cancers. In a companion study from this same laboratory (163) using immunohistochemistry, nuclear accumulation of p53 (detected by immunohistochemistry) in Caucasians but not African-Americans correlated with a poor prognosis.

Krieger et al. (85) using an immunohistochemical analysis and a cut-of-point of 5% reported p53 positive tumors in Caucasians (19.1%), African-Americans (16.7%) and Asians (28.6%). They found no statistically significant differences in p53 positivity based on immunostaining in any of these three racial groups. Also, Elledge et al. (29) reported no differences in the proportion of breast cancers with nuclear accumulation of p53. In contrast to Shiao et al. (162), Blaszyk et al. (164) reported an atypical excess of A:T to G:C transitions in p53 mutational events in African-American patients. Many of these studies with regard to racial differences in p53 mutational patterns and their correlation with prognosis are contradictory and these issues must be evaluated in larger, more controlled studies (also, please, see the contribution of F.O. Ikpatt and O.I. Olopade – Chapter 2 - in this book).

4.7.2 Bcl-2

Bcl-2 is a molecule sometimes classified as a protooncogene that inhibits apoptosis by interaction in the signal transduction pathway to inhibit the effects of Bax and of cytochrome C release from mitochondria, and the activation of procaspase 9 to caspase 9 via the interaction of Bax and cytochrome-C.

In addition, Bcl-2 has other cellular functions; for example, it can affect cellular levels of calcium and can modulate the proliferation of cells (165-166). Most importantly, it can interact with numerous cellular proteins such as the p53-binding protein designated Bp2.

Because Bcl-2 inhibits apoptosis, its expression at high levels would be expected to increase the aggressiveness of tumors. This is the case for follicular B cell lymphomas where increased levels of Bcl-2 are associated with a t (14:18) chromosomal transposition that causes Bcl-2 to be expressed highly and may block the apoptosis associated with high levels of Myc (167). High expression of Bcl-2 also correlates with aggressiveness of transitional cell carcinomas of the bladder, medullary carcinomas of the thyroid and adenocarcinomas of the prostate (168). However, in colorectal cancer (155), (169-171), and in breast cancer, the detection of increased levels of Bcl-2 by immunohistochemistry is associated with a good prognosis (172-174). The phenotypic expression of Bcl-2 has been correlated to ER positivity (175) in breast cancers.

In our preliminary study of expression of Bcl-2 in adenocarcinomas of the breast (Figure1, Tables 2 and 3) we noted that in general, the phenotypic expression of Bcl-2 seemed to be lower in African-American patients than in Caucasian patients. However, the tendency of $p53^+$ tumors (those with nuclear accumulation in 10% or more of the tumor cells) to have a lower phenotypic expression of Bcl-2 than $p53^-$ tumors was the same in both groups. Specifically for African-American patients, 16 of 33 (48%) $p53^-$ tumors had phenotypic expression of Bcl-2 of >1.0 while only 1 of 13 (8%) $p53^+$ tumors had a Bcl-2 value of >1.0. For Caucasians, 66 of 118 (56%) $p53^-$ tumors had values of Bcl-2 of 1.0 while 12 of 32 (38%) $p53^+$ tumors had values of Bcl-2 of >1.0. Racial differences in the prognostic usefulness of Bcl-2 have not been reported in adenocarcinomas of the breast to our knowledge.

4.7.3 Bax

Bax increases apoptosis via its interaction with Bcl-2 with which it can form heterodimers in the signal transduction pathway of apoptosis. In fact, it may be the Bax/Bcl-2 ratio that is important in initiating apoptosis. Other actions may be important; for example, apoptosis inducing events such as γ–radiation tend to increase levels of Bax in part because they increase the expression of p53 which stimulates the bax gene via binding to the regulatory components of the Bax pathway. Decreased phenotypic expression of Bax in the cells of breast ductal adenocarcinomas predicts a poor response to combination chemotherapy and a poor clinical outcome (176-177). We are unaware of any racial differences in phenotypic expression of Bax.

5. SUMMARY

In summary, several molecular features have been observed to change as neoplasia of the breast develops and pre-invasive neoplastic lesions progress to IDC and to metastatic ductal adenocarcinoma. Some of these molecular features such as the phenotypic expression of p185^{erbB-2}, ER, EGFr, Myc, and Bcl-2 and the nuclear accumulation of p53 have been associated directly or inversely with the aggressiveness of IDC, the responsiveness of IDC to specific therapies, and the clinical outcome of patients with tumors expressing these molecular features. Similarly, tumor characteristics such as cytomorphometric parameters (e.g., nuclear grade) and stage of the tumor and demographic parameters including age and race affect clinical outcome. Racial differences include the following: 1) ductal adenocarcinomas of African-Americans tend to be ER$^-$, to have a higher nuclear grade, to occur at younger ages, and to have a poorer clinical outcome. Also, the importance of prognostic parameters may vary among racial groups.

REFERENCES

(1) Radford, D.M., Phillips, N.J., Fair, K.L., Ritter, J.H., Holt, M., and Donis-Keller, H. (1995). Allelic loss and the progression of breast cancer.(erratum appears in Cancer Res 1996 Feb 15;56(4):935). Cancer Research. 55, 5180-5183.

(2) Brenner, A.J., and Aldaz, C.M. (1997). The genetics of sporadic breast cancer. In Etology of breast and gynecological cancers, C.M. Aldaz, M.N. Gould, J. Mclauchlan and J. Slaga, eds. (New York, NY: Wiley-Liss), pp. 63-82.

(3) Jensen, R.A., Dupont, W.D., and Page, D.L. (1993). Diagnostic criteria and cancer risk of proliferative breast lesions. Journal of Cellular Biochemistry - Supplement. 17G, 59-64.

(4) Marshall, L.M., Hunter, D.J., Connolly, J.L., Schnitt, S.J., Byrne, C., London, S.J., and Colditz, G.A. (1997). Risk of breast cancer associated with atypical hyperplasia of lobular and ductal types. Cancer Epidemiology, Biomarkers & Prevention. 6, 297-301.

(5) Couch, F.J., and Weber, B.L. (1998). Breast cancer. In The genetic basis of human cancer, B. Vogelstein and K.W. Kinzler, eds. (New York, NY: McGraw Hill), pp. 537-563.

(6) Devilee, P., and Cornelisse, C.J. (1994). Somatic genetic changes in human breast cancer. Biochimica et Biophysica Acta. 1198, 113-130.

(7) Kuukasjarvi, T., Karhu, R., Tanner, M., Kahkonen, M., Schaffer, A., Nupponen, N., Pennanen, S., Kallioniemi, A., Kallioniemi, O.P., and Isola, J. (1997). Genetic heterogeneity and clonal evolution underlying development of asynchronous metastasis in human breast cancer. Cancer Research. 57, 1597-1604.

(8) Kuukasjarvi, T., Tanner, M., Pennanen, S., Karhu, R., Kallioniemi, O.P., and Isola, J. (1997). Genetic changes in intraductal breast cancer detected by comparative genomic hybridization. American Journal of Pathology. 150, 1465-1471.

(9) Nishizaki, T., DeVries, S., Chew, K., Goodson, W.H., 3rd, Ljung, B.M., Thor, A., and Waldman, F.M. (1997). Genetic alterations in primary breast cancers and their

metastases: direct comparison using modified comparative genomic hybridization. Genes, Chromosomes & Cancer. 19, 267-272.
(10) Tsuda, H., Fukutomi, T., and Hirohashi, S. (1995). Pattern of gene alterations in intraductal breast neoplasms associated with histological type and grade. Clinical Cancer Research. 1, 261-267.
(11) Weinstat-Saslow, D., Merino, M.J., Manrow, R.E., Lawrence, J.A., Bluth, R.F., Wittenbel, K.D., Simpson, J.F., Page, D.L., and Steeg, P.S. (1995). Overexpression of cyclin D mRNA distinguishes invasive and in situ breast carcinomas from non-malignant lesions.(comment). Nature Medicine. 1, 1257-1260.
(12) Beckmann, M.W., Niederacher, D., Schnurch, H.G., Gusterson, B.A., and Bender, H.G. (1997). Multistep carcinogenesis of breast cancer and tumour heterogeneity. Journal of Molecular Medicine. 75, 429-439.
(13) Davidoff, A.M., Kerns, B.J., Iglehart, J.D., and Marks, J.R. (1991). Maintenance of p53 alterations throughout breast cancer progression. Cancer Research. 51, 2605-2610.
(14) Zhang, G.J., Kimijima, I., Abe, R., Kanno, M., Katagata, N., Hara, K., Watanabe, T., and Tsuchiya, A. (1997). Correlation between the expression of apoptosis-related bcl-2 and p53 oncoproteins and the carcinogenesis and progression of breast carcinomas. Clinical Cancer Research. 3, 2329-2335.
(15) Orr, J. (1958). The mechanisms of chemical carcinogenesis. Br Med Bull 14, 99-101.
(16) Orr, J., and Spencer, A. (1972). Transplantation studies of the role of the stroma in epidermal carcinogenesis. In Tissue interactions in carcinogenesis, T. D, ed. (London: Academic Press), pp. 291-304.
(17) Brues, A., and CJ, D. (1955). Radiation as a carcinogenic agent. Radiat Res 3, 272-281.
(18) Dawe, C., Morgan, W., and Slatic, M. (1996). Influence of epithelio-mesenchymal interactions of tumor induction by polyoma virus. Int J Cancer 1, 419-450.
(19) Sieweke, M.H., and Bissell, M.J. (1994). The tumor-promoting effect of wounding: a possible role for TGF-beta-induced stromal alterations. Crit Rev Oncog 5, 297-311.
(20) Walker, R.A. (2001). The complexities of breast cancer desmoplasia. Breast Cancer Res 3, 143-145.
(21) Weidner, N. (1998). Tumoural vascularity as a prognostic factor in cancer patients: the evidence continues to grow. Journal of Pathology 184, 119-122.
(22) Weidner, N., Semple, J.P., Welch, W.R., and Folkman, J. (1991). Tumor angiogenesis and metastasis--correlation in invasive breast carcinoma. New England Journal of Medicine 324, 1-8.
(23) Weidner, N., Folkman, J., Pozza, F., Bevilacqua, P., Allred, E.N., Moore, D.H., Meli, S., and Gasparini, G. (1992). Tumor angiogenesis: a new significant and independent prognostic indicator in early-stage breast carcinoma. Journal of the National Cancer Institute 84, 1875-1887.
(24) Elmore, J.G., Moceri, V.M., Carter, D., and Larson, E.B. (1998). Breast carcinoma tumor characteristics in black and white women. Cancer 83, 2509-2515.
(25) Newman, L.A., and Alfonso, A.E. (1997). Age-related differences in breast cancer stage at diagnosis between black and white patients in an urban community hospital. Ann Surg Oncol 4, 655-662.
(26) Kovi, J., Mohla, S., Norris, H.J., Sampson, C.C., and Heshmat, M.Y. (1989). Breast lesions in black women. Pathol Annu 24 Pt 1, 199-218.
(27) Perkins, P., Cooksley, C.D., and Cox, J.D. (1996). Breast cancer. Is ethnicity an independent prognostic factor for survival? Cancer 78, 1241-1247.

(28) Trock, B.J. (1996). Breast cancer in African American women: epidemiology and tumor biology. Breast Cancer Res Treat 40, 11-24.
(29) Elledge, R.M., Clark, G.M., Chamness, G.C., and Osborne, C.K. (1994). Tumor biologic factors and breast cancer prognosis among white, Hispanic, and black women in the United States. J Natl Cancer Inst 86, 705-712.
(30) Simon, M.S., and Severson, R.K. (1997). Racial differences in breast cancer survival: the interaction of socioeconomic status and tumor biology. Am J Obstet Gynecol 176, S233-239.
(31) Weiss, S.E., Tartter, P.I., Ahmed, S., Brower, S.T., Brusco, C., Bossolt, K., Amberson, J.B., and Bratton, J. (1995). Ethnic differences in risk and prognostic factors for breast cancer. Cancer 76, 268-274.
(32) Wojcik, B.E., Spinks, M.K., and Optenberg, S.A. (1998). Breast carcinoma survival analysis for African American and white women in an equal-access health care system. Cancer 82, 1310-1318.
(33) Shavers, V.L., Harlan, L.C., and Stevens, J.L. (2003). Racial/ethnic variation in clinical presentation, treatment, and survival among breast cancer patients under age 35. Cancer 97, 134-147.
(34) Aziz, H., Hussain, F., Sohn, C., Mediavillo, R., Saitta, A., Hussain, A., Brandys, M., Homel, P., and Rotman, M. (1999). Early onset of breast carcinoma in African American women with poor prognostic factors. Am J Clin Oncol 22, 436-440.
(35) Ownby, H.E., Frederick, J., Russo, J., Brooks, S.C., Swanson, G.M., Heppner, G.H., and Brennan, M.J. (1985). Racial differences in breast cancer patients. J Natl Cancer Inst 75, 55-60.
(36) Hunter, C.P., Redmond, C.K., Chen, V.W., Austin, D.F., Greenberg, R.S., Correa, P., Muss, H.B., Forman, M.R., Wesley, M.N., Blacklow, R.S., and et al. (1993). Breast cancer: factors associated with stage at diagnosis in black and white women. Black/White Cancer Survival Study Group. J Natl Cancer Inst 85, 1129-1137.
(37) Eley, J.W., Hill, H.A., Chen, V.W., Austin, D.F., Wesley, M.N., Muss, H.B., Greenberg, R.S., Coates, R.J., Correa, P., Redmond, C.K., and et al. (1994). Racial differences in survival from breast cancer. Results of the National Cancer Institute Black/White Cancer Survival Study. Jama 272, 947-954.
(38) Chen, V.W., Correa, P., Kurman, R.J., Wu, X.C., Eley, J.W., Austin, D., Muss, H., Hunter, C.P., Redmond, C., Sobhan, M., and et al. (1994). Histological characteristics of breast carcinoma in blacks and whites. Cancer Epidemiol Biomarkers Prev 3, 127-135.
(39) Mohla, S., Sampson, C.C., Khan, T., Enterline, J.P., Leffall, L., Jr., and White, J.E. (1982). Estrogen and progesterone receptors in breast cancer in Black Americans: Correlation of receptor data with tumor differentiation. Cancer 50, 552-559.
(40) Albain, K., Green, S., LeBlanc, M., Rivkin, S., O'Sullivan, J., and Osborne, C. (1992). Proportional hazards and recurrence partitioning and amalgamation analyses of the Southwest Oncology Group node-positive adjuvant CMFVP breast cancer database: a pilot study. Breast Cancer Res Treat 22, 273-284.
(41) Elledge, R.M., McGuire, W.L., and Osborne, C.K. (1992). Prognostic factors in breast cancer. Seminars in Oncology 19, 244-253.
(42) Albain, K.S., Allred, D.C., and Clark, G.M. (1994). Breast cancer outcome and predictors of outcome: are there age differentials? Journal of the National Cancer Institute. Monographs, 35-42.
(43) Furberg, H., Millikan, R., Dressler, L., Newman, B., and Geradts, J. (2001). Tumor characteristics in African American and white women. Breast Cancer Res Treat 68, 33-43.

(44) Kimmick, G., Muss, H.B., Case, L.D., and Stanley, V. (1991). A comparison of treatment outcomes for black patients and white patients with metastatic breast cancer. The Piedmont Oncology Association experience. Cancer 67, 2850-2854.

(45) Pierce, L., Fowble, B., Solin, L.J., Schultz, D.J., Rosser, C., and Goodman, R.L. (1992). Conservative surgery and radiation therapy in black women with early stage breast cancer. Patterns of failure and analysis of outcome. Cancer 69, 2831-2841.

(46) Miller, B.A., Hankey, B.F., and Thomas, T.L. (2002). Impact of sociodemographic factors, hormone receptor status, and tumor grade on ethnic differences in tumor stage and size for breast cancer in US women. Am J Epidemiol 155, 534-545.

(47) Dickson, R.B., and Lippman, M.E. (2001). Cancer of the Breast. In Cancer Principles & Practice of Oncology, Volume 2, 6th Edition, V.T. DeVita, S. Hellman and S. Rosenberg, eds. (Philadelphia: Lippincott Williams & Wilkins), pp. 1633-1651.

(48) Dickson, R.B., and Lippman, M.E. (2001). Molecular Basis of Breast Cancer in The Molecular Basis of Cancer, 2nd Edition (Philadelphia: W.B. Saunders Company).

(49) Olopade, O.I., Fackenthal, J.D., Dunston, G., Tainsky, M.A., Collins, F., and Whitfield-Broome, C. (2003). Breast cancer genetics in African Americans. Cancer 97, 236-245.

(50) Clark, G.M., Dressler, L.G., Owens, M.A., Pounds, G., Oldaker, T., and McGuire, W.L. (1989). Prediction of relapse or survival in patients with node-negative breast cancer by DNA flow cytometry. N Engl J Med 320, 627-633.

(51) Toikkanen, S., Joensuu, H., and Klemi, P. (1989). The prognostic significance of nuclear DNA content in invasive breast cancer--a study with long-term follow-up. Br J Cancer 60, 693-700.

(52) Beerman, H., Kluin, P.M., Hermans, J., van de Velde, C.J., and Cornelisse, C.J. (1990). Prognostic significance of DNA-ploidy in a series of 690 primary breast cancer patients. Int J Cancer 45, 34-39.

(53) Witzig, T.E., Ingle, J.N., Cha, S.S., Schaid, D.J., Tabery, R.L., Wold, L.E., Grant, C., Gonchoroff, N.J., and Katzmann, J.A. (1994). DNA ploidy and the percentage of cells in S-phase as prognostic factors for women with lymph node negative breast cancer. Cancer 74, 1752-1761.

(54) Shiao, Y.H., Chen, V.W., Lehmann, H.P., Wu, X.C., and Correa, P. (1997). Patterns of DNA ploidy and S-phase fraction associated with breast cancer survival in blacks and whites. Clin Cancer Res 3, 587-592.

(55) Multiple Authors. (1992). Systemic treatment of early breast cancer by hormonal, cytotoxic, or immune therapy. 133 randomised trials involving 31,000 recurrences and 24,000 deaths among 75,000 women. Early Breast Cancer Trialists' Collaborative Group. Lancet 339, 1-15.

(56) Multiple Authors (1992). Systemic treatment of early breast cancer by hormonal, cytotoxic, or immune therapy. 133 randomised trials involving 31,000 recurrences and 24,000 deaths among 75,000 women. Early Breast Cancer Trialists' Collaborative Group. Lancet 339, 71-85.

(57) Horwitz, K.B., and McGuire, W.L. (1978). Estrogen control of progesterone receptor in human breast cancer. Correlation with nuclear processing of estrogen receptor. J Biol Chem 253, 2223-2228.

(58) Fisher, B., Redmond, C., Fisher, E.R., and Caplan, R. (1988). Relative worth of estrogen or progesterone receptor and pathologic characteristics of differentiation as indicators of prognosis in node negative breast cancer patients: findings from National Surgical Adjuvant Breast and Bowel Project Protocol B-06. J Clin Oncol 6, 1076-1087.

(59) Clark, G.M., Sledge, G.W., Jr., Osborne, C.K., and McGuire, W.L. (1987). Survival from first recurrence: relative importance of prognostic factors in 1,015 breast cancer patients. J Clin Oncol 5, 55-61.

(60) Pertschuk, L.P., and Axiotis, C.A. (1999). Steroid Hormone Receptor Immunohistochemistry in Breast Cancer: Past, Present, and Future. Breast J 5, 3-12.

(61) Alexieva-Figusch, J., Van Putten, W.L., Blankenstein, M.A., Blonk-Van Der Wijst, J., and Klijn, J.G. (1988). The prognostic value and relationships of patient characteristics, estrogen and progestin receptors, and site of relapse in primary breast cancer. Cancer 61, 758-768.

(62) Foekens, J.A., Portengen, H., van Putten, W.L., Peters, H.A., Krijnen, H.L., Alexieva-Figusch, J., and Klijn, J.G. (1989). Prognostic value of estrogen and progesterone receptors measured by enzyme immunoassays in human breast tumor cytosols. Cancer Res 49, 5823-5828.

(63) Aamdal, S., Bormer, O., Jorgensen, O., Host, H., Eliassen, G., Kaalhus, O., and Pihl, A. (1984). Estrogen receptors and long-term prognosis in breast cancer. Cancer 53, 2525-2529.

(64) Hahnel, R., Woodings, T., and Vivian, A.B. (1979). Prognostic value of estrogen receptors in primary breast cancer. Cancer 44, 671-675.

(65) Hilsenbeck, S.G., Ravdin, P.M., de Moor, C.A., Chamness, G.C., Osborne, C.K., and Clark, G.M. (1998). Time-dependence of hazard ratios for prognostic factors in primary breast cancer. Breast Cancer Res Treat 52, 227-237.

(66) Meyer, J.S., Friedman, E., McCrate, M.M., and Bauer, W.C. (1983). Prediction of early course of breast carcinoma by thymidine labeling. Cancer 51, 1879-1886.

(67) Meyer, J.S., Prey, M.U., Babcock, D.S., and McDivitt, R.W. (1986). Breast carcinoma cell kinetics, morphology, stage, and host characteristics. A thymidine labeling study. Lab Invest 54, 41-51.

(68) Meyer, J.S., and Province, M.A. (1994). S-phase fraction and nuclear size in long term prognosis of patients with breast cancer. Cancer 74, 2287-2299.

(69) Gaffney, E.V., 2nd, Venz-Williamson, T.L., Hutchinson, G., Biggs, P.J., and Nelson, K.M. (1996). Relationship of standardized mitotic indices to other prognostic factors in breast cancer. Arch Pathol Lab Med 120, 473-477.

(70) Helin, M.L., Helle, M.J., Helin, H.J., and Isola, J.J. (1989). Proliferative activity and steroid receptors determined by immunohistochemistry in adjacent frozen sections of 102 breast carcinomas. Arch Pathol Lab Med 113, 854-857.

(71) Osborne, C.K. (1998). Steroid hormone receptors in breast cancer management. Breast Cancer Res Treat 51, 227 238.

(72) Schiff, R., and Suqua, S. (2002). The importance of estrogen receptor in breast cancer. In Breast Cancer: prognosis, treatment and prevention, J.R. Pasqualini, ed. (New York: Marcel Dekker, Inc.), pp. 149-173.

(73) Ravdin, P.M., Green, S., Dorr, T.M., McGuire, W.L., Fabian, C., Pugh, R.P., Carter, R.D., Rivkin, S.E., Borst, J.R., Belt, R.J., and et al. (1992). Prognostic significance of progesterone receptor levels in estrogen receptor-positive patients with metastatic breast cancer treated with tamoxifen: results of a prospective Southwest Oncology Group study. J Clin Oncol 10, 1284-1291.

(74) Rose, D.P., and Royak-Schaler, R. (2001). Tumor biology and prognosis in black breast cancer patients: a review. Cancer Detect Prev 25, 16-31.

(75) Natarajan, N., Nemoto, T., Mettlin, C., and Murphy, G.P. (1985). Race-related differences in breast cancer patients. Results of the 1982 national survey of breast cancer by the American College of Surgeons. Cancer 56, 1704-1709.

(76) Nemoto, T., Vana, J., Natarajan, N., Bedwani, R., and Mettlin, C. (1981). Observations on short-term and long-term surveys of breast cancer by the American College of Surgeons. I. Significance of the number of axillary nodes and II. Estrogen receptor assay in the U.S. in 1977. Int Adv Surg Oncol 4, 209-239.
(77) Stanford, J.L., Szklo, M., Boring, C.C., Brinton, L.A., Diamond, E.A., Greenberg, R.S., and Hoover, R.N. (1987). A case-control study of breast cancer stratified by estrogen receptor status. Am J Epidemiol 125, 184-194.
(78) Gapstur, S.M., Dupuis, J., Gann, P., Collila, S., and Winchester, D.P. (1996). Hormone receptor status of breast tumors in black, Hispanic, and non-Hispanic white women. An analysis of 13,239 cases. Cancer 77, 1465-1471.
(79) Joslyn, S.A., and West, M.M. (2000). Racial differences in breast carcinoma survival. Cancer 88, 114-123.
(80) Joslyn, S.A. (2002). Hormone receptors in breast cancer: racial differences in distribution and survival. Breast Cancer Res Treat 73, 45-59.
(81) Chu, K.C., and Anderson, W.F. (2002). Rates for breast cancer characteristics by estrogen and progesterone receptor status in the major racial/ethnic groups. Breast Cancer Res Treat 74, 199-211.
(82) Li, C.I., Malone, K.E., and Daling, J.R. (2002). Differences in breast cancer hormone receptor status and histology by race and ethnicity among women 50 years of age and older. Cancer Epidemiol Biomarkers Prev 11, 601-607.
(83) Li, C.I., Malone, K.E., and Daling, J.R. (2003). Differences in breast cancer stage, treatment, and survival by race and ethnicity. Arch Intern Med 163, 49-56.
(84) Ansell, D., Whitman, S., Lipton, R., and Cooper, R. (1993). Race, income, and survival from breast cancer at two public hospitals. Cancer 72, 2974-2978.
(85) Krieger, N., van den Eeden, S.K., Zava, D., and Okamoto, A. (1997). Race/ethnicity, social class, and prevalence of breast cancer prognostic biomarkers: a study of white, black, and Asian women in the San Francisco bay area. Ethn Dis 7, 137-149.
(86) Silvestrini, R., Daidone, M.G., Luisi, A., Boracchi, P., Mezzetti, M., Di Fronzo, G., Andreola, S., Salvadori, B., and Veronesi, U. (1995). Biologic and clinicopathologic factors as indicators of specific relapse types in node-negative breast cancer. J Clin Oncol 13, 697-704.
(87) Muss, H.B. (1992). Endocrine therapy for advanced breast cancer: a review. Breast Cancer Res Treat 21, 15-26.
(88) Silvestrini, R., and Daidone, M.G. (1993). Review of proliferative variables and their predictive value. Recent Results Cancer Res 127, 71-76.
(89) Wenger, C.R., Beardslee, S., Owens, M.A., Pounds, G., Oldaker, T., Vendely, P., Pandian, M.R., Harrington, D., Clark, G.M., and McGuire, W.L. (1993). DNA ploidy, S-phase, and steroid receptors in more than 127,000 breast cancer patients. Breast Cancer Res Treat 28, 9-20.
(90) Lyman, G.H., Lyman, S., Balducci, L., Kuderer, N., Reintgen, D., Cox, C., Baekey, P., Greenberg, H., and Horton, J. (1996). Age and the Risk of Breast Cancer Recurrence. Cancer Control 3, 421-427.
(91) Remvikos, Y., Gerbault-Seureau, M., Magdelenat, H., Prieur, M., and Dutrillaux, B. (1992). Proliferative activity of breast cancers increases in the course of genetic evolution as defined by cytogenetic analysis. Breast Cancer Res Treat 23, 43-49.
(92) Veronesi, U., Marubini, E., Del Vecchio, M., Manzari, A., Andreola, S., Greco, M., Luini, A., Merson, M., Saccozzi, R., Rilke, F., and et al. (1995). Local recurrences and distant metastases after conservative breast cancer treatments: partly independent events. J Natl Cancer Inst 87, 19-27.

(93) Tynninen, O., von Boguslawski, K., Aronen, H.J., and Paavonen, T. (1999). Prognostic value of vascular density and cell proliferation in breast cancer patients. Pathol Res Pract 195, 31-37.

(94) Stillman, B. (1996). Cell cycle control of DNA replication. Science 274, 1659-1664.

(95) Catzavelos, C., Bhattacharya, N., Ung, Y.C., Wilson, J.A., Roncari, L., Sandhu, C., Shaw, P., Yeger, H., Morava-Protzner, I., Kapusta, L., Franssen, E., Pritchard, K.I., and Slingerland, J.M. (1997). Decreased levels of the cell-cycle inhibitor p27Kip1 protein: prognostic implications in primary breast cancer. Nat Med 3, 227-230.

(96) Chu, J.S., Huang, C.S., and Chang, K.J. (1999). p27 expression as a prognostic factor of breast cancer in Taiwan. Cancer Lett 141, 123-130.

(97) .Porter, P.L., Malone, K.E., Heagerty, P.J., Alexander, G.M., Gatti, L.A., Firpo, E.J., Daling, J.R., and Roberts, J.M. (1997). Expression of cell-cycle regulators p27Kip1 and cyclin E, alone and in combination, correlate with survival in young breast cancer patients. Nat Med 3, 222-225.

(98) Tan, P., Cady, B., Wanner, M., Worland, P., Cukor, B., Magi-Galluzzi, C., Lavin, P., Draetta, G., Pagano, M., and Loda, M. (1997). The cell cycle inhibitor p27 is an independent prognostic marker in small (T1a,b) invasive breast carcinomas. Cancer Res 57, 1259-1263.

(99) Joe, A.K., Arber, N., Bose, S., Heitjan, D., Zhang, Y., Weinstein, I.B., and Hibshoosh, H. (2001). Cyclin D1 overexpression is more prevalent in non-Caucasian breast cancer. Anticancer Research. 21, 3535-3539.

(100) Talley, L.I., Grizzle, W.E., Waterbor, J.W., Brown, D., Weiss, H., and Frost, A.R. (2002). Hormone receptors and proliferation in breast carcinomas of equivalent histologic grades in pre- and postmenopausal women. Int J Cancer 98, 118-127.

(101) Bartkova, J., Lukas, J., Strauss, M., and Bartek, J. (1994). Cell cycle-related variation and tissue-restricted expression of human cyclin D1 protein. Journal of Pathology. 172, 237-245.

(102) Bates, S., and Peters, G. (1995). Cyclin D1 as a cellular proto-oncogene. Seminars in Cancer Biology. 6, 73-82.

(103) Lukas, J., Bartkova, J., Rohde, M., Strauss, M., and Bartek, J. (1995). Cyclin D1 is dispensable for G1 control in retinoblastoma gene-deficient cells independently of cdk4 activity. Molecular & Cellular Biology. 15, 2600-2611.

(104) Fantl, V., Smith, R., Brookes, S., Dickson, C., and Peters, G. (1993). Chromosome 11q13 abnormalities in human breast cancer.(comment). Cancer Surveys. 18, 77-94.

(105) Gillett, C., Fantl, V., Smith, R., Fisher, C., Bartek, J., Dickson, C., Barnes, D., and Peters, G. (1994). Amplification and overexpression of cyclin D1 in breast cancer detected by immunohistochemical staining. Cancer Research. 54, 1812-1817.

(106) Seshadri, R., Lee, C.S., Hui, R., McCaul, K., Horsfall, D.J., and Sutherland, R.L. (1996). Cyclin DI amplification is not associated with reduced overall survival in primary breast cancer but may predict early relapse in patients with features of good prognosis. Clinical Cancer Research. 2, 1177-1184.

(107) Pelosio, P., Barbareschi, M., Bonoldi, E., Marchetti, A., Verderio, P., Caffo, O., Bevilacqua, P., Boracchi, P., Buttitta, F., Barbazza, R., Dalla Palma, P., and Gasparini, G. (1996). Clinical significance of cyclin D1 expression in patients with node-positive breast carcinoma treated with adjuvant therapy. Annals of Oncology. 7, 695-703.

(108) van Diest, P.J., Michalides, R.J., Jannink, L., van der Valk, P., Peterse, H.L., de Jong, J.S., Meijer, C.J., and Baak, J.P. (1997). Cyclin D1 expression in invasive breast cancer. Correlations and prognostic value. American Journal of Pathology. 150, 705-711.

(109) Barnes, D.M. (1997). Cyclin D1 in mammary carcinoma. Journal of Pathology. 181, 267-269.

(110) Toi, M., Kashitani, J., and Tominaga, T. (1993). Tumor angiogenesis is an independent prognostic indicator in primary breast carcinoma. International Journal of Cancer 55, 371-374.

(111) Barbareschi, M., Pelosio, P., Caffo, O., Buttitta, F., Pellegrini, S., Barbazza, R., Dalla Palma, P., Bevilacqua, G., and Marchetti, A. (1997). Cyclin-D1-gene amplification and expression in breast carcinoma: relation with clinicopathologic characteristics and with retinoblastoma gene product, p53 and p21WAF1 immunohistochemical expression. International Journal of Cancer. 74, 171-174.

(112) Dickson, C., Fantl, V., Gillett, C., Brookes, S., Bartek, J., Smith, R., Fisher, C., Barnes, D., and Peters, G. (1995). Amplification of chromosome band 11q13 and a role for cyclin D1 in human breast cancer. Cancer Letters. 90, 43-50.

(113) Michalides, R., Hageman, P., van Tinteren, H., Houben, L., Wientjens, E., Klompmaker, R., and Peterse, J. (1996). A clinicopathological study on overexpression of cyclin D1 and of p53 in a series of 248 patients with operable breast cancer. British Journal of Cancer. 73, 728-734.

(114) Zhang, S.Y., Caamano, J., Cooper, F., Guo, X., and Klein-Szanto, A.J. (1994). Immunohistochemistry of cyclin D1 in human breast cancer. American Journal of Clinical Pathology. 102, 695-698.

(115) Zukerberg, L.R., Yang, W.I., Gadd, M., Thor, A.D., Koerner, F.C., Schmidt, E.V., and Arnold, A. (1995). Cyclin D1 (PRAD1) protein expression in breast cancer: approximately one-third of infiltrating mammary carcinomas show overexpression of the cyclin D1 oncogene. Modern Pathology. 8, 560-567.

(116) McIntosh, G.G., Anderson, J.J., Milton, I., Steward, M., Parr, A.H., Thomas, M.D., Henry, J.A., Angus, B., Lennard, T.W., and Horne, C.H. (1995). Determination of the prognostic value of cyclin D1 overexpression in breast cancer. Oncogene. 11, 885-891.

(117) Gillett, C., Smith, P., Gregory, W., Richards, M., Millis, R., Peters, G., and Barnes, D. (1996). Cyclin D1 and prognosis in human breast cancer. International Journal of Cancer. 69, 92-99.

(118) Nielsen, N.H., Emdin, S.O., Cajander, J., and Landberg, G. (1997). Deregulation of cyclin E and D1 in breast cancer is associated with inactivation of the retinoblastoma protein. Oncogene. 14, 295-304.

(119) Blood, C.H., and Zetter, B.R. (1990). Tumor interactions with the vasculature: angiogenesis and tumor metastasis. Biochimica et Biophysica Acta 1032, 89-118.

(120) Folkman, J. (1990). What is the evidence that tumors are angiogenesis dependent? Journal of the National Cancer Institute 82, 4-6.

(121) Bosari, S., Lee, A.K., DeLellis, R.A., Wiley, B.D., Heatley, G.J., and Silverman, M.L. (1992). Microvessel quantitation and prognosis in invasive breast carcinoma. Human Pathology 23, 755-761.

(122) Gasparini, G., Weidner, N., Bevilacqua, P., Maluta, S., Dalla Palma, P., Caffo, O., Barbareschi, M., Boracchi, P., Marubini, E., and Pozza, F. (1994). Tumor microvessel density, p53 expression, tumor size, and peritumoral lymphatic vessel invasion are relevant prognostic markers in node-negative breast carcinoma. Journal of Clinical Oncology 12, 454-466.

(123) Fox, S.B., Leek, R.D., Smith, K., Hollyer, J., Greenall, M., and Harris, A.L. (1994). Tumor angiogenesis in node-negative breast carcinomas--relationship with epidermal growth factor receptor, estrogen receptor, and survival. Breast Cancer Research & Treatment 29, 109-116.

(124) Horak, E.R., Leek, R., Klenk, N., LeJeune, S., Smith, K., Stuart, N., Greenall, M., Stepniewska, K., and Harris, A.L. (1992). Angiogenesis, assessed by platelet/endothelial cell adhesion molecule antibodies, as indicator of node metastases and survival in breast cancer. Lancet 340, 1120-1124.

(125) Adams, J., Carder, P.J., Downey, S., Forbes, M.A., MacLennan, K., Allgar, V., Kaufman, S., Hallam, S., Bicknell, R., Walker, J.J., Cairnduff, F., Selby, P.J., Perren, T.J., Lansdown, M., and Banks, R.E. (2000). Vascular endothelial growth factor (VEGF) in breast cancer: comparison of plasma, serum, and tissue VEGF and microvessel density and effects of tamoxifen. Cancer Research 60, 2898-2905.

(126) Gasparini, G., Toi, M., Gion, M., Verderio, P., Dittadi, R., Hanatani, M., Matsubara, I., Vinante, O., Bonoldi, E., Boracchi, P., Gatti, C., Suzuki, H., and Tominaga, T. (1997). Prognostic significance of vascular endothelial growth factor protein in node-negative breast carcinoma. Journal of the National Cancer Institute 89, 139-147.

(127) Lee, A.H., Dublin, E.A., Bobrow, L.G., and Poulsom, R. (1998). Invasive lobular and invasive ductal carcinoma of the breast show distinct patterns of vascular endothelial growth factor expression and angiogenesis. Journal of Pathology 185, 394-401.

(128) Linderholm, B., Grankvist, K., Wilking, N., Johansson, M., Tavelin, B., and Henriksson, R. (2000). Correlation of vascular endothelial growth factor content with recurrences, survival, and first relapse site in primary node-positive breast carcinoma after adjuvant treatment. Journal of Clinical Oncology 18, 1423-1431.

(129) Obermair, A., Kucera, E., Mayerhofer, K., Speiser, P., Seifert, M., Czerwenka, K., Kaider, A., Leodolter, S., Kainz, C., and Zeillinger, R. (1997). Vascular endothelial growth factor (VEGF) in human breast cancer: correlation with disease-free survival. International Journal of Cancer 74, 455-458.

(130) Toi, M., Kondo, S., Suzuki, H., Yamamoto, Y., Inada, K., Imazawa, T., Taniguchi, T., and Tominaga, T. (1996). Quantitative analysis of vascular endothelial growth factor in primary breast cancer. Cancer 77, 1101-1106.

(131) Brown, L.F., Berse, B., Jackman, R.W., Tognazzi, K., Guidi, A.J., Dvorak, H.F., Senger, D.R., Connolly, J.L., and Schnitt, S.J. (1995). Expression of vascular permeability factor (vascular endothelial growth factor) and its receptors in breast cancer. Human Pathology 26, 86-91.

(132) Tischer, E., Mitchell, R., Hartman, T., Silva, M., Gospodarowicz, D., Fiddes, J.C., and Abraham, J.A. (1991). The human gene for vascular endothelial growth factor. Multiple protein forms are encoded through alternative exon splicing. Journal of Biological Chemistry 266, 11947-11954.

(133) Ferrara, N., Heinsohn, H., Walder, C.E., Bunting, S., and Thomas, G.R. (1995). The regulation of blood vessel growth by vascular endothelial growth factor. Annals of the New York Academy of Sciences 752, 246-256.

(134) Toi, M., Inada, K., Suzuki, H., and Tominaga, T. (1995). Tumor angiogenesis in breast cancer: its importance as a prognostic indicator and the association with vascular endothelial growth factor expression. Breast Cancer Research & Treatment 36, 193-204.

(135) Gasparini, G., Toi, M., Miceli, R., Vermeulen, P.B., Dittadi, R., Biganzoli, E., Morabito, A., Fanelli, M., Gatti, C., Suzuki, H., Tominaga, T., Dirix, L.Y., and Gion, M. (1999). Clinical relevance of vascular endothelial growth factor and thymidine phosphorylase in patients with node-positive breast cancer treated with either adjuvant chemotherapy or hormone therapy. Cancer Journal From Scientific American. 5, 101-111.

(136) Wu, Y., Saldana, L., Chillar, R., and Vadgama, J.V. (2002). Plasma vascular endothelial growth factor is useful in assessing progression of breast cancer post

surgery and during adjuvant treatment. International Journal of Oncology. 20, 509-516.

(137) Eppenberger, U., Kueng, W., Schlaeppi, J.M., Roesel, J.L., Benz, C., Mueller, H., Matter, A., Zuber, M., Luescher, K., Litschgi, M., Schmitt, M., Foekens, J.A., and Eppenberger-Castori, S. (1998). Markers of tumor angiogenesis and proteolysis independently define high- and low-risk subsets of node-negative breast cancer patients. Journal of Clinical Oncology. 16, 3129-3136.

(138) Beguinot, L., Lyall, R.M., Willingham, M.C., and Pastan, I. (1984). Down-regulation of the epidermal growth factor receptor in KB cells is due to receptor internalization and subsequent degradation in lysosomes. Proc Natl Acad Sci U S A 81, 2384-2388.

(139) Dunn, W.A., and Hubbard, A.L. (1984). Receptor-mediated endocytosis of epidermal growth factor by hepatocytes in the perfused rat liver: ligand and receptor dynamics. J Cell Biol 98, 2148-2159.

(140) Piyathilake, C.J., Frost, A.R., Manne, U., Weiss, H., Bell, W.C., Heimburger, D.C., and Grizzle, W.E. (2002). Differential expression of growth factors in squamous cell carcinoma and precancerous lesions of the lung. Clin Cancer Res 8, 734-744.

(141) Liu, X.H., Wiley, H.S., and Meikle, A.W. (1993). Androgens regulate proliferation of human prostate cancer cells in culture by increasing transforming growth factor-alpha (TGF-alpha) and epidermal growth factor (EGF)/TGF-alpha receptor. J Clin Endocrinol Metab 77, 1472-1478.

142) Schuurmans, A.L., Bolt, J., and Mulder, E. (1988). Androgens stimulate both growth rate and epidermal growth factor receptor activity of the human prostate tumor cell LNCaP. Prostate 12, 55-63.

(143) Myers, R.B., Brown, D., Oelschlager, D.K., Waterbor, J.W., Marshall, M.E., Srivastava, S., Stockard, C.R., Urban, D.A., and Grizzle, W.E. (1996). Elevated serum levels of p105(erbB-2) in patients with advanced-stage prostatic adenocarcinoma. Int J Cancer 69, 398-402.

(144) Myers, R.B., Oelschlager, D., Manne, U., Coan, P.N., Weiss, H., and Grizzle, W.E. (1999). Androgenic regulation of growth factor and growth factor receptor expression in the CWR22 model of prostatic adenocarcinoma. Int J Cancer 82, 424-429.

(145). Nicholson, S., Halcrow, P., Sainsbury, J.R., Angus, B., Chambers, P., Farndon, J.R., and Harris, A.L. (1988). Epidermal growth factor receptor (EGFr) status associated with failure of primary endocrine therapy in elderly postmenopausal patients with breast cancer. Br J Cancer 58, 810-814.

(146) Toi, M., Osaki, A., Yamada, H., and Toge, T. (1991). Epidermal growth factor receptor expression as a prognostic indicator in breast cancer. Eur J Cancer 27, 977-980.

(147) Fox, S.B., Smith, K., Hollyer, J., Greenall, M., Hastrich, D., and Harris, A.L. (1994). The epidermal growth factor receptor as a prognostic marker: results of 370 patients and review of 3009 patients. Breast Cancer Res Treat 29, 41-49.

(148) Muss, H.B., Thor, A.D., Berry, D.A., Kute, T., Liu, E.T., Koerner, F., Cirrincione, C.T., Budman, D.R., Wood, W.C., Barcos, M., and et al. (1994). c-erbB-2 expression and response to adjuvant therapy in women with node-positive early breast cancer. N Engl J Med 330, 1260-1266.

(149) Slamon, D.J., Godolphin, W., Jones, L.A., Holt, J.A., Wong, S.G., Keith, D.E., Levin, W.J., Stuart, S.G., Udove, J., Ullrich, A., and et al. (1989). Studies of the HER-2/neu proto-oncogene in human breast and ovarian cancer. Science 244, 707-712.

(150) Slamon, D.J., Clark, G.M., Wong, S.G., Levin, W.J., Ullrich, A., and McGuire, W.L. (1987). Human breast cancer: correlation of relapse and survival with amplification of the HER-2/neu oncogene. Science 235, 177-182.

(151) Paik, S., Hazan, R., Fisher, E.R., Sass, R.E., Fisher, B., Redmond, C., Schlessinger, J., Lippman, M.E., and King, C.R. (1990). Pathologic findings from the National Surgical Adjuvant Breast and Bowel Project: prognostic significance of erbB-2 protein overexpression in primary breast cancer. J Clin Oncol 8, 103-112.

(152) Urban, D., Irwin, W., Kirk, M., Markiewicz, M.A., Myers, R., Smith, M., Weiss, H., Grizzle, W.E., and Barnes, S. (2001). The effect of isolated soy protein on plasma biomarkers in elderly men with elevated serum prostate specific antigen. J Urol 165, 294-300.

(153) Wu, Y., Khan, H., Chillar, R., and Vadgama, J.V. (1999). Prognostic value of plasma HER-2/neu in African American and Hispanic women with breast cancer. Int J Oncol 14, 1021-1037.

(154) Hollstein, M., Rice, K., Greenblatt, M.S., Soussi, T., Fuchs, R., Sorlie, T., Hovig, E., Smith-Sorensen, B., Montesano, R., and Harris, C.C. (1994). Database of p53 gene somatic mutations in human tumors and cell lines. Nucleic Acids Res 22, 3551-3555.

(155) Manne, U., Myers, R.B., Moron, C., Poczatek, R.B., Dillard, S., Weiss, H., Brown, D., Srivastava, S., and Grizzle, W.E. (1997). Prognostic significance of Bcl-2 expression and p53 nuclear accumulation in colorectal adenocarcinoma. Int J Cancer 74, 346-358.

(156) Manne, U., Weiss, H.L., Myers, R.B., Danner, O.K., Moron, C., Srivastava, S., and Grizzle, W.E. (1998). Nuclear accumulation of p53 in colorectal adenocarcinoma: prognostic importance differs with race and location of the tumor. Cancer 83, 2456-2467.

(157) Isola, J., Visakorpi, T., Holli, K., and Kallioniemi, O.P. (1992). Association of overexpression of tumor suppressor protein p53 with rapid cell proliferation and poor prognosis in node-negative breast cancer patients. J Natl Cancer Inst 84, 1109-1114.

(158) Thor, A.D., Moore, D.H., II, Edgerton, S.M., Kawasaki, E.S., Reihsaus, E., Lynch, H.T., Marcus, J.N., Schwartz, L., Chen, L.C., Mayall, B.H., and et al. (1992). Accumulation of p53 tumor suppressor gene protein: an independent marker of prognosis in breast cancers. J Natl Cancer Inst 84, 845-855.

(159). Allred, D.C., Clark, G.M., Elledge, R., Fuqua, S.A., Brown, R.W., Chamness, G.C., Osborne, C.K., and McGuire, W.L. (1993). Association of p53 protein expression with tumor cell proliferation rate and clinical outcome in node-negative breast cancer. J Natl Cancer Inst 85, 200-206.

(160) Silvestrini, R., Benini, E., Daidone, M.G., Veneroni, S., Boracchi, P., Cappelletti, V., Di Fronzo, G., and Veronesi, U. (1993). p53 as an independent prognostic marker in lymph node-negative breast cancer patients. J Natl Cancer Inst 85, 965-970.

(161) Borresen Dale, A.L. (2003). TP53 and breast cancer. Hum Mutat 21, 292-300.

(162) Shiao, Y.H., Chen, V.W., Scheer, W.D., Wu, X.C., and Correa, P. (1995). Racial disparity in the association of p53 gene alterations with breast cancer survival. Cancer Res 55, 1485-1490.

(163) Shiao, Y.H., Chen, V.W., Wu, X.C., Scheer, W.D., Lehmann, H.P., Malcom, G.T., Boudreau, D.A., Ruiz, B., and Correa, P. (1996). Racial comparison of p53 alterations in breast cancer: difference in prognostic value. In Vivo 10, 169-173.

(164) Blaszyk, H., Vaughn, C.B., Hartmann, A., McGovern, R.M., Schroeder, J.J., Cunningham, J., Schaid, D., Sommer, S.S., and Kovach, J.S. (1994). Novel pattern of p53 gene mutations in an American black cohort with high mortality from breast cancer. Lancet 343, 1195-1197.

(165) Upadhyay, S., Li, G., Liu, H., Chen, Y.Q., Sarkar, F.H., and Kim, H.R. (1995). bcl-2 suppresses expression of p21WAF1/CIP1 in breast epithelial cells. Cancer Res 55, 4520-4524.

(166) Vairo, G., Innes, K.M., and Adams, J.M. (1996). Bcl-2 has a cell cycle inhibitory function separable from its enhancement of cell survival. Oncogene 13, 1511-1519.

(167) Piris, M.A., Pezzella, F., Martinez-Montero, J.C., Orradre, J.L., Villuendas, R., Sanchez-Beato, M., Cuena, R., Cruz, M.A., Martinez, B., Pezella, F., and et al. (1994). p53 and bcl-2 expression in high-grade B-cell lymphomas: correlation with survival time. Br J Cancer 69, 337-341.

(168). Bilim, V., Tomita, Y., Kawasaki, T., Katagiri, A., Imai, T., Takeda, M., and Takahashi, K. (1996). Prognostic value of Bcl-2 and p53 expression in urinary tract transitional cell cancer. J Natl Cancer Inst 88, 686-688.

(169) Sinicrope, F.A., Hart, J., Michelassi, F., and Lee, J.J. (1995). Prognostic value of bcl-2 oncoprotein expression in stage II colon carcinoma. Clin Cancer Res 1, 1103-1110.

(170) Manne, U., Weiss, H.L., and Grizzle, W.E. (2000). Bcl-2 expression is associated with improved prognosis in patients with distal colorectal adenocarcinomas. Int J Cancer 89, 423-430.

(171) Grizzle, W.E., Manne, U., Weiss, H.L., Jhala, N., and Talley, L. (2002). Molecular staging of colorectal cancer in African-American and Caucasian patients using phenotypic expression of p53, Bcl-2, MUC-1 AND p27(kip-1). Int J Cancer 97, 403-409.

(172) Joensuu, H., Pylkkanen, L., and Toikkanen, S. (1994). Bcl-2 protein expression and long-term survival in breast cancer. Am J Pathol 145, 1191-1198.

(173) Lipponen, P., Pietilainen, T., Kosma, V.M., Aaltomaa, S., Eskelinen, M., and Syrjanen, K. (1995). Apoptosis suppressing protein bcl-2 is expressed in well-differentiated breast carcinomas with favourable prognosis. J Pathol 177, 49-55.

(174) Silvestrini, R., Veneroni, S., Daidone, M.G., Benini, E., Boracchi, P., Mezzetti, M., Di Fronzo, G., Rilke, F., and Veronesi, U. (1994). The Bcl-2 protein: a prognostic indicator strongly related to p53 protein in lymph node-negative breast cancer patients. J Natl Cancer Inst 86, 499-504.

(175) Bhargava, V., Kell, D.L., van de Rijn, M., and Warnke, R.A. (1994). Bcl-2 immunoreactivity in breast carcinoma correlates with hormone receptor positivity. Am J Pathol 145, 535-540.

(176) Krajewski, S., Blomqvist, C., Franssila, K., Krajewska, M., Wasenius, V.M., Niskanen, E., Nordling, S., and Reed, J.C. (1995). Reduced expression of proapoptotic gene BAX is associated with poor response rates to combination chemotherapy and shorter survival in women with metastatic breast adenocarcinoma. Cancer Res 55, 4471-4478.

(177) Reed, J.C. (1996). Balancing cell life and death: bax, apoptosis, and breast cancer. J Clin Invest 97, 2403-2404.

Chapter 4

SCREENING AND EARLY DETECTION OF BREAST CANCER IN WOMEN IN AFRICA AND THE MIDDLE EAST

B. L. Rapoport MD, M. Med. (Int. Med.)
The Medical Oncology Centre of Rosebank, Johannesburg, South Africa

1. OBJECTIVE

Historically, breast cancer has been diagnosed when a woman seeks medical attention for a breast symptom, such as a palpable mass or soreness. Breast cancer is usually diagnosed by pathologic review of a fixed specimen of breast tissue. The biopsy can be obtained by surgically removing part of a palpable mass, or by surgically excising an abnormal area identified by mammogram, with surgical needle localization under x-ray guidance. Alternatively, a core needle biopsy of a mammographically suspicious area can be obtained with use of stereotactic x-ray or ultrasound.

Reduction of breast mortality is the primary and fundamental objective of breast screening. Breast screening should also have the potential to reduce breast cancer morbidity by allowing less extensive surgical procedures, with less need for postoperative radiotherapy and adjuvant chemotherapy, all of which may result in an increased quality of life for breast cancer patients (1).

2. INCIDENCE AND MORTALITY

An important paradox is the difference in breast cancer incidence and mortality rates between white and black women. Incidence rates are 20% higher in white than in black women, yet among women diagnosed in the USA between 1983 and 1989, the 5-year relative survival rate is 16% lower

C. Williams et al. (eds.), Breast Cancer in Women of African Descent, 71-80.

for black than for white women. Among breast cancer diagnosed between 1992 and 1998, 64% of white women but only 53% of black women had localised disease. The 5-year survival for localised disease was 97.4% for white women and 88.9% for black women; for regional disease it was 80.2% and 65.4%; and for distant metastasis, it was 24% and 14.7% respectively. This higher mortality in black women has primarily been attributed to inadequate screening practices in this population, which lead to delayed diagnosis and later stage at diagnosis. However, even when black and white women are compared stage for stage, the mortality in black women is higher. The reasons for this are unknown. (2-5).

Only three countries in sub-Saharan Africa (Benin, Botswana and Mauritius) provided mortality data of adequate quality to the WHO in 1996. Thus, no data on cancer exist in most countries in Africa. To address this gap the South African National Cancer Registry was established in 1986. Between 1993 and 1995, an annual average of 3785 new cases of breast cancer was diagnosed in South Africa. The crude incidence was 18.5/100 000 and the Age Standardized Incidence Rates 25.1/100 000. The lifetime risk was 1 in 36 overall, but varied from 1 in 81 in black to 1 in 13 in white females, a 6-fold difference. The age standardized rates (ASR) of 11.3 in black South African women compare well with rates from central Africa (Harare ASR = 20.4/100 000, Kampala 16.4/100 000). Breast cancer is even rarer in Gambia (3.4/100 000) (6).

Table 1 summarizes the breast cancer incidence and prevalence in the different regions of Africa and some African Countries.

Table 1. Breast Cancer Incidence and Prevalence in Certain African Countries.

	Cases	1 –year prevalence	5 year prevalence
Eastern Africa	**13615**	**9970**	**39776**
Ethiopia	4286	3089	12329
Kenya	1533	1114	4347
Mauritius	182	128	497
Tanzania	1577	1064	4372
Zambia	307	236	934
Zimbabwe	582	406	1554
Middle Africa	**3902**	**2850**	**11258**
Angola	877	638	2514
Cameroon	626	462	1839
Congo	1564	1136	4482
Northern Africa	**18724**	**15120**	**59214**
Algeria	3507	3018	11894
Egypt	7921	6470	25305
Morocco	3207	2556	9978
Southern Africa	**5537**	**4442**	**17690**
Botswana	74	53	215
South Africa	5130	4146	1650
Western Africa	**17389**	**13165**	**51516**
Burkino Fasco	942	671	26001
Cote D'Ivoire	755	571	2330
Ghana	16332	1165	4540
Nigeria	10528	8174	31955
Senegal	451	322	1252

3. SCREENING METHODS

The cornerstones of screening are breast self-examination (BSE), clinical breast examination (CBE), and screening mammography (7-10). Specific recommendations depend on the woman's personal risk of breast cancer. Before screening is initiated, any severe co-morbid conditions that limit life expectancy or therapeutic intervention should be considered. Screening is not appropriate in women with a life expectancy of less than 10 years or in women in whom therapeutic intervention is not planned.

4. INTERNATIONAL RECOMMENDATIONS

Although breast cancer screening is well established in the developed world, its value continues to be debated (9). Several trials have shown that the breast cancer mortality was reduced by screening mammography (11-12). Mammography is an expensive test that requires great care and expertise both to perform and in the interpretation of results. Although there is inadequate evidence to suggest that physical examination of the breast as a single screening modality reduces mortality from breast cancer, there are indications that good clinical breast examinations by specially trained health workers could have an important role (8).

5. CURRENT PRACTICE

No organized screening programs exist for breast cancer in African and middle Eastern countries. In advocating screening programs as part of early detection it is important to avoid imposing the "high technology" of the developed world on countries that lack the infrastructure and resources to use the technology appropriately or to achieve adequate coverage of the population.

6. CRITICAL ISSUES

Before embarking on general screening programs, several other issues also need to be considered. It is critical to recognize the role of socio-economic status and racial/ethnic factors, access to health care and compliance with recommended protocols, all of which influence disease prognosis.

6.1 Socio-economic Status

Numerous population-based studies demonstrate that, despite dramatic improvement over time in life expectancy and death rates for all major subgroups of the population, socio-economic status is still a major determinant of health status. Low is consistently associated with adverse health consequences and appears to be largely responsible for the differences in health status between whites and other racial/ethnic groups especially black. Social inequity represents a compelling challenge.

The total number of people in the world living in absolute poverty is steadily increasing despite the general increase in per capita income in most parts of the world. Poverty in Africa is a harsh reality. Half the population lives on less than one dollar a day. More than half the population has no access to safe drinking water. More than 2 million infants die annually before reaching their first birthday.

A notable exception is the inverse gradient between socio-economic status and disease is the fact that the incidence of breast cancer is higher in upper socio-economic status.

6.2 Healthcare System and Funding

It is unrealistic to expect that health care changes made in the USA and Europe could be directly imported into a country of limited resources in Africa or the Middle East. Different countries have vastly different social issues and financial frameworks for change and improvement in medical healthcare delivered. Currently existing guidelines are most useful for countries with fully equipped healthcare systems. As such, currently existing guidelines do not address the need for developing regions of the world with major limitations to healthcare access.

Lack of adequate health insurance is a major problem for many women, poor low-income women, and women in reproductive age. Women in general are more likely than men to have low paying part-time jobs that don't provide health insurance. Women are also more likely than men to lose insurance as a consequence of being divorced or widowed.

South Africa has by far the best-developed health care system in the region and many patients from the other African countries, e.g. Angola, Congo, Kenya and Zambia, are travelling to South Africa for medical treatment.

South Africa has a two-tiered healthcare system comprising of the Public Health Services, that provides the bulk of care to between 75% and 80% of the population, and the Private Sector that provides services to the remaining 20% to 25% of the population. Traditionally the organization of public sector services resulted in inequities in access based upon race. In an attempt to address these problems the South African government has initiated a major reform process of all South Africa public services that promotes a Primary Health Care philosophy. The commitment to Primary Healthcare has resulted in a shift in resources from tertiary to primary health care.

6.3 HIV/AIDS and Life Expectancy in Africa

HIV is continuing to devastate Africa's economies, communities and development. Sixty million Africans have been touched by AIDS. They are either living with HIV, have died of AIDS or have lost their parents to AIDS.

In several Southern African countries the prevalence rate among pregnant women has now reached 30%.

It is estimated that by 2010 AIDS will account for 28-58% of all deaths. HIV/AIDS prevalence is strongly linked to adverse socio-economic conditions of individuals (13).

6.4 Tuberculosis

Tuberculosis remains a major problem, compounded by the emergence of multi-drug resistant organisms and the association between tuberculosis and HIV/AIDS. As the incidence of HIV increases so will the prevalence of tuberculosis (14).

6.5 Cultural Beliefs

For African patients traditional medicine is the oldest most tried and tested form of medicine. There is a belief by African individuals that African medicine in fact is the original medicine. Traditional methods of treating illness and diseases were used in Africa long before the coming of Europeans and western medicine. It is difficult to separate African medicine from African religion. The two main reasons for this are the following: firstly, the African general theory of illness is very broad; it includes African theology. In other words, the theory not only attempts to explain illness and disease but also the relations between God and the universe. The second reason, related to the previous one, is that many traditional healers are also religious leaders and vice versa. In fact, many Africans who became Christians found it difficult to abandon their religion and medicine completely.

It is important to mention the traditional medical sector has continued to grow despite the attempts by western cultures to suppress it; and it has continued to grow because traditional healers are successful in curing a large number of illnesses as well as controlling disease symptoms. African patients trust traditional healers. Traditional healers use both scientific and non-scientific or subjective knowledge. Scientific medicines are obtained mainly from plants. Many plant medicines recommended by traditional healers are correct even when judged by modern scientific methods.

African patients perceive traditional medicine as more holistic in its approach than allopathic medicine. Traditional medicine has been criticized by western cultures as emphasizing the cure of symptoms rather than the underlying causes. Traditional healers hold an esteemed and powerful position in southern African societies. Their role is that of physician, counsellor, psychiatrist, and priest. In Africa there is knowledge of the potential effectiveness of traditional healers as primary health givers and their importance in the fight against HIV and AIDS. Many organizations such as WHO also supports the integration of western medicine and traditional healing, encouraging referrals between the two groups. Years ago, people had the traditional belief that sickness and diseases such as malaria and other misfortunes in life are a result of spiritual forces. These spiritual forces may be some kind of punishment from gods or from their enemies to harm them. Most traditional doctors maintain their practices in rural areas because of the expensive costs of running a business in the city (15-16).

Indigenous Southern African people have practiced natural medicine for centuries. Africa has different categories of practitioners mainly the sangoma, inyanga and witchdoctor. Each category specializes in its own field.

6.5.1 Sangoma

Western culture refers to sangomas as diviners. Sangomas are primarily diagnosticians. Several different kinds of sangomas exist, each with their own method of divination. Sangomas are mainly female and their gift is said to be hereditary. They serve as councillors and advisers to their local communities and its leaders. It could be said that they perform the task of marriage councillors, medical general practitioners, advisors and sometimes jurors. They have a general knowledge of medicinal plants and its uses.

6.5.2 Inyanga

An inyanga could be called a healer or doctor. The true inyanga has a deep sense of dedication to his calling. Inyangas are generally men. Primarily the inyanga is a herbalist. He gathers the raw material for his *mutis* (medicines) on errands through the countryside. Like the western herbalist, the inyanga may prescribe medicine to be taken whole, as an infusion, decoction or inhalant. The inyanga is generally highly skilled in its art and has a detailed knowledge and acquaintance with botany. Most inyangas do not pretend all-round knowledge but specialize in a particular disease or a particular area of the body such as the liver, lungs or gastrointestinal tract.

6.5.3 Witchdoctors

A witchdoctor is a native healer who practices among rural people in his or her own country.

For African peoples, the belief and ideas about ancestors to form an essential part of the effort to inculcate, mobilize, and promote the community ideal of harmonious living in society. African people share a common understanding of the importance of ancestors in daily life. Traditional Africans believe that spiritual beings, especially ancestral spirits, guarantee and legitimate the ethical code. When they have lost touch with their ancestors, illness may result or bad luck. Then a traditional healer, or sangoma, is sought out who may prescribe herbs, changes in lifestyle, a career change, or changes in relationships.

While it is true that the traditional religion still has considerable influence in the life and culture of many African peoples, it no longer enjoys exclusive dominance and control over the life of the vast majority of the population. The rate of displacement of the traditional religion by the forces of radical social change in Africa is generally slower in rural areas than in urban cities.

7. RECOMMENDATIONS

Given the present level of evidence breast cancer screening by mammography should not be recommended. Emphasis should rather be placed on early diagnosis of breast cancer by creating awareness in the communities or by offering clinical breast examinations to those women concerned about their breasts who attend primary healthcare centres or other healthcare facilities for other reasons. Mammography should not be introduced for screening unless adequate resources are available to ensure effective and reliable screening. It is important to avoid imposing the high technology of the developed world on countries with limited resources like Africa and the Middle East. The success of screening depends on having sufficient numbers of trained personnel to perform screening tests and on the availability of facilities that can undertake the subsequent diagnosis, treatment and follow-up of these women. It is important to avoid over-diagnosing. Over-diagnosing may result in over-servicing and therefore increase in costs at the expense of other essential services.

Information about breast screening must be improved if women are to understand fully both the benefits and potential harms in order to make an informed decision.

7.1 Recommendations for Prevention of Advanced Disease

A special category of patients refers to the high proportion of African patients presenting with very advanced disease, either locally advanced or metastatic breast cancer.

Proper studies have not yet been conducted addressing the prevention of locally advanced breast cancer. Education and breast self examination may play a very important role in preventing African women to present with locally advanced disease. Mammography is not an adequate screening test in these patients. Breast physical examination by experienced medical personnel may be more effective and less costly. Prevention of locally advanced breast cancer may result in decreased usage of neo-adjuvant chemotherapy, radiotherapy and extensive surgery. Prevention of locally advanced disease may result in an increase in survival, cost reduction, and an increase in quality of life in African women.

REFERENCES

(1) Harris R, Leininger L: Clinical strategies for breast cancer screening: weighing and using the evidence. Ann Intern Med 122 : 539-547, 1995

(2) Ries LA, Eisner MP, Kosary CL, et al.: SEER Cancer Statistics Review, 1973-1999. Bethesda, Md: National Cancer Institute, 2002 early detection of cancer. J Natl Cancer Inst 91: 1409-1415, 1999

(3) Roetzheim RG, Pal N, Tennant C, et al.: Effects of health insurance and race on

(4) Bradley CJ, Given CW, Roberts C: Race, socioeconomic status, and breast cancer treatment and survival. J Natl Cancer Inst 94: 490-496, 2002

(5) McCarthy EP, Burns RB, Coughlin SS, et al.: Mammography use helps to explain differences in breast cancer stage at diagnosis between older black and white women. Ann Intern Med 128: 729-736, 1998

(6) Cancer in South Africa, 1993 -1995

(7) Miller AB, To T, Baines CJ, et al.: Canadian National Breast Screening Study-2: 13-year results of a randomized trial in women aged 50-59 years. J Natl Cancer Inst 92 : 1490-1499, 2000

(8) Baines CJ, Miller AB, Bassett AA: Physical examination. Its role as a single screening modality in the Canadian National Breast Screening Study. Cancer 63: 1816-1822, 1989

(9) Olsen O, Gøtzsche PC: Screening for breast cancer with mammography. Cochrane Database Syst Rev: CD001877, 2001

(10) Nyström L, Andersson I, Bjurstam N, et al.: Long-term effects of mammography screening: updated overview of the Swedish randomized trials. Lancet 359: 909-919, 2002

(11) Kerlikowske K: Efficacy of screening mammography among women aged 40 to 49 years and 50 to 69 years: comparison of relative and absolute benefit. J Natl Cancer Inst Monogr 22: 79-86, 1997.

(12) Duffy SW, Tabár L, Chen HH, et al.: The impact of organized mammography service screening on breast carcinoma mortality in seven Swedish counties. Cancer 95: 458-469, 2002
(13) UNAIDS. Aids epidemic update. December 2002
(14) Corbett EL, Watt CJ, Walker N, Maher D, Williams BG, Raviglione MC, Dye C. The growing burden of tuberculosis: global trends and interactions with the HIV epidemic. Arch Intern Med. 163:1009-1021, 2003
(15) Miller NN. Traditional medicine in East Africa: the search for a synthesis. Am Univ Field Staff Rep Afr. 22:1-15, 1980
(16) Chipfakacha VG. STD/HIV/AIDS knowledge, beliefs and practices of traditional healers in Botswana. AIDS Care. 9:417-425, 1997

Chapter 5

BREAST CANCER AT DIAGNOSIS IN WOMEN OF AFRICA AND THE MIDDLE EAST

Hussein M. Khaled, MD
Medical Oncology Department, National Cancer Institute, Cairo University, Cairo, Egypt

1. INTRODUCTION

The myth behind the aetiology and efforts towards treatment and prevention of breast cancer were always present since antiquity. Two Egyptian papyruses, written approximately 1600 B.C. are concerned with the disease, the Edwin Smith Papyrus and the Ebers Papyrus. The first includes a case (no. 45) of "bulging tumour of the breast recorded in lines 9 to 20 of the reproduction:

> "If thou examinest a man having bulging tumours on his breast, (and) thou findest that (swellings) have spread over the breast; if thou puttest thy hand upon his breast upon these tumours, (and) thou findest them very cool, there being no fever at all therein when thy hand touches him, they have no granulation, they form no fluid, and they are bulging to thy hand, thou should say concerning him one having bulging tumours. An ailment with which I will attend. There is no (treatment)".

From the Ebers Papyrus, there is the following case description:

> "When thou meetest a large tumour of the God Xensu in any part of the limb of a person, it is loathsome and suffers many pustules to come forth; something arises therein as though wind were in it , causing irritation. The tumour calls with a loud voice to thee: It is a tumour of the God Xensu. Do thou nothing there against".

The Arabic physician Ibn Sina or Avicenna (980–1037 AC) in his Qanoun of Medicine differentiated between cancerous swelling and

C. Williams et al. (eds.), Breast Cancer in Women of African Descent, 81-90.

induration. Ramazzini, in his De Morbis Artificuum of 1700, recorded the high occurrence of breast cancer among nuns giving the first report on association of occupation and breast cancer. Hunter (1728- 1793) later stated that the predisposing factors to be mainly age, partly hereditary predisposition, and perhaps the climate. Then information concerning breast cancer and its spread grew markedly during the late 18th and early 19th centuries. It was only in the 1970s that systematic use of mammography as a screening method for breast cancer was tested.

2. BREAST CANCER BURDEN IN AFRICA AND THE MIDDLE EAST

The large variation in breast cancer incidence and mortality rates, from one part of the world to another, has been a subject of considerable interest to many scientists. Breast cancer is considered as the most common cause of cancer death among women worldwide.

Incidence rates are high in more developed countries, whereas rates in less developed countries and in Japan are relatively low, but increasing. So, in 2000 worldwide there were over 10 million new cases of cancer, nearly 5.5 million, of which were in the less developed countries. For women, the most common cancers worldwide are breast and cervical cancer, although cervical cancer is primarily seen in less developed countries. Thus in 2000 breast cancer accounted for 10% of all new cancers in males and females, which made it the second most common site of malignant neoplasm after the lung (*1*). Twenty-two percent of all new cancer cases in women are those of the breast, thus making it by far the most common cancer in females (For additional information on this subject, please, see the contribution of Max Parkin – Chapter 1 – in this book). In less developed countries, approximately 47,000 new cases of breast cancer are diagnosed per year representing also the most common cancer among females.

It is known that, the risk of breast cancer is lower in the low-income regions including Africa and the Middle East, where the possibility of developing breast cancer by the age of 75 is one third that of other high-income regions where an average of 6% of women in high-income regions develop invasive breast cancer before the age of 75 (*2*).

Clear increases in the incidence of and mortality from breast cancer were reported until the early 1980s in both high and low-income countries. However, substantial improvements in survival have been observed in many western countries since late 1970s and early 1980s. Improvements in treatment and possibly breast screening by mammography are possible factors responsible for the reduced mortality. In low-income countries, survival from breast cancer is generally poorer than that in high-income regions reflecting mainly late presentation of cases (*3*). According to WHO in 2000, non-communicable diseases including cancer accounted for half of all deaths in the Middle East and Northern Africa and less than 25% of all deaths in sub-Saharan Africa.

Worldwide, however, the risk continues to increase (see the contribution of Dr. Max Parkin in this book) and with this observed variance in incidence in different regions, the underlying causes must be largely different among different countries (*4*). The environmental factors, low parity, late age at first pregnancy, early menarche and late menopause are all factors that are consistently associated with an increased risk. Also, and as for most epithelial tumours, the risk for breast cancer increases steadily with age.

According to the WHO world cancer report (*5*), breast cancer age standardized incidence rates are 28.2 and 22/100,000 population for north and sub-Saharan African populations respectively. In Egypt, there was a lack of true cancer incidence rates. Most of the published data were hospital-based frequency rates. However, recent population – based cancer registry (*6*) has shown that breast cancer incidence rate among Egyptian population represented by Gharbiah governorate is 29.5/100,000 population. At the Cairo National Cancer Institute, the largest cancer centre in the region, the relative frequency of breast cancer among females is 35.1 % (www.nci.edu.eg). Likewise, in a case-control study conducted over a 4 years period in Western Cape of South Africa (*7*) the overall incidence rate of breast cancer was 23.1/100,000 women per year, being 25.6/100,000 for coloured women. It was 26.6/100,000 in urban areas vs. 16.3/100,000 in rural areas.

3. DIAGNOSTIC WORKUP

The first sign of breast cancer is still a palpable lump although a sizable proportion of all breast cancers are non palpable and visible only at mammography or ultrasound. Physical examination usually has a low specificity and further examinations are required to decrease the risk of missing cancers. Mammography is the first choice examination with 60%-75% specificity and 85%-95% sensitivity in most series. These variations depend mainly on age where dense breasts in young patients have lower specificity and sensitivity. Ultrasound of the breast is invaluable adjunct to mammography especially in dense breasts (*8*). The primary role of sonography in breast imaging is to determine if a focal mass is cystic or solid. If the mass is solid, breast ultrasound may differentiate benign from malignant lesions. Fine-needle aspiration or core biopsy provides the highest evidence for malignancy. The diagnostic accuracy of these examinations has been shown to be near 99%.

Other promising diagnostic techniques for breast cancer are currently under evaluation. These include breast MRI, computer aided diagnosis and PET scan. Although several of these techniques hold promise, a systematic review showed that few have been used for either screening or diagnosis, and the studies were generally small and of poor quality, so that the evidence for their routine use is weak.

A series of studies were done to evaluate the accuracy, sensitivity, and specificity of fine needle aspiration cytology (FNAC) during a 3-year period (1994-1996) at Cairo National Cancer Institute. In one study 999 FNAC samples were included (*9*). The cytological diagnosis was compared with the results of the histological examination of the same lesions. The overall accuracy rate was 70 % with sensitivity and specificity rate of 92.4 % and 100 % respectively. In another study, the diagnostic results of another more simplified fine needle sampling technique without aspiration was compared to that of the routine classic FNAC technique in 115 female patients having both benign and malignant breast tumours. With this technique, trauma was reduced and better tumour sampling was obtained. Moreover, insufficient cellular yield was reported in 6.8% of lesions examined with the fine needle alone, compared to 8% of these obtained by aspiration technique (*10*).

The role of mammography and echography in the diagnosis of breast cancer in Dakar was studied in 25 cases (*11*). The authors reported mammographic sensitivity of 80%, false positive rate of 28% and a false positive rate of 8% against respectively 96%, 32% and 12% for echography. They, thus, recommended the use of both mammography and echography in diagnosis of breast cancer in Senegal.

Also, the role of MRI in breast cancer diagnosis was assessed in a series of 24 Egyptian cases with pathologically proven breast cancer(*12*). Results were compared to those of mammography and ultrasonography of the breast and showed an overall sensitivity of 100% in tumour detection. In another study for breast imaging, Tc^{-99m}MIBI scintimammography was evaluated and showed a good diagnostic accuracy of detecting malignant lesions specially those larger than 1 cm in diameter with high-grade subtypes and with positive ER status (*13*).

4. CLINICO-PATHOLOGIC DIAGNOSTIC PROFILE

For breast cancer, as well as for many other cancers, the medically underserved are understudied and are not well understood by many in the field of oncology. This is mainly because, in developing countries, health care institutions are usually under-funded and do not have access to state-of-the-art cancer screening, clinical follow-up, diagnosis, and treatment. Medically underserved women usually present with late-stage disease, and have the greatest mortality rates. The cause is multifactorial and may include inaccessibility of screening services for high risk women, many barriers to effective follow up, diagnosis, and treatment as well as complex underlying biological, psychological, behavioural, and socio-cultural processes and pathways (Please, see the contribution of Christopher Williams – Chapter 14 in this book – for additional information on this subject).

Evidence in support of different tumour biologic characteristics of patients in Africa and the Middle East when compared to those of patients in western parts of the world rests essentially on higher incidence of hormone receptor negative tumours and poor tumour grade as well as on prevalence of special types e.g. inflammatory breast cancer in certain localities. However, data on these differences in prognostic markers and tumour biologic criteria are too few for any clear conclusion (Please, see the contribution of Francis Ikpatt and Olufunmilayo I. Olopapade – Chapter 2 in this book – for additional discussion of this subject).

Numerous studies on epidemiologic, clinical, and pathologic features of breast cancer patients in different African Nations and those of African origin have been published in last years. It is clear that in these studies late presentation of cases is the salient feature.

Fifty cases were recruited and analysed in 1997 at Muhimbili Medical Centre in Tanzania. There was not a single case in stage I disease. One case presented with stage II, 7 cases with stage IIIA, 37 cases with stage IIIB, and 5 cases with stage IV disease. The majority of patients were premenopausal and below the age of 50 years. In a subsequent study from the same centre in Tanzania (*14*), the characteristics of 63 benign and 184 malignant lesions of the female breast in a sub-Saharan African population were reported. Fibrocystic disease with no proliferative lesions and fibroadenomas were the most frequent benign lesions, while the vast majority of breast carcinomas were invasive. As a striking feature, the majority of those studied (66%) were of the non-special type displaying a more aggressive behaviour than the remaining tumours of the special type. In the latter group, cribriform type constitutes 41% of cases, which may be a special feature in African females with breast cancer.

The expression of oestrogen and progesterone receptors was assessed by immunohistochemistry in 60 indigenous sub-Saharan females with breast cancer, from the same centre in Dakar. A lower frequency of ER (33%), and PR (18%) as compared to literature reports was observed (*15*) further indicating that this malignancy might have a different biology from that in western females.

The demographic characteristics and clinical profile of breast cancer in Ethiopian patients was reported by a group from Addis Ababa University (*16*). The study included 72 patients, 62 females and 10 males. The age range was 21-82 years with a mean of 41 years for females and 52 years for males. The median time interval between the onset of symptoms to diagnosis was 12 months. Infiltrating ductal and lobular carcinoma accounted for 85% and 11% of cases respectively. Seventy-two percent of female cases were premenopausal and 76% had stage III and IV disease at presentation.

In Nigeria (*17*), patients with breast cancer comprised 30% of all patients with breast disease seen in 4 hospitals located in Eastern Nigeria. The mean age was 44 years, 69% were premenopausal, and 64% had advanced disease at presentation (stages III and IV).

In the northern part of the continent, 689 new cases of mammary cancers with proven pathologic or cytological diagnosis presenting in Tunisia during the year 1994 were analysed (*18*). The average age was 50 years, with an age-adjusted incidence rate of 16.7/100,000 women. The average tumour size was 49.5 mm with the infiltrating and grade II lesions being the most frequent (53.6%). According to TNM classification of 1988, 7.2% of tumours were T1, 48.9% T2, 18.5% T3, and 23.4% T4 (6.2% T4a, and 16.1% T4b). About 22% of tumours were M1 and 3.3% were in situ carcinoma.

Breast cancer in Egyptian patients has a younger age distribution with the majority of cases between 30 – 60 years. The median age is 46 years, one decade younger than the corresponding age in western countries and most of the patients are premenopausal (60.5%). Late presentation of most patients is a characteristic feature and the inflammatory subtype is relatively more frequent. At the National Cancer Institute, clinical T1, T2, T3, and T4 lesions were 1.2%, 30%, 26.4%, and 42.4% respectively. The mean tumour size was 4.5 cm. The frequency of axillary lymph node metastases was 75% with the number of positive nodes being 1-3 in 23%, 4-10 in 22%, and more than 10 positive nodes in 17% of patients. Most tumours were invasive duct carcinoma (83.4%), while intraductal carcinoma was present in 1.5% of cases. Invasive lobular, medullary and mucoid carcinoma was detected in 7.12%, 1.63%, and 2.27% of cases respectively. Pathologic grading showed a low incidence of grade I (5.4%). Grades II, and III tumours were 66% and 28.6% respectively. Multiple tumours formed 1.5% and were mostly of the lobular subtype and bilateral presentation occurred in 0.6% of cases (*19*). The profile of hormone receptor status as determined by immunohistochemistry was positive for ER in 40.9%, and for PR 31.4%, and for both receptors in 27.2% of patients.

Lastly, the pattern of breast cancer among white American, African-American, and non-immigrant West African women were reviewed by Ijaduola and Smith (*20*). Among the three populations, breast cancer appears to be least common in non-immigrant West African women. The peak incidence in African American and West African occurs around the premenopausal period while it occurs in postmenopausal period in whites. Also, the first two groups present late for treatment with a greater cancer burden and consequently lower survival rates. The highest percentage (33%) of infiltrating poorly differentiated anaplastic carcinoma occurs in the West African patients.

5. SPECIAL TYPES AND ASSOCIATIONS

5.1 Inflammatory Breast Cancer

Inflammatory carcinoma (IBC) is one of the most aggressive types of breast cancer. This type of cancer has been considered a clinico-pathological entity characterized by diffuse brawny induration of the skin of the breast with an erysipeloid edge, usually without an underlying palpable mass. A strikingly high incidence of such form of breast cancer has been described in Tunisia and Egypt compared to western reports. It is unclear, however, whether such a difference is related to basic biologic characteristics of the disease or variability in diagnostic criteria. Two types are usually distinguished: (1) primary type, in which inflammatory changes appear simultaneously with the carcinoma, and (2) secondary type: in which the inflammatory manifestations appear in breast with long-standing carcinoma.

The PEV staging system for tumour evolution is usually adopted for staging patients with inflammatory breast cancer (suggest also to see the chapter in this book on "Breast Cancer Aggressiveness in Women of African Descent" by Paul H Levine and Carmela Veneroso). In a study (*21*) of 73 patients classified according to this staging system and conducted at Cairo NCI in the 1980s, 48 cases were diagnosed as inflammatory (PEV2 and PEV3), and 25 were non-inflammatory (PEV0 and PEV1). Forty-five cases, which were verified histopathologically, consisted of 35 having the primary and 10 the secondary types. Twenty-nine per cent of the inflammatory cases were postmenopausal. The median age of such cases was 42 with a peak incidence in fifth decade of life (44%). Bilateral breast involvement was encountered in 4 cases. All the IBC cases had lymph node involvement at presentation, 75 had axillary, and 25% had both axillary and supraclavicular nodes. A great number of cases occurred during pregnancy (27%). A bigger tumour size at presentation (mean 7.2 cm) was also observed. The majority of IBC cases were ER negative (73%). The relation between evolutionary phase and 5-year relapse-free survival are presented in Table 1.

Table 1. Rrelation between evolutionary phase and 5-year relapse-free survival ($p<0.001$).

Evolution Phase	*No. of Patients*	*Survival*
PEV0	233	46.0%
PEV1	56	26.0%
PEV2	65	18.5%
PEV3	14	21.4%
Total	368	

5.2 Breast Cancer and HIV Epidemic in African Populations

The incidence and some other changes of different cancers, including breast cancer among patients with HIV/AIDS infection have been observed and suggested. In one study (*22*), a statistically significant decrease in the incidence of breast cancer was observed during AIDS epidemic period in both males and females. However, this suggested protective effect of HIV infection against breast cancer was challenged in a subsequent study where the authors concluded that it is the AIDS associated mortality which commonly occurs in the second and third decades of life that probably have changed the demographic of the disease in an African population (*23*).

5.3 Frequency and Genome Load of EBV in Breast Cancer Cases

Few data have explored a possible association between EBV and breast cancer with conflicting results. In one of these studies, PCR of a sub region of EBV BamHIC encoding to EBERs demonstrated that 31.8% of the tumours contained the viral genome. However, the presence of EBV genome was not correlated with age, menopausal status, tumour size, nodal status, or histological grade (*24*).

5.4 Factors Responsible for Late Presentation of the Disease in the African Population

Because late presentation of patients with breast cancer is common in most African nations, it is crucial that this phenomenon be addressed as a research objective. Early diagnosis is an important determining factor in lowering treatment cost and in improving quality of life and survival of breast cancer patients.

One study from Mali (*25*) has suggested that the medical itinerary of the patients from traditional medicine to conventional medicine was the most frequently observed factor for late diagnosis. Furthermore, few patients would have received appropriate information about their illness before their specialized consultation. A cross-sectional survey conducted among nurses in Nigeria (*26*) has found that only 30% had had a clinical breast examination, and 8% a mammogram within the past three years. Twenty eight percent did not know how to estimate the risk of cancer and 61% believed they were not at risk (Please, see Christopher Williams' contribution on "Barriers to successful management of breast cancer" – Chapter 14 in this book – for further discussion of this subject).

Thus many different determinants including socio-economic, cultural, geographic accessibility to medical centres with oncologic services, availability of traditional healers and other factors affect patients with breast cancer in their decision to obtain early medical help, as well as to refrain from the proposed therapeutic methods (surgery, radiotherapy and chemotherapy).

REFERENCES

(1) Parkin DM, Bray F, Ferlay J, Pisani P: Estimating the world cancer burden: Globocan 2000. International Journal of Cancer. 2001;94:153-6.

(2) Ferlay J, Pisani P, Parkin DM, editors: Globocan 2000: IARC Press; 2001.

(3) Sankaranarayanan R, Black RJ, Parkin DM: An overview of cancer survival in developing countries. In Sankaranarayanan R, Black RJ, Swaminathan R, Parkin DM, editors. Cancer survival in developing countries. Lyon: IARC Press, 1998, p. 135-75.

(4) Parkin DM, Whelan SL, Ferlay J, Raymond L, Young JL, editors: Cancer Incidence in Five Continents. Lyon: IARC Press; 1997.

(5) Stewart WB, Kleihues P, editors: World Cancer Report. Lyon. IARC Press, 2003.

(6) Ibrahim AS, Hussein H, Ismail K, Hablas A, Abdel BI, Ramadan M, editors: Cancer profile in Gharbiah-Egypt; Methodology and results; 1999.

(7) Hoffman M, de Pinho H, Cooper D, Sayed R, Dent DM, Gudgeon A, et al: Breast cancer incidence and determinants of cancer stage in the Western Cape. South African Medical Journal. 2000;90:1212-6.

(8) Stavros AT, Thickman D, Rapp CL, Dennis MA, Parker SH, Sisney GA: Solid breast nodules: use of sonography to distinguish between benign and malignant lesions.[comment]. Radiology. 1995;196:123-34.

(9) El Gemeie E, El Didi M, Nouh A: A fine needle aspiration of breast lesions: the experience of the Egyptian National Cancer Inst. J. Egyptian National Cancer Institute 1997;9:239-46.

(10) Salih AM, El Morsy B, Mohammed NH: Cytopathological diagnosis of breast lesions among Egyptian by fine needle sampling without aspiration: using only the needle. J. Egyptian National Cancer Institute 1996;8:203-11.

(11) Dipo-Ba S, Badian M, Niang EH, Ly Ba A, Alfidja A, Favre Y: Contribution of mammography and echography to the diagnosis of breast cancer in Dakar: apropos of 25 cases. Dakar Medical. Dakar Medical 1997;42:114-17.

(12) El Badawi M, Zaki I, Tantawi S: Magnetic response mammography diagnostic potentials in breast cancer. J. Egyptian National Cancer Institute 1999;11:91-98.

(13) Saber RA: Assessment of carcinoma by 99mTC-Sestamibi scintimammography: Effect of lesion site, size, and histopathological characteristics. J. Egyptian National Cancer Institute 2000;12:183-89.

(14) Mbonde MP, Amir H, Mbembati NA, Holland R, Schwartz-Albiez R, Kitinya JN: Characterisation of benign lesions and carcinomas of the female breast in a sub-Saharan African population. Pathology, Research & Practice. 1998;194:623-9.

(15) Mbonde MP, Amir H, Schwartz-Albiez R, Akslen LA, Kitinya JN: Expression of estrogen and progesterone receptors in carcinomas of the female breast in Tanzania. Oncology Reports. 2000;7:277-83.

(16) Gebremedhin A, Shamebo M: Clinical profile of Ethiopian patients with breast cancer. East African Medical Journal. 1998;75:640-3.

(17) Anyanwu SN: Breast cancer in eastern Nigeria: a ten year review. West African Journal of Medicine. 2000;19:120-5.

(18) Maalej M, Frikha H, Ben Salem S, Daoud J, Bouaouina N, Ben Abdallah M, et al: [Breast cancer in Tunisia: clinical and epidemiological study]. Bulletin du Cancer. 1999;86:302-6.

(19) El Bolkainy MN: Breast cancer. Cairo: The National Cancer Institute, Cairo University; 2000.

(20) Ijaduola TG, Smith EB: Pattern of breast cancer among white-American, African-American, and nonimmigrant west-African women. Journal of the National Medical Association. 1998;90:547-51.

(21) Omar SH, El Aaser A, Eissa S, al. e: End results of breast cancer therapy. II: Clinical immunological and hormonal profile of inflammatory carcinoma of the breast. A.: Clinicopathological aspects. Cancer Prevention & therapy 1989;1.

(22) Amir H, Kaaya EE, Kwesigabo G, Kiitinya JN: Breast cancer before and during the AIDS epidemic in women and men: a study of Tanzanian Cancer Registry Data 1968 to 1996. Journal of the National Medical Association. 2000;92:301-5.

(23) Amir H, Makwaya C, Mhalu F, Mbonde MP, Schwartz-Albiez R: Breast cancer during the HIV epidemic in an African population. Oncology Reports. 2001;8:659-61.

(24) Fina F, Romain S, Ouafik L, Palmari J, Ben Ayed F, Benharkat S, et al: Frequency and genome load of Epstein-Barr virus in 509 breast cancers from different geographical areas. British Journal of Cancer. 2001;84:783-90.

(25) Ly M, Diop S, Sacko M, Baby M, Diop CT, Diallo DA: [Breast cancer: factors influencing the therapeutic itinerary of patients in a medical oncology unit in Bamako (Mali)]. Bulletin du Cancer. 2002;89:323-6.

(26) Odusanya OO, Tayo OO: Breast cancer knowledge, attitudes and practice among nurses in Lagos, Nigeria. Acta Oncologica. 2001;40:844-8.

Chapter 6

BREAST CANCER AGGRESSIVENESS IN WOMEN OF AFRICAN DESCENT

Paul H. Levine, MD and Carmela Veneroso, MPH
The George Washington University School of Public Health and Health Services, Washington, DC, USA

1. INTRODUCTION

While the incidence of breast cancer is higher in White women (115.5/100,000) than African-American women (101.5/100,000), the mortality pattern is just the opposite (1). Not only do African-American women have a higher mortality (31.0/100,000) than White women (24.3/100,000) but the mortality rates are falling more rapidly in White women. There are many factors that may contribute to these disparities, such as inequalities in access to healthcare and poverty (2-4) and lower education levels (5). These factors create barriers to health care access contributing to African-Americans being diagnosed at a later stage of disease (6-11). Some studies have shown that socio-economic factors are associated with a poorer outcome and may account for some of the difference between African-Americans and Whites (12-17). However, even when African-Americans have equal access to medical care, there are still racial differences in outcome (18-19). In the Jatoi et al.'s 2003 study of breast cancer survival in the U.S. Department of Defence's Healthcare System, an "equal access system", (18) they found, after adjusting for age and stage, that not only was there a disparity in survival between African-American women as compared to Whites, but that this disparity has increased since 1980. Other studies show that factors other than delay in presentation and socio-economic status explain some of the disparity and that African-American women still have poorer survival after controlling for these factors (20-22).

C. Williams et al. (eds.), Breast Cancer in Women of African Descent, 91-118.

There is increasing evidence that breast cancer is more aggressive in African-Americans and that African-Americans have more biologically aggressive tumours defined by specific markers that are associated with a worse prognosis or worse survival. One marker of aggressive breast cancer on which there is general agreement is tumour grade. Many studies have shown that histological grade is a statistically significant prognostic factor for disease free survival and overall survival (23-34). Grade is evaluated at time of diagnosis and therefore reflects events occurring in the tumour before diagnosis and treatment. Grade provides measurements of differentiation, nuclear grade, and mitotic count, important parameters in the aggressiveness of the tumour. Other markers associated with worse prognosis and more aggressive disease are negative hormone receptors (35-36), aneuploid tumours (37-38), high s-phase (39-40) , and increased microvessel density (41). Many studies have shown that a larger percentage of African-American women as compared to white women have these markers (6-7), (42-55).

One form of breast cancer that offers an excellent opportunity to identify aggressive breast cancer is inflammatory breast cancer (IBC), one of the most aggressive forms of breast malignancies. IBC reportedly comprises only 1-6% breast cancer cases has the greatest impact on African-American women. Some investigators categorize IBC as a subgroup of locally advanced breast cancer (LABC)(56-57), but as noted by Wolff and Davidson (58) despite the inclusion of IBC in most classifications of LABC, it has a distinct clinical behaviour and worse prognosis. As noted below, the importance of chemotherapy as primary treatment for IBC is based on its early dissemination of micrometastases which are more susceptible to destruction before they have a chance to develop resistance.

Analyses of data from Surveillance, Epidemiology, and End Results (SEER) Program (59-60) document the greater impact of IBC on African-American women than any other racial/ethnic group, but the extent of this impact depends on the case definition. In our earlier report (59), where we analysed SEER data between 1975 and 1981 involving 56,683 cases of invasive breast cancer in women, (51,030 White, 3,834 Black, and 1,819 other non-White), we concentrated on analysis of data from White patients because of the larger number of cases. In this group, patients were classified as having clinical features of IBC without pathologic confirmation (1,181 patients), pathologic features of IBC without clinical features (38 patients), both clinical and pathologic features of IBC (62 patients), and no evidence of IBC (25,089 patients). Using the broadest definition of IBC (all three IBC groups), 10.1% of African-American women with breast cancer had evidence of IBC as compared to 6.2% in White patients and 5.1% in other non-Caucasians. With the requirement for pathologic confirmation, the percentage dropped to only 0.7% of African-American breast cancer patients having IBC vs. 0.5% in Caucasians (59). In a follow-up study encompassing 1975-1992 (60), there was confirmation that the incidence of IBC in African-American women is significantly higher than White women (1.1 per 100,000 person years as compared to 0.7 per 100,000 person-years), the relatively small number resulting from the exclusion of those cases with only clinical features of IBC (in our earlier study there were 11 times the number of patients with only clinical features as those with pathological evidence of IBC (59)). A second intriguing finding in Chang's study is that the incidence in both African-American and White women doubled between 1975-77 and 1990-92. Whether this represents a true increase in IBC or a greater awareness of the need for skin biopsies in IBC patients to document invasion of the dermal lymphatic system remains to be demonstrated. In our experience with the Inflammatory Breast Cancer Registry (IBCR) (61), a recently initiated project designed to obtain uniform clinical and epidemiological information as well as biospecimens from IBC patients throughout the United States and Canada, we see that a higher proportion of women are getting multiple skin biopsies to document pathologic involvement, and in one cluster of patients in California, the surgeon had to take more than ten biopsies before "pathologic proof" of IBC could be found (Levine, unpublished data).

In this chapter, we will review some of the pertinent studies that have shed light on the problem of aggressiveness and emphasize the emerging data on risk factors for aggressive breast cancer, which actually cross into all racial/ethnic groups and are the target of intensive research in our University.

2. HISTORICAL ASPECTS

A focus on aggressive breast cancer has generally been attributed to an English surgeon, Sir Charles Bell, who noted in 1814 that an enlarged purplish painful breast was a poor prognostic sign (62). Several authors have noted breast cancer associated with pregnancy also tends to be more aggressive with a poor prognosis (63-64) but this has not been universally accepted (64). Until recently, however, a consistent focus on aggressive breast cancer was not possible because the tools for detecting these cases were not readily available. Only when there were dramatic clinical signs, as with IBC, could a poor survival be predicted. The emphasis on diagnosing IBC as a clinical entity continued when Taylor and Meltzer in 1938 emphasized the clinical manifestations but noted that invasion of the dermal lymphatic system should be considered as "pathologic proof" (65). As noted above, this is the approach currently adopted by AJCC (66).

The move towards a pathologic definition began with the 1974 report of Ellis and Teitelbaum (67) based on their examination of skin biopsies in five long term IBC survivors noting that "none of these patients had dermal lymphatic metastases." Further support of IBC as a pathologic entity was provided by Saltzstein (68) who described the opposite end of the spectrum, noting dermal lymphatic invasion in four patients with rapid progression of breast cancer but no clinical evidence of IBC. He used the term "clinically occult inflammatory breast cancer," which we identified in the SEER database as Group III and which appears to have a worse prognosis than those with only clinical manifestations (59). In 1978, Lucas and Perez-Mesa (69) documented a poor survival in their 58 patients with clinical IBC and 15 patients with "occult" inflammatory cancer, thus indicating that either clinical or pathologic features were sufficient to support a poor prognosis.

In France's Institut Gustav Roussy, Denoix developed a terminology to investigate aggressive breast relying more on the rapidity of tumour growth than any other characteristic (70). Using the term "*pousée évolutive*" (PEV) to designate rapidly progressing breast cancer, he defined four forms: PEV-0 is a designation given to patients without inflammatory signs and no history of rapid tumour growth; PEV-1 is a designation given to patients who describe rapid tumour growth but who show no inflammatory signs; PEV-2 is a designation given to patients with inflammatory signs involving less than half of the breast; PEV-3 is a designation given to patients with inflammatory signs involving more than half of the breast. PEV-3 would be recognized as inflammatory breast cancer readily by clinicians worldwide. Investigators at the Institut Salah Azaiz in Tunisia noted that 58.5% of the 581 cases of breast cancer seen there between 1969-1974 were PEV positive (PEV-1, PEV-2, or PEV-3); of the 581 cases, 48.5% were PEV-3 (the IBC equivalent) and 10% were PEV-1 and PEV-2. (71). This finding launched a series of studies that have proven to be highly relevant to current studies of IBC in North America. Among the more important findings were the dramatic improvement in survival with neoadjuvant therapy (72-73), the observation that the risk factors for breast cancer aggressiveness differed from those for developing breast cancer (74) (see below), and the indication that there was a higher percentage of patients with evidence for a human breast cancer virus (75). More recently, we examined biopsies from 45 Tunisian patients using molecular techniques and found that microvessel density was significantly higher in those with clinical features of IBC compared to those without (41), indicating increased angiogenesis in the Tunisian IBC patients.

3. THE IMPACT OF AGGRESSIVE BREAST CANCER ON AFRICAN-AMERICAN WOMEN

As discussed above there are many studies that indicate that African-Americans have more aggressive disease (6-7), (42-55), (59). While data from the SEER program of the National Cancer Institute (NCI) document a worse tumour grade (43), an indicator of more aggressive cancer, and a poorer survival in African-American women, the strongest data for the importance of tumour aggressiveness come from studies of breast cancer in the "equal access" health care system of the military (18-19). While these studies looked at survival and not directly at tumour aggressiveness, the implication is that all patients in the military had essentially the same treatment for their disease and therefore the poorer survival in African-American women was not due to unequal access to care. Therefore, since many other studies show that African-Americans have tumours that are more biologically aggressive, aggressiveness may be the key factor in the survival difference in the military population. However, not all studies are in agreement. English et al. (76) found that there was no difference in overall survival or survival by stage in a study of 585 African-American and white women treated in their university teaching hospital between 1990 and 1999, despite the fact that the African-American patients were younger, presented with higher-stage tumours, were more often to have positive axillary lymph nodes, were more often to have negative oestrogen and progesterone receptors, and were more often premenopausal.

4. IMPORTANT RESEARCH QUESTIONS IN AFRICAN AMERICAN WOMEN

4.1 What is the Relative Impact of Tumour Aggressiveness vs. Access to Care on Mortality Rates in African-American Women?

At the present time, there are no available data that address the question of the relative impact of tumour aggressiveness vs. access to care on the poorer survival of African-American women with breast cancer. However, the poorer survival of African-American women in "equal access" studies (18-19), suggest that tumour aggressiveness may have a major impact on African-American women. A number of studies have noted that there is a major impact from socio-economic status, access to care, co-morbidities, quality of insurance and different treatment strategies (5), (77) but these issues are likely to be exacerbated by tumour aggressiveness. More recently we have investigated this issue on a nationwide basis using SEER data comparing tumour grade in African-American and Caucasian women by stage at presentation (43). We found that regardless of disease stage, the histological grade of the tumour was significantly higher in the African-American women. There may well be better markers of aggressiveness, such as molecular markers, but tumour grade is useful since it is determined at the time of diagnosis before treatment. While lymph node involvement is an important prognostic indicator, it does not distinguish between slow growing tumours that have been present for a considerable time and rapidly growing tumours of recent onset.

4.2 What are the Risk Factors for Breast Cancer Aggressiveness?

There is evidence that the development of aggressive breast cancer depends more heavily on environment than genetics (see below). International comparisons of aggressive breast cancer are difficult due to differences in Registry procedures and case definitions. However, one source of data comes from the study of aggressive breast cancer cases at the Institut Salah Azaiz in Tunisia (as discussed above) where 48.5% of all breast cancer patients presented as PEV-3 (the equivalent of IBC), compared to the United States where 1-6 % of all breast cancer cases patients are reported as IBC. Within Tunisia the proportion of PEV positive patients was more in the rural than urban population and there was a suggestion that pregnancy at an early age was a risk factor (74). Other studies discussed below also seem to indicate an environmental influence.

While the aetiology of breast cancer has been studied extensively and many risk factors have been established, risk factors for aggressive breast cancer have not been well studied. The few studies that have been done have looked at these factors in relation to survival and have yielded contradictory results. These contradictory results may be due to the difficulty in controlling for treatment that has an impact on survival. There have been even fewer studies that have looked at these factors in relation to the aggressiveness of the tumour. In some of these survival and aggressiveness studies, factors that are known to be protective of developing breast cancer have been associated with worse survival and/or a more aggressive form of breast cancer. Mourali et al.(74) found that late age at menarche, an established protective factor associated with a decreased risk of developing breast cancer (78), was associated with an increased risk of developing aggressive breast cancer. Korzeniowski et al.(79) found that reproductive factors known to decrease risk, specifically late menarche and parity, were associated with an adverse impact on survival. Kroman et al.(80) found that the mother's age when she gave birth to her first child (age at first birth), a very well established risk factor for decreased risk for developing breast cancer, was associated with a worse prognosis. This finding is compatible with the observation that 14/15 Tunisian women who had their first child under the age of 18 were PEV positive (74). Other risk factors that have been associated with more aggressive tumours are young age at diagnosis, oral contraceptive (OC) use, exposure to organochlorines, and obesity (see below).

4.2.1 Early Age at First Pregnancy

A number of studies have found that women who had their first child at an early age had a poorer prognosis (74), (80-83). In Schouten et al.'s 1997 study of 866 breast cancer patients, they found that young age at first full-term pregnancy was related to decreased survival (82). In Kroman et al.'s 1998 study of the prognosis of reproductive factors in 10,703 women with primary breast cancer in the Danish Cancer Registry, they found that women who had their first child before the age of 20 had a higher risk of dying than women who had their first child at age 20 and above (80). Supportive evidence was provided by Chang's study on IBC, where it was found that IBC patients were younger at the time of their first live birth than non-inflammatory breast cancer patients and non-breast cancer patients (81). And finally in an early study on pre-menopausal women, Greenberg found that women who were older when they had their first child had a better prognosis (83).

However, some studies have not supported a poorer prognosis for young age at first birth. In Lund's study of breast cancer mortality of 800,814 Norwegian married women, women who had their first birth after age 35 had a 2.6 higher risk of mortality than women who had their first birth before age 20 (84). This discordant study is difficult to explain, but could be due to population differences and some factor associated with late age at first pregnancy. In a Northern Alberta study, age at first birth was not found to have a significant effect on survival but this analysis was performed using age at first birth as a continuous variable; direct comparison of women whose first birth was at less than age 20 was not made with those whose first birth was after age 20 (85).

4.2.2 Young Age at Diagnosis

Many studies have found an association between young age at diagnosis and a poor prognosis (86-97). Some of these studies have shown that patients diagnosed at a young age have more aggressive tumours. In Maggard et al.'s 2003 study (94) of 24,935 invasive breast cancer patients using the SEER database (1992-1998), they found that young breast cancer patients had poorer survival as compared with an older cohort and that the younger women presented at a more advanced stage disease and had more aggressive tumour characteristics, that is, higher grade tumours and more oestrogen- and progesterone receptor-negative tumours. Kollias' study (96) found that patients who were less than 35 years old presented more frequently with high grade tumours. In Winchester et al.'s 1996 study of 508,724 breast cancer patients diagnosed from 1985 to 1993 from hospital cancer registries throughout the United Sates, they found that women younger than age 35 had higher grade primary tumours, more advanced disease, and poorer 5-year survival compared to older premenopausal women (98). Walker et al. (97) found that women aged under 35 years had a significantly high incidence of having poorly differentiated tumours, higher proliferation rates, and a significantly high incidence of p53 protein staining. Bonnier et al. (91) found a higher frequency of high grade and undifferentiated tumours, microscopic lymph-node involvement, and negative hormonal receptor status was observed in patients under 35 years.

4.2.3 Oral Contraceptive Use

A complex relationship between OC use and breast cancer prognosis has been evident in many of the studies on OC use and prognosis. A number of early studies showed no association with OC use with prognosis (83, 99-101). However, some of these studies were very small and had few users before their first full-term pregnancy. On the other hand, other studies found use of OC was associated with a poorer prognosis although with conflicting results. Some of these studies found that pre-menopausal patients with a history of OC use had larger tumours, more metastases, lower PR and ER receptors, higher S-phase, frequency of aneuploidy, and poorer survival (102-105). And in Brinton's study (106) on OC use and breast cancer risk among younger women, they found that OC associations were stronger for more advanced tumours. In contrast, however, some studies found a beneficial effect. A more recent study done by Sauerbrei et al. found no effect on survival and found that OC users had smaller tumours (107). In Vessey's study published in 1983 they found that women who never used OC presented with more advanced tumours; however, only small numbers of cases and controls had prolonged OC use before their first pregnancy (108). The conflicting data regarding the effect of OC use on tumour aggressiveness and survival may be the result of the different approaches to evaluating the length of time of OC use as well as incomplete information on dosage of oral OCs.

Other studies also found that the effect of OC use varied, depending on the length of use. A study published in 1994 by Holmberg suggested that short duration of use had a favourable effect on the prognosis (109). Holmberg found that 5-year survival estimates for users of 1-3 years (short-term users) had a significantly better prognosis than never users, while users of four years or more had a non-significant worse prognosis. Yet, in Schonborn's study also published in 1994, they found that long-term use of OC had a significantly increased 5 year survival time, but only significantly for those who used OC for greater than 4 years (110). They also found that long-term OC use increased survival for patients with poor histopathological prognostic factors (number of positive nodes, large tumours, low ER, low PR, histological grade). Of note, was that they found statistical significance on all of the factors except tumour grade. They found that long-term users had a statistically significantly higher number of poorly differentiated tumours, perhaps suggesting an effect of OC use on tumour biology. This poor tumour biology should suggest a worse prognosis. Interestingly, they also found a significant correlation between long-term use and current use. Perhaps this strong correlation of long-term users with current users suggests current OC use (hormonal influences) has an effect on the behaviour of the tumour in the subclinical phase. The results may have been different if they did not include current users. Finally, in a more recent study published in 1997, Schouten et al. looked at the association between OC use and survival (82) and found no association with prior use of OCs.

Other studies looked at the effect of the age at which the woman started her use of OC. A couple of studies showed that survival was worse in women who had started OC use before the age of 20 (103, 111). In a study by Olsson et al. in 1991 of primary tumour specimens from 72 premenopausal women, they found that amplification of Her-2/neu, which is associated with more aggressive tumours (112-113), was much more common among OC users who started using OC before the age of 20 (114). They found no significant associations between amplification and the variables of parity, age at first full-term pregnancy, or late abortion, suggesting that the higher rate of Her-2/neu amplification among early OC users is an effect of the OC use per se rather than of the relative youth of the users. However, Holmberg found no evidence of a worse prognosis for women who used OCs at an early age (109).

4.3 Other Risk Factors

Other risk factors that have been associated with more aggressive tumours are exposure to organochlorines and obesity. Although Demers et al. (115) found no relation between organochlorines and the risk of developing breast cancer, they found that some organochlorines and especially p,p'-DDE was associated with breast cancer aggressiveness. Specifically they found a probability of lymph-node invasion among breast cancer cases with increased exposure to 1,1-dichloro-2,2-bis (4-chlorophenyl) ethylene and that p,p'-DDE exposure was associated with a dose-related increased relative risk of exhibiting both lymph-node involvement and a large tumour. Similar associations were noted with beta-hexachlorocyclohexane, oxychlordane, and transnonachlor. Woolcott et al.'s study (116) found that many polychlorinated biphenyls (PCB) were more strongly associated with tumours of poor prognosis, that is tumours which were larger and higher-grade and oestrogen receptor negative.

Finally, several studies have found that obesity was associated with a worse prognosis (81), (117-118). Daling et al. (118) found that the women younger than 45 years of age in the highest quartile of BMI were more likely to be oestrogen receptor negative and have a high S-phase fraction, a high histological grade, a high mitotic cell count, and large tumour size compared with the tumours of women whose BMI was in the first quartile. Relative to the large tumours in women in the lowest BMI quartile, the large tumours in women in the highest BMI quartile were more likely to express markers of high proliferation, indicating they may have grown faster than similar size tumours of the thinnest women. Finally Chang et al. found that high BMI was significantly associated with an increased risk of IBC (81).

5. HOW SHOULD AGGRESSIVE BREAST CANCER AND IBC BE DEFINED?

There are a number of studies that have investigated racial differences using various markers of aggressiveness and as noted earlier many of those studies looked at the differences in tumour grade. African American women have been observed to have higher grade tumours compared with white women (43, 45, 47, 119). More recently, a study by Henson et al.(43) used SEER registry data from 1992-1999 and looked at the correlation between survival and histological grade, stage of disease, and tumour size for African-American and white women. This study found that for nearly every combination of stage and grade, regardless of age, African-American women presented with proportionally more Grade III and fewer Grade II tumours.

Higher grade was associated with a less favourable 6-year cause specific survival. (The difference was not statistically significant for every combination of grade and stage, but it was observed in 12 of the 13 combinations analysed).

There have been a number of studies that looked at differences in different markers of aggressiveness between African-American and white female breast cancer patients. Many of these studies have shown that African-American women are more likely to present with oestrogen receptor negative tumours (42), (46), (54-55) and high s-phase (46). Research may not only benefit from better classification of aggressive breast cancer, but also a more acceptable and consistent definition of IBC, the prototype of aggressive breast cancer.

The current definition of IBC continues to be problematic. Not only do individual clinicians differ in their criteria for diagnosis of IBC but national organizations also disagree. The American Joint Committee on Cancer (AJCC) emphasizes the clinical features and states that "Inflammatory carcinoma is a clinicopathologic entity characterized by diffuse erythema and oedema (peau d'orange) of the breast (which) should involve the majority of the skin of the breast." "It is important to remember that inflammatory carcinoma is primarily a clinical diagnosis. Involvement of the dermal lymphatic system alone does not indicate inflammatory carcinoma in the absence of clinical findings"(66). In contrast, the National Cancer Institute's Surveillance, Epidemiology and End Results (SEER) Program defines IBC as a clinical diagnosis verified by biopsy of the tumour and overlying skin (120). The two major reviews of IBC using SEER data (59-60) note that although a relatively small group, SEER does include cases defined pathologically without any clinical evidence of disease. The implications of these disparate definitions and approaches are considerable and explain notable differences in estimates of the incidence of IBC, even using the same source of data. Our first review of SEER data used cases fitting clinical or pathologic criteria identified between 1975 and 1981 (59). In an updated analysis including patients identified by SEER through 1992, Chang et al. evaluated only women who had pathologically defined IBC because of their concern that the clinical classification could include cases of neglected breast cancer (60). Regardless of whether the criteria are primarily clinical or pathological, it is clear that the prognosis is generally much worse than for any other form of breast cancer (59).

Cristofanilli et al. (56) report that IBC patients "usually present with a rapid onset of swelling of the involved breast." They add that "The classic criteria established by Haagenson (121) include diffuse erythema, oedema involving more than two-thirds of the breast, peau d'orange, tenderness, induration, warmth, enlargement and diffuseness of the tumour on palpation." While this is indeed the classic definition, more often patients are being diagnosed at a much earlier stage where the redness may be far more limited and there may initially be peau d'orange without erythema or the converse. Waiting until the breast shows the classic findings can seriously diminish the chances of a cure, which is now possible in a significant percentage of patients.

6. CURRENT RESEARCH

6.1 Case Definition

The investigation of a disease, whether related to aetiology, pathogenesis or control, requires a tight case definition. As discussed above, there is no universally accepted case definition for aggressive breast cancer or for IBC. Investigators often rely on tumour grade but other criteria, such as hormone receptor and Her2-neu status have been used. Similarly, there is disagreement as to the precise case definition of IBC. The AJCC (66), which defines IBC as predominantly a clinical disease involving more than half of the breast, may well be inadequate since the diagnosis is made with far less clinical involvement, and this early diagnosis appears to be appropriate.

In 2001 we initiated the Inflammatory Breast Cancer Registry to describe the variations in the diagnosis of IBC and to attempt to determine if molecular diagnostic tools could be identified to bring some improved classification to a disease defined so differently by diverse organizations and clinicians. Registries for relatively rare diseases have been useful in the study of other rare malignancies, where existing case definitions masked the presence of multiple entities within the same category. For example, the American Burkitt's lymphoma (BL) Registry (122-124) helped to clarify the fact that a single pathologically defined entity actually consisted of at least two biologically distinct diseases, one characterized by the presence of the Epstein-Barr virus within the tumour cells and different responses to chemotherapy. The Epstein Barr virus (EBV) –associated BL has a consistent chromosome translocation (125), is predominant in sub-Saharan Africa and other areas of holoendemic malaria and appears to respond to less aggressive chemotherapy than the non-EBV-associated BL, which is the predominant form in the United States (124).

To improve the evaluation of American Burkitt's Lymphoma, we divided our cases into subgroups based on the quality of the diagnostic pathology material. A similar approach is being taken with the IBCR, which divides patients with IBC into subgroups according to the degree of clinical and pathologic criteria.

In the first report from the IBCR (61), among the intriguing findings are the large number of patients who have clinical findings involving less than half of the breast, not the AJCC definition of IBC. In fact, most of the patients in our IBC Registry diagnosed by practicing physicians do not present with this classical form. In a significant number of our Caucasian patients, the first symptom was the appearance of a small pink spot, with no obvious peau d'orange or noticeable breast swelling at the beginning. This early manifestation may not be noticed in African-American women, leading to a later diagnosis.

Unfortunately, the diagnosis of IBC is delayed considerably in a large percentage of women because the clinician does not consider IBC, probably because of inexperience with the disease. This problem is not restricted to primary care physicians as we have had patients referred to surgeons with the possible diagnosis of IBC and the patient has been placed on weeks to months of antibiotics. Approximately one third of our patients were given antibiotics for an infection for up to five months for some before the diagnosis of IBC was made. Young women with a painful breast, a common presenting feature in IBC, were often told that they could not have breast cancer because they were too young and breast cancer is not painful. In our series, however 34% of our IBC patients presented with breast pain. Furthermore, mammography is often not helpful; in our IBCR series to date only 30% of the diagnostic mammograms showed a discrete mass. Our data are similar to those of Kushwaha et al (126) who reported that a mass could be detected in only 15% of their cases. In contrast is the report by Dershaw et al. (127) who observed discrete masses in 21/22 of their IBC patients. In general, however, not only the diffuseness of the tumour but its frequent occurrence in young women with dense breasts interferes with the diagnosis.

Studies are now in progress to test the tissues from the patients in this Registry by a number of molecular techniques to determine if there are different identifiable subgroups of IBC.

6.2 Molecular Biology

A number of molecular approaches are currently being pursued to understand more thoroughly the aetiology, pathogenesis and control of aggressive breast cancer and IBC. There are numerous examples of molecular markers being important tools to identify sub-groups of disease, which could have important etiologic and prognostic implications. Non-Hodgkin's lymphoma (NHL) provides many such examples, with B and T-cell markers now being used extensively in classification. In one form of

NHL the identification of human T-lymphotropic virus type-I (128) is used to distinguish classic adult T-cell lymphoma from morphologically similar tumours, a critical factor in understanding the aetiology of this tumour. In another form of NHL, Burkitt's lymphoma, at least two subtypes have been defined with the detection of Epstein-Barr virus in the tumour cells of some patients being associated with specific chromosome translocations and other genetic markers (125). It is hoped that similar molecular efforts can be utilized to better define aggressive breast in general and IBC in particular.

As noted in the Introduction, "locally advanced breast cancer" is a completely inappropriate term to be used for IBC because it is apparent that the disease is systemic when first detected. In addition to the invasion of the dermal lymphatic system, microemboli are another hallmark of IBC and the spread of the these tumour cells systemically explains why successful treatment of IBC relies primarily on neoadjuvant therapy which is more likely to destroy tumour cells before they have had an opportunity to establish their defences. It is likely that the increased invasiveness of IBC as compared to non-IBC breast cancer has some molecular counterparts. Several that have been suggested include increased angiogenesis (41), the loss of expression of a novel gene initially called LIBC and but later identified as WISP3 (129), and the increased expression of e-cadherin (130). Reasonable mechanisms have been proposed for each of these observations. Angiogenesis as identified by increased microvessel density (MVD) was identified in Tunisian breast cancer patients with objective signs of IBC as compared to other Tunisian breast cancer patients without these signs. Increased angiogenesis is associated with rapid tumour growth as the increased vasculature helps to nourish the tumour. (131). Loss of LIBC was found in a study by van Golen et al. (129) where they investigated 29 IBC and 19 non-IBC stage III archival breast samples and they found a significant difference in the expression of the LIBC gene which was expressed in only 20% of the IBC tumours and in 79% of the non-IBC tumours. They also found that transcript T6, RhoC GTPase was overexpressed in 90% of the IBC samples, in comparison with only 38% of the non-IBC samples. When comparing the concordance of having both of these genes altered, they found that the loss of LIBC and the overexpression of RhoC occurred in 91% of the IBC tumours whereas concordance was not seen in any of the non-IBC samples. In Kleer et. al.'s study of 20 IBC and 22 non-IBC matched by stage, they found a strong association between E-cadherin and IBC. All the IBC patients' tumours expressed E-cadherin, whereas only 68% of the non-IBC patients' tumours expressed the protein, and the intralymphatic tumour emboli in the IBC cases also expressed E-cadherin (130). Using a human/mouse model of IBC (where human breast carcinoma was grafted in scid/nude mice), Alpaugh et al. (132) found a 10-

20 fold overexpression of E-cadherin in the IBC xenografts as compared to the non-IBC xenographs, and in a later study (133) they found that E-cadherin was involved in the passive dissemination of tumour emboli in IBC.

Among the recent genetic studies, Lerebours et al. (134) reported more genetic alterations in IBC patients compared to non-IBC patients. Specifically they found loss of heterozygosity (LOH) patterns in IBC patients that were less frequent in non-IBC patients and that LOH patterns differed between patients with localized and extensive breast inflammation. They also found that extensive breast inflammation at the first clinical examination was associated with a poorer outcome and the overall frequency of LOH was also higher in this group. While the progress being made in the laboratory is highly encouraging, much remains to be done.

Another interesting tool that is being applied to IBC is the investigation of viral footprints. Viral studies in breast cancer have a long history (for a review, see Robert-Guroff M, Buehring GC (135). Early virologic techniques (including electron microscopy) had their basis in the study of the mouse mammary tumour virus (MMTV) as a model for a human breast cancer virus. A focus on the relationship of MMTV-associated antigens and molecular sequences to aggressive breast cancer began in 1984 when we applied the findings of Sol Spiegelman and his colleagues to our studies of aggressive breast cancer in Tunisia. Spiegelman's laboratory had noted that human breast cancers contained an antigen that cross-reacted with the gp-52 of MMTV (136-137) and in our initial applications to the Tunisian study, a far higher proportion of cases (70%) were noted to have this antigen than had been found in U.S. cases (30%) (75). These findings have gained support in preliminary studies using more recent molecular techniques (138-139) with a tendency for more MMTV-related antigenic and molecular expression in the more aggressive PEV cases than the non-PEV controls (140). At the present time, studies are in progress to investigate further the geographic patterns of these MMTV-like sequences but currently there is no definite relationship to aggressiveness. However, in view of the apparent increase in aggressiveness when a breast cancer arises during pregnancy, and the increased incidence of MMTV-related sequences in breast cancer associated with pregnancy and lactation (62% vs. 30-38% in U.S. cases) (141), the possible relationship between viral footprints and breast cancer aggressiveness needs to be pursued. Whether or not these sequences prove to be truly related to a human breast cancer virus or to aggressive breast cancer, the definition of subgroups of breast cancer by current laboratory methods is a promising field since such approaches have been useful in classifying other malignancies, such as Burkitt's lymphoma and other non-Hodgkin's lymphoma.

6.3 Epidemiologic Studies

6.3.1 International Patterns

International comparisons regarding the incidence of aggressive breast cancer and IBC are extremely difficult because of differences in case definition and the quality of the data in different registries. Based on the data available, however, current research into the patterns in different countries is extremely important. Several studies, for example, indicate that Africa has a higher proportion of cases of aggressive breast cancer compared to the United States (71), (142-146). The studies in Sub-Saharan Africa are not as population-based as in North Africa and one of the major concerns in case definition, as in all countries, is distinguishing IBC from neglected or locally advanced breast cancer. As the methods for defining aggressive breast cancer on a pathologic and molecular basis evolve, however, international comparisons should become more feasible.

The early reports indicating that north Africa had a significantly higher proportion of patients with aggressive breast cancer (PEV) and particularly those with clinical signs of IBC (PEV-2 and PEV-3) than virtually any other country is an observation that is now being confirmed and extended by standardized methods. The earlier study at the ISA in Tunisia went beyond the well documented clinical findings and showed that the aggressive breast cancer patients had notable differences in pathologic features (147), hormonal patterns (148), and molecular patterns (primarily micro-vessel density (41) than the non-aggressive cases, confirming the validity of the PEV-2 and PEV-3 classification at that time. Another intriguing finding was the identification of an antigen indistinguishable from the gp-52 of the mouse mammary tumour virus in 70% of Tunisian breast cancer cases vs approximately 30% of U.S. cases (75). The viral footprints were somewhat more apparent in the PEV cases than the non-PEV cases (140), but the most important observation was the overall increase in all Tunisian cases. These findings are currently being confirmed by comparable studies using current molecular techniques (149). A more recent report from Tunisia by another group of ISA investigators provided an interesting follow-up through a national survey of breast cancer patients (142). This study included breast cancer patients throughout Tunisia between Jan. 1, 1994 through Dec. 31, 1994 and compared their findings with the report by Tabbane et al. focused on patients at the Institut Salah Azaiz, the major cancer centre, between 1969-1974. The current study found the mean clinical tumour size to decrease 5-6 mm every 10 years from 63.9 mm in 1969-74, 55.8 mm in 1981-5, and 49.5 mm in 1994. Concomitantly, the percentage of patients with any objective clinical findings of IBC (including PEV 2, which involves less than half of the breast) represented only 23.2% of the 1994 cases vs. 55.2% in the earlier ISA series. The percentage of PEV 3 or T4d cases, comparable to IBC In the AJCC classification, declined from 48.7% in the early series to 6.2% in the 1994 series. These data suggest that the proportion of cases with aggressive breast cancer is decreasing, providing strong support for the importance of environmental factors on the aetiology of aggressive breast cancer.

The difficulty in dissecting neglected or locally advanced breast cancer from aggressive breast cancer is described in a Nigerian study (144) , where a series of 116 Nigerian women seen at the University of Benin Teaching Hospital from 1974-79 was reported. Slightly over 10% (12 patients) of the Nigerian patients were either pregnant or lactating and 99 (85.3%) of the study group presented with TNM Stages III and IV disease. Evidence for tumour aggressiveness is provided by the pathologic observation that 50% of the patients had anaplastic carcinomas.

6.3.2 U.S. Patterns

As the SEER Registry improves its identification of IBC and we learn how to better use the existing data, it is possible to re-examine the earlier reports (59-60) on IBC patterns in the United States. We are again analysing and updating the trends in the incidence of IBC and survival with this disease using new as well as old SEER classifications. We have analysed the incidence of IBC and survival with this disease using new as well as old SEER classifications. The SEER Registry has had several modifications of its identification of IBC as a clinical entity, the latest being a new code (998) established in 2002 which is included under Extent of Disease. This code is used for diffuse tumour involving more than ¾ of the breast or inflammatory breast cancer. In our current re-evaluation of the SEER data (150) using comparable methods to the original report (59) to identify clinically as well as pathologically identified IBC, we found that between the three-year intervals of 1975-1977 and 1998-2000, the incidence of IBC in both African-American and Caucasian women has more than doubled with the incidence in African-American women being 50% higher (1.7/100,000 vs. 1.1/100,000 in Caucasians). Survival from this disease in African-American women was also significantly shorter than for Caucasian women, approximately 51 months vs. 113 months (150).

6.3.3 Risk Factors

As noted above, there is evidence that the risk factors for developing IBC and other aggressive breast cancers differ significantly from the risk factors for breast cancer in general. Studies have shown that reproductive factors known to decrease the risk of breast cancer have an adverse effect on prognosis. Mourali et al. (74) found that late age at menarche, an established risk factor for decreased risk of developing breast cancer, was associated with a increased risk of developing PEV, and they observed that for the patients for whom they had information on date of first pregnancy, 14 of the 15 patients who had their first births at the age of 18 or younger were diagnosed as PEV positive. And in Korzeniowski et al.'s study they found that reproductive factors known to decrease risk, specifically late menarche and parity, were associated with an adverse impact on survival (79).

Based on the results of a pilot study done by Veneroso et al in 1997 (151) we are presently conducting a study on risk factors for aggressive breast cancer. The pilot study was a case-case study of 215 breast cancer patients seen at the George Washington University Medical Faculty Associates. 215 patients were eligible for the study. Tumour aggressiveness was defined by tumour grade and breast cancer patients with tumours that were not aggressive were compared to breast cancer patients with aggressive tumours. The data showed women who had their first child before the age of 20 had about a 3 times greater odds of having an aggressive cancer, and that women who had ever used OC (ever users) had lower odds of aggressive cancer than never users, but the longer they used OC, the worse their odds for an aggressive cancer. Women who were diagnosed at an early age had 4% greater odds for each year younger at diagnosis. The identification of risk factors for aggressive breast cancer in general and IBC in particular is likely to be enhanced by the identification of better and more specific case definitions.

6.4 Treatment

A number of treatment trials are being carried out at large institutions such as the National Cancer Institute in Bethesda and MD Anderson Hospital in Texas, which involve various approaches such as inhibition of angiogenesis, vaccines, bone marrow transplants and new agents or combination of agents. Some of these new approaches as well as the current standard approach to the management of IBC have been summarized recently by Cristofanilli et al. (56).

7. SUMMARY

Aggressive breast cancer is a well-recognized but poorly understood phenomenon that has a particularly important impact on women of African descent. The poor survival of African-American women compared to any other U.S. racial/ethnic group is well documented, and this chapter describes the evidence that this adverse outcome is not solely related to barriers to care. The current weight of epidemiological evidence indicates that tumour aggressiveness results primarily from environmental rather than genetic factors, leading to the possibility that more detailed studies will provide opportunities to reduce the risk of developing an aggressive breast malignancy. The growing success in molecular epidemiology, which is enhancing the opportunity to compare a wide variety of patients in countries throughout the world, should greatly improve our ability to understand the aetiology and the possibility of control of aggressive breast cancer.

ACKNOWLEDGEMENTS

Supported in part by a grant with the Susan G. Komen Foundation, and the Department of Defense of the Government of USA.

REFERENCES

(1) American Cancer Society. Breast Cancer Facts and Figures, 2001-2002.

(2) Bradley, C. J., Given, C. W., and Roberts, C. Race, socioeconomic status, and breast cancer treatment and survival. J Natl Cancer Inst, 94: 490-6, 2002.

(3) Yood, M. U., Johnson, C. C., Blount, A., Abrams, J., Wolman, E., McCarthy, B. D., Raju, U., Nathanson, D. S., Worsham, M., and Wolman, S. R. Race and differences in breast cancer survival in a managed care population. J Natl Cancer Inst, 91: 1487-91, 1999.

(4) Blendon, R. J., Aiken, L. H., Freeman, H. E., and Corey, C. R. Access to medical care for black and white Americans. A matter of continuing concern. Jama, 261: 278-81, 1989.

(5) The unequal burden of cancer: an assessment of NIH research and programs for ethnic minorities and the medically underserved. Institute of Medicine: National Academy Press, 1999.

(6) Natarajan, N., Nemoto, T., Mettlin, C., and Murphy, G. P. Race-related differences in breast cancer patients. Results of the 1982 national survey of breast cancer by the American College of Surgeons. Cancer, 56: 1704-9, 1985.

(7) Eley, J. W., Hill, H. A., Chen, V. W., Austin, D. F., Wesley, M. N., Muss, H. B., Greenberg, R. S., Coates, R. J., Correa, P., Redmond, C. K., and et al. Racial differences in survival from breast cancer. Results of the National Cancer Institute Black/White Cancer Survival Study. Jama, 272: 947-54, 1994.

(8) Roach, M., 3rd, and Alexander, M. The prognostic significance of race and survival from breast cancer: a model for assessing the reliability of reported survival differences. J Natl Med Assoc, 87: 214-9, 1995.

(9) Farley, T. A., and Flannery, J. T. Late-stage diagnosis of breast cancer in women of lower socioeconomic status: public health implications. Am J Public Health, 79: 1508-12, 1989.

(10) Mandelblatt, J., Andrews, H., Kerner, J., Zauber, A., and Burnett, W. Determinants of late stage diagnosis of breast and cervical cancer: the impact of age, race, social class, and hospital type. Am J Public Health, 81: 646-9, 1991.

(11) El-Tamer, M. B., Homel, P., and Wait, R. B. Is race a poor prognostic factor in breast cancer? J Am Coll Surg, 189: 41-5, 1999.

(12) Dayal, H. H., Power, R. N., and Chiu, C. Race and socio-economic status in survival from breast cancer. J Chronic Dis, 35: 675-83, 1982.

(13) Gordon, N. H., Crowe, J. P., Brumberg, D. J., and Berger, N. A. Socioeconomic factors and race in breast cancer recurrence and survival. Am J Epidemiol, 135: 609-18, 1992.

(14) Berg, J. W., Ross, R., and Latourette, H. B. Economic status and survival of cancer patients. Cancer, 39: 467-77, 1977.

(15) Bassett, M. T., and Krieger, N. Social class and black-white differences in breast cancer survival. Am J Public Health, 76: 1400-3, 1986.

(16) Cella, D. F., Orav, E. J., Kornblith, A. B., Holland, J. C., Silberfarb, P. M., Lee, K. W., Comis, R. L., Perry, M., Cooper, R., Maurer, L. H., and et al. Socioeconomic status and cancer survival. J Clin Oncol, 9: 1500-9, 1991.

(17) Lannin, D. R., Mathews, H. F., Mitchell, J., Swanson, M. S., Swanson, F. H., and Edwards, M. S. Influence of socioeconomic and cultural factors on racial differences in late-stage presentation of breast cancer. Jama, 279: 1801-7, 1998.

(18) Jatoi, I., Becher, H., and Leake, C. R. Widening disparity in survival between white and African-American patients with breast carcinoma treated in the U. S. Department of Defense Healthcare system. Cancer, 98: 894-9, 2003.

(19) Wojcik, B. E., Spinks, M. K., and Optenberg, S. A. Breast carcinoma survival analysis for African American and white women in an equal-access health care system. Cancer, 82: 1310-8, 1998.

(20) Vernon, S. W., Tilley, B. C., Neale, A. V., and Steinfeldt, L. Ethnicity, survival, and delay in seeking treatment for symptoms of breast cancer. Cancer, 55: 1563-71, 1985.

(21) Bain, R. P., Greenberg, R. S., and Whitaker, J. P. Racial differences in survival of women with breast cancer. J Chronic Dis, 39: 631-42, 1986.

(22) Simon, M. S., and Severson, R. K. Racial differences in survival of female breast cancer in the Detroit metropolitan area. Cancer, 77: 308-14, 1996.

(23) Bloom, H. Prognosis in carcinoma of the breast. Brit J Cancer, 4: 259-288, 1950.

(24) Bloom, H., and Richardson, W. Histological grading and prognosis in breast cancer. Brit J Cancer, 11: 359-377, 1957.

(25) Fisher, E., Redmond, C., Fisher, B., and Bass, G. Pathologic findings from the National Surgical Adjuvant Breast and Bowel Projects (NSABP). Cancer, 65: 2121-2128, 1990.

(26) Contesso, G., Jotti, G. S., and Bonadonna, G. Tumor grade as a prognostic factor in primary breast cancer. Eur J Cancer Clin Oncol, 25: 403-9, 1989.

(27) Davis, B. W., Gelber, R. D., Goldhirsch, A., Hartmann, W. H., Locher, G. W., Reed, R., Golouh, R., Save-Soderbergh, J., Holloway, L., Russell, I., and et al. Prognostic significance of tumor grade in clinical trials of adjuvant therapy for breast cancer with axillary lymph node metastasis. Cancer, 58: 2662-70, 1986.

(28) Fisher, B., Redmond, C., Fisher, E. R., and Caplan, R. Relative worth of estrogen or progesterone receptor and pathologic characteristics of differentiation as indicators of

prognosis in node negative breast cancer patients: findings from National Surgical Adjuvant Breast and Bowel Project Protocol B-06. J Clin Oncol, 6: 1076-87, 1988.
(29) Pinder, S. E., Murray, S., Ellis, I. O., Trihia, H., Elston, C. W., Gelber, R. D., Goldhirsch, A., Lindtner, J., Cortes-Funes, H., Simoncini, E., Byrne, M. J., Golouh, R., Rudenstam, C. M., Castiglione-Gertsch, M., and Gusterson, B. A. The importance of the histologic grade of invasive breast carcinoma and response to chemotherapy. Cancer, 83: 1529-39, 1998.
(30) Seshadri, R., Horsfall, D. J., McCaul, K., and Leong, A. S. A simple index to predict prognosis independent of axillary node information in breast cancer. Aust N Z J Surg, 67: 765-70, 1997.
(31) Carriaga, M. T., and Henson, D. E. The histologic grading of cancer. Cancer, 75: 406-21, 1995.
(32) Freedman, L. S., Edwards, D. N., McConnell, E. M., and Downham, D. Y. Histological grade and other prognostic factors in relation to survival of patients with breast cancer. Br J Cancer, 40: 44-55, 1979.
(33) Galea, M. H., Blamey, R. W., Elston, C. E., and Ellis, I. O. The Nottingham Prognostic Index in primary breast cancer. Breast Cancer Res Treat, 22: 207-19, 1992.
(34) Saimura, M., Fukutomi, T., Tsuda, H., Sato, H., Miyamoto, K., Akashi-Tanaka, S., and Nanasawa, T. Prognosis of a series of 763 consecutive node-negative invasive breast cancer patients without adjuvant therapy: analysis of clinicopathological prognostic factor. J Surg Oncol, 71: 101-5, 1999.
(35) Ewers, S. B., Attewell, R., Baldetorp, B., Borg, A., Ferno, M., Langstrom, E., and Killander, D. Prognostic significance of flow cytometric DNA analysis and estrogen receptor content in breast carcinomas--a 10 year survival study. Breast Cancer Res Treat, 24: 115-26, 1993.
(36) Parl, F. F., Schmidt, B. P., Dupont, W. D., and Wagner, R. K. Prognostic significance of estrogen receptor status in breast cancer in relation to tumor stage, axillary node metastasis, and histopathologic grading. Cancer, 54: 2237-42, 1984.
(37) Tsutsui, S., Ohno, S., Murakami, S., Hachitanda, Y., and Oda, S. Prognostic value of DNA ploidy in 653 Japanese women with node-negative breast cancer. Int J Clin Oncol, 6: 177-82, 2001.
(38) Bracko, M., Us-Krasovec, M., Cufer, T., Lamovec, J., Zidar, A., and Goehde, W. Prognostic significance of DNA ploidy determined by high-resolution flow cytometry in breast carcinoma. Anal Quant Cytol Histol, 23: 56-66, 2001.
(39) Mirza, A., Mirza, N., Vlastos, G., and Singletary, S. Prognostic factors in node-negative breast cancer: a review of studies with sample size more than 200 and follow-up more than 5 years. Am Surg, 235: 10-26, 2002.
(40) Tolentino, R. S., Ang, S. D., Cajacom, C. C., Eufemio, G. G., Talla, V., and Lo, R. W. The prognostic significance of the S-phase fraction and DNCA ploidy in node-negative estrogen receptor-positive breast cancer. Philipp J Surg Spec, 53: 25-30, 1998.
(41) McCarthy, N. J., Yang, X., Linnoila, I. R., Merino, M. J., Hewitt, S. M., Parr, A. L., Paik, S., Steinberg, S. M., Hartmann, D. P., Mourali, N., Levine, P. H., and Swain, S. M. Microvessel density, expression of estrogen receptor alpha, MIB-1, p53, and c-erbB-2 in inflammatory breast cancer. Clin Cancer Res, 8: 3857-62, 2002.
(42) Joslyn, S. A. Racial differences in survival from breast cancer. Jama, 273: 1000, 1995.
(43) Henson, D. E., Chu, K. C., and Levine, P. H. Histologic grade, stage, and survival in breast carcinoma: comparison of African American and Caucasian women. Cancer, 98: 908-17, 2003.
(44) Elmore, J. G., Moceri, V. M., Carter, D., and Larson, E. B. Breast carcinoma tumor characteristics in black and white women. Cancer, 83: 2509-15, 1998.

(45) Simon, M. S., and Severson, R. K. Racial differences in breast cancer survival: the interaction of socioeconomic status and tumor biology. Am J Obstet Gynecol, 176: S233-9, 1997.
(46) Elledge, R. M., Clark, G. M., Chamness, G. C., and Osborne, C. K. Tumor biologic factors and breast cancer prognosis among white, Hispanic, and black women in the United States. J Natl Cancer Inst, 86: 705-12, 1994.
(47) Chen, V. W., Correa, P., Kurman, R. J., Wu, X. C., Eley, J. W., Austin, D., Muss, H., Hunter, C. P., Redmond, C., Sobhan, M., and et al. Histological characteristics of breast carcinoma in blacks and whites. Cancer Epidemiol Biomarkers Prev, 3: 127-35, 1994.
(48) Mohla, S., Sampson, C. C., Khan, T., Enterline, J. P., Leffall, L., Jr., and White, J. E. Estrogen and progesterone receptors in breast cancer in Black Americans: Correlation of receptor data with tumor differentiation. Cancer, 50: 552-9, 1982.
(49) Mohla, S., Enterline, J. P., Sampson, C. C., Khan, T., Leffall, L., Jr., and White, J. E. A predominance of poorly differentiated tumors among black breast cancer patients: management and screening implications. Prog Clin Biol Res, 83: 249-58, 1982.
(50) Crowe, J. P., Jr., Gordon, N. H., Hubay, C. A., Pearson, O. H., Marshall, J. S., and McGuire, W. L. The interaction of estrogen receptor status and race in predicting prognosis for stage I breast cancer patients. Surgery, 100: 599-605, 1986.
(51) Chu, K. C., Anderson, W. F., Fritz, A., Ries, L. A., and Brawley, O. W. Frequency distributions of breast cancer characteristics classified by estrogen receptor and progesterone receptor status for eight racial/ethnic groups. Cancer, 92: 37-45, 2001.
(52) Stanford, J. L., and Greenberg, R. S. Breast cancer incidence in young women by estrogen receptor status and race. Am J Public Health, 79: 71-3, 1989.
(53) Dignam, J. J., Redmond, C. K., Fisher, B., Costantino, J. P., and Edwards, B. K. Prognosis among African-American women and white women with lymph node negative breast carcinoma: findings from two randomized clinical trials of the National Surgical Adjuvant Breast and Bowel Project (NSABP). Cancer, 80: 80-90, 1997.
(54) Newman, L. A., Bunner, S., Carolin, K., Bouwman, D., Kosir, M. A., White, M., and Schwartz, A. Ethnicity related differences in the survival of young breast carcinoma patients. Cancer, 95: 21-7, 2002.
(55) Siegel, R., Blacklow, B., and Schwartz, A. Survival of black women with stage I and stage II breast cancer is inferior to survival among white women when treates in the same way at a single institution. Proc. ASCO, 1994.
(56) Cristofanilli, M., Buzdar, A. U., and Hortobagyi, G. N. Update on the management of inflammatory breast cancer. Oncologist, 8: 141-8, 2003.
(57) Kokal, W. A., Hill, L. R., Porudominsky, D., Beatty, J. D., Kemeny, M. M., Riihimaki, D. U., and Terz, J. J. Inflammatory breast carcinoma: a distinct entity? J Surg Oncol, 30: 152-5, 1985.
(58) Wolff, A. C., and Davidson, N. E. Preoperative therapy in breast cancer: lessons from the treatment of locally advanced disease. Oncologist, 7: 239-45, 2002.
(59) Levine, P. H., Steinhorn, S. C., Ries, L. G., and Aron, J. L. Inflammatory breast cancer: the experience of the surveillance, epidemiology, and end results (SEER) program. J Natl Cancer Inst, 74: 291-7, 1985.
(60) Chang, S., Parker, S. L., Pham, T., Buzdar, A. U., and Hursting, S. D. Inflammatory breast carcinoma incidence and survival: the surveillance, epidemiology, and end results program of the National Cancer Institute, 1975-1992. Cancer, 82: 2366-72, 1998.

(61) Levine, P., and Zolfaghari, L. The Inflammatory Breast Cancer Registery: Lessons learned from the first 50 Patients. 26th Annual San Antonio Breast Cancer Symposium 2003, 2003.
(62) Bell, C. A system of operative surgery, volume 2. Hartford, CT: Hale and Homser: 136, 1814.
(63) Schumann, E. A study of carcinoma mastitoides. Ann Surg, 54: 69-77, 1911.
(64) Swain, S. M., and Lippman, M. E. Treatment of patients with inflammatory breast cancer. Important Adv Oncol: 129-50, 1989.
(65) Taylor, G., and Meltzer, A. "Inflammatory carcinoma" of the breast. Am J Cancer, 33: 33-49, 1938.
(66) Singletary, S. E., Allred, C., Ashley, P., Bassett, L. W., Berry, D., Bland, K. I., Borgen, P. I., Clark, G. M., Edge, S. B., Hayes, D. F., Hughes, L. L., Hutter, R. V., Morrow, M., Page, D. L., Recht, A., Theriault, R. L., Thor, A., Weaver, D. L., Wieand, H. S., and Greene, F. L. Staging system for breast cancer: revisions for the 6th edition of the AJCC Cancer Staging Manual. Surg Clin North Am, 83: 803-19, 2003.
(67) Ellis, D. L., and Teitelbaum, S. L. Inflammatory carcinoma of the breast. A pathologic definition. Cancer, 33: 1045-7, 1974.
(68) Saltzstein, S. L. Clinically occult inflammatory carcinoma of the breast. Cancer, 34: 382-8, 1974.
(69) Lucas, F. V., and Perez-Mesa, C. Inflammatory carcinoma of the breast. Cancer, 41: 1595-605, 1978.
(70) Denoix, P. Treatment of malignant breast cancer. Recent results in cancer research 31, pp. 92-94. Berlin: Springer/Verlag, 1970.
(71) Tabbane, F., Muenz, L., Jaziri, M., Cammoun, M., Belhassen, S., and Mourali, N. Clinical and prognostic features of a rapidly progressing breast cancer in Tunisia. Cancer, 40: 376-82, 1977.
(72) Mourali, N., Tabbane, F., Muenz, L. R., Bahi, J., Belhassen, S., Kamaraju, L. S., and Levine, P. H. Preliminary results of primary systemic chemotherapy in association with surgery or radiotherapy in rapidly progressing breast cancer. Br J Cancer, 45: 367-74, 1982.
(73) Mourali, N., Tabbane, F., Muenz, L. R., Behi, J., Ben Moussa, F., Jaziri, M., and Levine, P. H. Ten-year results utilizing chemotherapy as primary treatment in nonmetastatic, rapidly progressing breast cancer. Cancer Invest, 11: 363-70, 1993.
(74) Mourali, N., Muenz, L. R., Tabbane, F., Belhassen, S., Bahi, J., and Levine, P. H. Epidemiologic features of rapidly progressing breast cancer in Tunisia. Cancer, 46: 2741 6, 1980.
(75) Levine, P. H., Mesa-Tejada, R., Keydar, I., Tabbane, F., Spiegelman, S., and Mourali, N. Increased incidence of mouse mammary tumor virus-related antigen in Tunisian patients with breast cancer. Int J Cancer, 33: 305-8, 1984.
(76) English, W. P., Cleveland, K. E., and Barber, W. H. There is no difference in survival between African-American and white women with breast cancer. Am Surg, 68: 594-7, 2002.
(77) Newman, L. A., Mason, J., Cote, D., Vin, Y., Carolin, K., Bouwman, D., and Colditz, G. A. African-American ethnicity, socioeconomic status, and breast cancer survival: a meta-analysis of 14 studies involving over 10,000 African-American and 40,000 White American patients with carcinoma of the breast. Cancer, 94: 2844-54, 2002.
(78) Kelsey, J. L., Gammon, M. D., and John, E. M. Reproductive factors and breast cancer. Epidemiol Rev, 15: 36-47, 1993.
(79) Korzeniowski, S., and Dyba, T. Reproductive history and prognosis in patients with operable breast cancer. Cancer, 74: 1591-4, 1994.

(80) Kroman, N., Wohlfahrt, J., Andersen, K. W., Mouridsen, H. T., Westergaard, T., and Melbye, M. Parity, age at first childbirth and the prognosis of primary breast cancer. Br J Cancer, 78: 1529-33, 1998.
(81) Chang, S., Buzdar, A. U., and Hursting, S. D. Inflammatory breast cancer and body mass index. J Clin Oncol, 16: 3731-5, 1998.
(82). Schouten, L. J., Hupperets, P. S., Jager, J. J., Volovics, L., Wils, J. A., Verbeek, A. L., and Blijham, G. H. Prognostic significance of etiological risk factors in early breast cancer. Breast Cancer Res Treat, 43: 217-23, 1997.
(83) Greenberg, E. R., Vessey, M. P., McPherson, K., Doll, R., and Yeates, D. Body size and survival in premenopausal breast cancer. Br J Cancer, 51: 691-7, 1985.
(84) Lund, E. Breast cancer mortality and the change in fertility risk factors at menopause: a prospective study of 800,000 married Norwegian women. Epidemiology, 2: 285-8, 1991.
(85) Lees, A. W., Jenkins, H. J., May, C. L., Cherian, G., Lam, E. W., and Hanson, J. Risk factors and 10-year breast cancer survival in northern Alberta. Breast Cancer Res Treat, 13: 143-51, 1989.
(86) Earley, T. K., Gallagher, J. Q., and Chapman, K. E. Carcinoma of the breast in women under thirty years of age. Am J Surg, 118: 832-4, 1969.
(87) Brightmore, T. G., Greening, W. P., and Hamlin, I. An analysis of clinical and histopathological features in 101 cases of carcinoma of breast in women under 35 years of age. Br J Cancer, 24: 644-69, 1970.
(88) Noyes, R. D., Spanos, W. J., Jr., and Montague, E. D. Breast cancer in women aged 30 and under. Cancer, 49: 1302-7, 1982.
(89) Adami, H. O., Malker, B., Holmberg, L., Persson, I., and Stone, B. The relation between survival and age at diagnosis in breast cancer. N Engl J Med, 315: 559-63, 1986.
(90) Host, H., and Lund, E. Age as a prognostic factor in breast cancer. Cancer, 57: 2217-21, 1986.
(91) Bonnier, P., Romain, S., Charpin, C., Lejeune, C., Tubiana, N., Martin, P. M., and iana, L. Age as a prognostic factor in breast cancer: relationship to pathologic and biologic features. Int J Cancer, 62: 138-44, 1995.
(92) Sant, M., Gatta, G., Micheli, A., Verdecchia, A., Capocaccia, R., Crosignani, P., and Berrino, F. Survival and age at diagnosis of breast cancer in a population-based cancer registry. Eur J Cancer, 27: 981-4, 1991.
(93) de la Rochefordiere, A., Asselain, B., Campana, F., Scholl, S. M., Fenton, J., Vilcoq, J. R., Durand, J. C., Pouillart, P., Magdelenat, H., and Fourquet, A. Age as prognostic factor in premenopausal breast carcinoma. Lancet, 341: 1039-43, 1993.
(94) Maggard, M. A., O'Connell, J. B., Lane, K. E., Liu, J. H., Etzioni, D. A., and Ko, C. Y. Do young breast cancer patients have worse outcomes? J Surg Res, 113: 109-13, 2003.
(95) Marcus, J. N., Watson, P., Page, D. L., and Lynch, H. T. Pathology and heredity of breast cancer in younger women. J Natl Cancer Inst Monogr: 23-34, 1994.
(96) Kollias, J., Elston, C. W., Ellis, I. O., Robertson, J. F., and Blamey, R. W. Early-onset breast cancer--histopathological and prognostic considerations. Br J Cancer, 75: 1318-23, 1997.
(97) Walker, R. A., E., L., MB., W., and SJ., D. Breast cancer occurring in young women (< 35 years) are different. Br J Cancer, 74: 1796-1800, 1996.
(98) Winchester, D. P., Osteen, R. T., and Menck, H. R. The National Cancer Data Base report on breast carcinoma characteristics and outcome in relation to age. Cancer, 78: 1838-43, 1996.

(99) Rosner, D., Lane, W. W., and Perez Brett, R. Influence of oral contraceptives on the prognosis of breast cancer in young women. Cancer, 55: 1556-62, 1985.
(100) Rosner, D. H., Joy, J. V., and Lane, W. W. Oral contraceptives and prognosis of breast cancer in women aged 35 to 50. J Surg Oncol, 30: 52-9, 1985.
(101) Rosner, D., and Lane, W. W. Oral contraceptive use has no adverse effect on the prognosis of breast cancer. Cancer, 57: 591-6, 1986.
(102). Kay, C. R., and Hannaford, P. C. Breast cancer and the pill--a further report from the Royal College of General Practitioners' oral contraception study. Br J Cancer, 58: 675-80, 1988.
(103) Olsson, H., Moller, T. R., Ranstam, J., Borg, A., and Ferno, M. Early oral contraceptive use as a prognostic factor in breast cancer. Anticancer Res, 8: 29-32, 1988.
(104) Olsson, H., Borg, A., Ferno, M., Moller, T. R., and Ranstam, J. Early oral contraceptive use and premenopausal breast cancer--a review of studies performed in southern Sweden. Cancer Detect Prev, 15: 265-71, 1991.
(105) Olsson, H., Ranstam, J., Baldetorp, B., Ewers, S. B., Ferno, M., Killander, D., and Sigurdsson, H. Proliferation and DNA ploidy in malignant breast tumors in relation to early oral contraceptive use and early abortions. Cancer, 67: 1285-90, 1991.
(106) Brinton, L. A., Daling, J. R., Liff, J. M., Schoenberg, J. B., Malone, K. E., Stanford, J. L., Coates, R. J., Gammon, M. D., Hanson, L., and Hoover, R. N. Oral contraceptives and breast cancer risk among younger women. J Natl Cancer Inst, 87: 827-35, 1995.
(107) Sauerbrei, W., Blettner, M., Schmoor, C., Bojar, H., and Schumacher, M. The effect of oral contraceptive use on the prognosis of node positive breast cancer patients. German Breast Cancer Study Group. Eur J Cancer, 34: 1348-51, 1998.
(108) Vessey, M., Baron, J., Doll, R., McPherson, K., and Yeates, D. Oral contraceptives and breast cancer: final report of an epidemiological study. Br J Cancer, 47: 455-62, 1983.
(109) Holmberg, L., Lund, E., Bergstrom, R., Adami, H. O., and Meirik, O. Oral contraceptives and prognosis in breast cancer: effects of duration, latency, recency, age at first use and relation to parity and body mass index in young women with breast cancer. Eur J Cancer, 30A: 351-4, 1994.
(110) Schonborn, I., Nischan, P., and Ebeling, K. Oral contraceptive use and the prognosis of breast cancer. Breast Cancer Res Treat, 30: 283-92, 1994.
(111) Ranstam, J., Olsson, H., Garne, J. P., Aspegren, K., and Janzon, L. Survival in breast cancer and age at start of oral contraceptive usage. Anticancer Res, 11: 2043-6, 1991.
(112) Han, S., Yun, I. J., Noh, D. Y., Choe, K. J., Song, S. Y., and Chi, J. G. Abnormal expression of four novel molecular markers represents a highly aggressive phenotype in breast cancer. Immunohistochemical assay of p53, nm23, erbB-2, and cathepsin D protein. J Surg Oncol, 65: 22-7, 1997.
(113) Revillion, F., Bonneterre, J., and Peyrat, J. P. ERBB2 oncogene in human breast cancer and its clinical significance. Eur J Cancer, 34: 791-808, 1998.
(114) Olsson, H., Borg, A., Ferno, M., Ranstam, J., and Sigurdsson, H. Her-2/neu and INT2 proto-oncogene amplification in malignant breast tumors in relation to reproductive factors and exposure to exogenous hormones. J Natl Cancer Inst, 83: 1483-7, 1991.
(115) Demers, A., Ayotte, P., Brisson, J., Dodin, S., Robert, J., and Dewailly, E. Risk and aggressiveness of breast cancer in relation to plasma organochlorine concentrations. Cancer Epidemiol Biomarkers Prev, 9: 161-6, 2000.
(116) Woolcott, C. G., Aronson, K. J., Hanna, W. M., SenGupta, S. K., McCready, D. R., Sterns, E. E., and Miller, A. B. Organochlorines and breast cancer risk by receptor status, tumor size, and grade (Canada). Cancer Causes Control, 12: 395-404, 2001.
(117) Cui, Y., Whiteman, M. K., Flaws, J. A., Langenberg, P., Tkaczuk, K. H., and Bush, T. L. Body mass and stage of breast cancer at diagnosis. Int J Cancer, 98: 279-83, 2002.

(118) Daling, J. R., Malone, K. E., Doody, D. R., Johnson, L. G., Gralow, J. R., and Porter, P. L. Relation of body mass index to tumor markers and survival among young women with invasive ductal breast carcinoma. Cancer, 92: 720-9, 2001.

(119) Ownby, H. E., Frederick, J., Russo, J., Brooks, S. C., Swanson, G. M., Heppner, G. H., and Brennan, M. J. Racial differences in breast cancer patients. J Natl Cancer Inst, 75: 55-60, 1985.

(120) National Cancer Institute's Surveillance, Epidemiology and End Results (SEER) Program. web site.

(121) Haagensen, C. Diseases of the breast, second edition. Philadelphia: Saunders: 576-584, 1971.

(122) Levine, P. H., Connelly, R. R., Berard, C. W., O'Connor, G. T., Dorfman, R. F., Easton, J. M., and DeVita, V. T. The American Burkitt Lymphoma Registry: a progress report. Ann Intern Med, 83: 31-6, 1975.

(123) Levine, P. H., Connelly, R. R., and McKay, F. W. Burkitt's lymphoma in the USA: cases reported to the American Burkitt Lymphoma Registry compared with population-based incidence and mortality data. IARC Sci Publ: 217-24, 1985.

(124) Levine, P. H., Kamaraju, L. S., Connelly, R. R., Berard, C. W., Dorfman, R. F., Magrath, I., and Easton, J. M. The American Burkitt's Lymphoma Registry: eight years' experience. Cancer, 49: 1016-22, 1982.

(125) Shiramizu, B., Barriga, F., Neequaye, J., Jafri, A., Dalla-Favera, R., Neri, A., Guttierez, M., Levine, P., and Magrath, I. Patterns of chromosomal breakpoint locations in Burkitt's lymphoma: relevance to geography and Epstein-Barr virus association. Blood, 77: 1516-26, 1991.

(126) Kushwaha, A. C., Whitman, G. J., Stelling, C. B., Cristofanilli, M., and Buzdar, A. U. Primary inflammatory carcinoma of the breast: retrospective review of mammographic findings. AJR Am J Roentgenol, 174: 535-8, 2000.

(127) Dershaw, D. D., Moore, M. P., Liberman, L., and Deutch, B. M. Inflammatory breast carcinoma: mammographic findings. Radiology, 190: 831-4, 1994.

(128) Poiesz, B. J., Ruscetti, F. W., Gazdar, A. F., Bunn, P. A., Minna, J. D., and Gallo, R. C. Detection and isolation of type C retrovirus particles from fresh and cultured lymphocytes of a patient with cutaneous T-cell lymphoma. Proc Natl Acad Sci U S A, 77: 7415-9, 1980.

(129) van Golen, K. L., Davies, S., Wu, Z. F., Wang, Y., Bucana, C. D., Root, H., Chandrasekharappa, S., Strawderman, M., Ethier, S. P., and Merajver, S. D. A novel putative low-affinity insulin-like growth factor-binding protein, LIBC (lost in inflammatory breast cancer), and RhoC GTPase correlate with the inflammatory breast cancer phenotype. Clin Cancer Res, 5: 2511-9, 1999.

(130) Kleer, C. G., van Golen, K. L., Braun, T., and Merajver, S. D. Persistent E-cadherin expression in inflammatory breast cancer. Mod Pathol, 14: 458-64, 2001.

(131) Folkman, J. Fundamental concepts of the angiogenic process. Curr Mol Med, 3: 643-51, 2003.

(132) Alpaugh, M. L., Tomlinson, J. S., Shao, Z. M., and Barsky, S. H. A novel human xenograft model of inflammatory breast cancer. Cancer Res, 59: 5079-84, 1999.

(133) Alpaugh, M. L., Tomlinson, J. S., Kasraeian, S., and Barsky, S. H. Cooperative role of E-cadherin and sialyl-Lewis X/A-deficient MUC1 in the passive dissemination of tumor emboli in inflammatory breast carcinoma. Oncogene, 21: 3631-43, 2002.

(134) Lerebours, F., Bertheau, P., Bieche, I., Plassa, L. F., Champeme, M. H., Hacene, K., Toulas, C., Espie, M., Marty, M., and Lidereau, R. Two prognostic groups of inflammatory breast cancer have distinct genotypes. Clin Cancer Res, 9: 4184-9, 2003.

(135) Guroff, M., and Buehring, G. In Pursuit of a human breast cancer virus, from mouse to human. In: J. Goedert (ed.), Infectious Causes of Cancer: Targets for intervention. Totowa, NJ: Humana Press Inc.
(136) Mesa-Tejada, R., Keydar, I., Ramanarayanan, M., Ohno, T., Fenoglio, C., and Spiegelman, S. Detection in human breast carcinomas of an antigen immunologically related to a group-specific antigen of mouse mammary tumor virus. Proc Natl Acad Sci U S A, 75: 1529-33, 1978.
(137) Mesa-Tejada, R., Keydar, I., Ramanarayanan, M., Ohno, T., Fenoglio, C., and Spiegelman, S. Immunohistochemical detection of a cross-reacting virus antigen in mouse mammary tumors and human breast carcinomas. J Histochem Cytochem, 26: 532-41, 1978.
(138) Pogo, B. G., Melana, S. M., Holland, J. F., Mandeli, J. F., Pilotti, S., Casalini, P., and Menard, S. Sequences homologous to the mouse mammary tumor virus env gene in human breast carcinoma correlate with overexpression of laminin receptor. Clin Cancer Res, 5: 2108-11, 1999.
(139) Wang, Y., Pelisson, I., Melana, S. M., Go, V., Holland, J. F., and Pogo, B. G. MMTV-like env gene sequences in human breast cancer. Arch Virol, 146: 171-80, 2001.
140. Levine PH, Coronel SM, Pogeo BG-T, Klouj A, Holland JF, Mourali N, and K., W. Increasing evidence for a human breast cancer virus. 25th Annual San Antonio Breast Cancer Symposium 2002, 2002.
(141) Wang, Y., Melana, S. M., Baker, B., Bleiweiss, I., Fernandez-Cobo, M., Mandeli, J. F., Holland, J. F., and Pogo, B. G. High prevalence of MMTV-like env gene sequences in gestational breast cancer. Med Oncol, 20: 233-6, 2003.
142. Maalej, M., Frikha, H., Ben Salem, S., Daoud, J., Bouaouina, N., Ben Abdallah, M., and Ben Romdhane, K. Breast cancer in Tunisia: clinical and epidemiological study]. Bull Cancer, 86: 302-6, 1999.
(143) Ijaduola, T. G., and Smith, E. B. Pattern of breast cancer among white-American, African-American, and nonimmigrant west-African women. J Natl Med Assoc, 90: 547-51, 1998.
(144) Chiedozi, L. C. Breast cancer in Nigeria. Cancer, 55: 653-7, 1985.
(145) Chiedozi, L. C. Rapidly progressing breast cancer in Nigeria. Eur J Surg Oncol, 13: 505-9, 1987.
(146) Mbonde, M. P., Amir, H., Mbembati, N. A., Holland, R., Schwartz-Albiez, R., and Kitinya, J. N. Characterisation of benign lesions and carcinomas of the female breast in a sub-Saharan African population. Pathol Res Pract, 194: 623-9, 1998.
(147) Costa, J., Webber, B. L., Levine, P. H., Muenz, L., O'Conor, G. T., Tabbane, F., Belhassen, S., Kamaraju, L. S., and Mourali, N. Histopathological features of rapidly progressing breast carcinoma in Tunisia: a study of 94 cases. Int J Cancer, 30: 35-7, 1982.
(148) Levine, P. H., Tabbane, F., Muenz, L. R., Kamaraju, L. S., Das, S., Polivy, S., Belhassen, S., Scholl, S. M., Bekesi, J. G., and Mourali, N. Hormone receptors in rapidly progressing breast cancer. Cancer, 54: 3012-6, 1984.
(149) Levine PH, Sherman M, and C, V. The Inflammatory Breast Cancer Registry: An approach to standarization. 25th Annual San Antonio Breast Cancer Symposium 2002, 2002.
(150) Hance, K., Anderson, W., Devesa, S., and Levine, P. Trends in Inflammatory Breast Cancer Incidence and Survival: The Surveillance, Epidemiology, and End Results Program at the National Cancer Institute. 26th Annual San Antonio Breast Cancer Symposium 2003.

(151) Veneroso, C., Levine, P., Biggar, H., Skourtis, S., and Siegel, R. Risk factors for Breast Cancer Aggressiveness. 24th Annual San Antonio Breast Cancer Symposium, 2001.

Chapter 7

RADIOTHERAPY OF BREAST CANCER

Olusegun Agboola, MB, BS (IB); DABR (T); FRCPC
Ottawa Regional Cancer Centre, and University of Ottawa, Ottawa, Ontario, Canada

1. INTRODUCTION

Radiation therapy plays an integral role in the management of breast cancer at all stages, as definitive, adjuvant and as palliative treatment for metastatic disease.

2. MECHANISM OF ACTION

X-ray generated from radiotherapy equipment deposit their energy as they transverse through cells. Deposition of this energy leads to cell death by two mechanisms, direct and indirect mechanism. By direct mechanism, the X-ray causes direct single or double strand break in the DNA strands of the cell chromosome. Significant number of the single strand breaks is successfully repaired. Attempts to repair double strand breaks result in chromosomal aberrations and cell death during replication. In indirect mechanism, the x-rays interact with the milieu in which the cells are suspended, that is the tissue fluid. The deposited energy ionises water thereby releasing radicals ($OH^·$) that are very reactive resulting in the formation of toxic substances i.e. hydrogen peroxide.

$H^2O^+ + H^2O \quad H_3O^+ + OH^·; \quad OH^· + OH^· \quad H_2O_2.$

Both malignant and normal cells are equally affected by ionising radiation, however the normal cell is capable of repair of radiation damage better than the malignant cell as long as the radiation damage is not too severe to kill the cell outright or destroy its ability to repair the damage. By

C. Williams et al. (eds.), Breast Cancer in Women of African Descent, 119-152.

administering small doses on a daily basis, the normal cell retains the ability to repair the damage scored to a significant degree. The accumulation of the daily damage eventually leads to the death of the malignant cell.

The difference in the ability of the normal and malignant cell to repair damage is the rational for the use of radiotherapy in the treatment of malignant tumours. This difference is referred to as therapeutic ratio, and it is a factor of tumour sensitivity and the sensitivity of the tissue in which the tumour is located. The higher the therapeutic ratio, the better the ability of radiotherapy to destroy the malignant lesion leaving the normal tissue intact.

3. EQUIPMENT

Radiotherapy equipment varies according to the type of radiation the machine generates, the source of the radiation either from isotope or electrical power and the energy of the beam. There are two types of radiation treatment: teletherapy and brachytherapy. The teletherapy is referred to as external beam radiation treatment, which involves the use of γ-rays, X-rays or electrons radiation beams, which are collectively called ionising radiation. The difference is the mode of generation of the each type of radiation; the biologic effects are the same.

3.1 Teletherapy

Gamma (γ) rays are generated from an isotope of Cobalt, i.e. Cobalt ^{60} that is the radioactive source of Co^{60} machines. The generated γ-rays are equivalent to 1.3 mega-voltage of generated X-rays from linear accelerators. The linear accelerator machines are high-energy X-ray machines, with energy ranging from 4 to 40 million volts. Accelerated electrons striking a platinum target in a vacuum tube generate the X-rays. Usually energy beams with the range of 4 to 10 MV are used for the treatment of breast cancer. In the developing countries, the Co^{60} machines are more in use, because they require less maintenance, and are less expensive to acquire. Orthovoltage machines: these generate X-rays on the same principle as the linear accelerators but at a considerably reduced energy. Its maximum energy is 500KV. It is still utilized in some parts of the world because it is less expensive than the Cobalt ^{60} machine, but it is less ideal than the latter for treatment of breast cancer because of increased skin reaction. The drawback in using Cobalt ^{60} or Orthovoltage machine for the treatment of breast cancer is their decreased ability to deliver adequate dose of radiation treatment to the desired depth of the breast tissue without excessively radiating the skin of the breast especially when treating fairly large breasts. (Figures 1&2) show typical Cobalt ^{60} and Linac Accelerator machines.

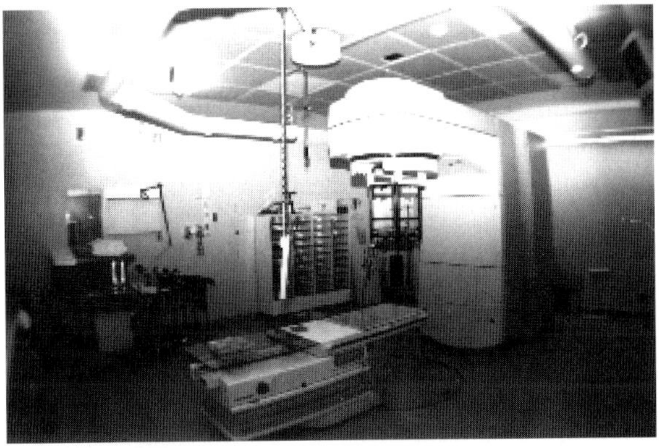

Figure 1. Linac Accelerator Machine.

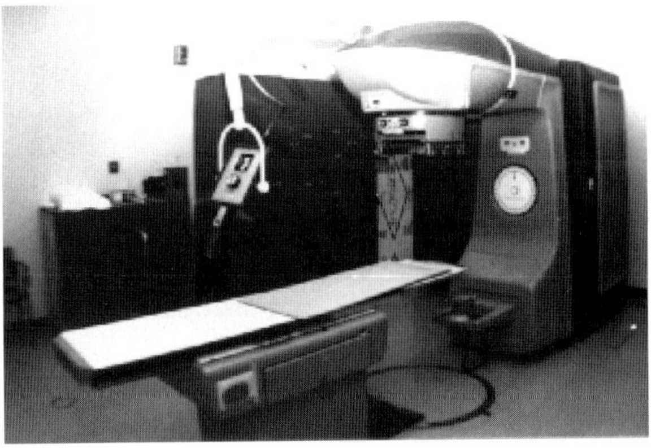

Figure 2. Cobalt ^{60} Machine.

Electron beams are also generated by Linac Accelerators. They have a well-defined depth of travel in the body tissue, which is dependent on the energy of the beam. They are useful in treating areas in which the underlying tissue is being spared from high dose of radiation. It is used to treat the chest wall following mastectomy, for boost treatment to the surgical scar and the tumour bed.

3.2 Brachytherapy (Short Treatment)

This is a treatment in which radioactive source is introduced directly into body tissue usually by the use of radioactive needles or wires e.g. Ir^{192} (Figure 3). It deposits its radiation energy in the immediate vicinity in which it is introduced and sparing surrounding tissues few centimetres away from high dose of radiation. The advantage is that a small area of the breast can be treated to a higher dose without significantly increasing the dose to the whole breast. This type of radiation treatment is used when the surgical scar or the tumour bed is given extra treatment to prevent recurrence. Studies are being conducted to determine if this type of radiation treatment could be used as the sole treatment in patients with early stage breast cancer usually detected at mammography and about 1 cm in size (1).

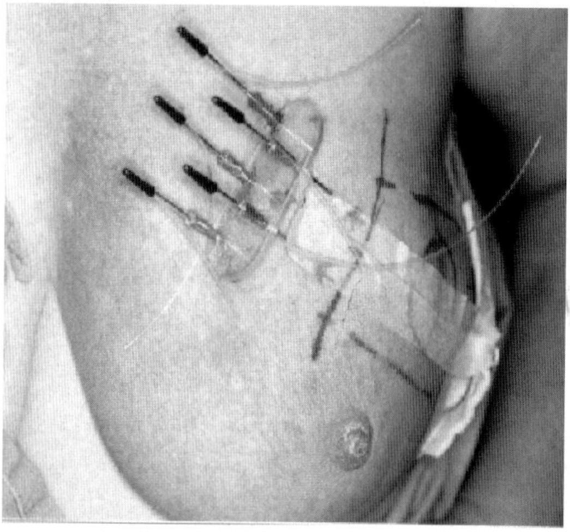

Figure 3. A patient with hollow needles inserted into the breast for brachytherapy.

4. TECHNIQUES OF TREATMENT

The body parts that need to be treated in breast cancer depend on the stage of the disease, the type and extent of surgery that has been performed. These regions are:

1. The intact breast following segmental resection.
2. The chest wall following mastectomy.
3. The axillary lymph nodes.
4. The supraclavicular lymph nodes.

5. The intramammary node.
    a. The general principle of treatment involves immobilization of the patient that allows accurate and reproducible set up on a daily basis.
    b. The choice of equipment that will adequately deliver the required treatment dose to the target area with acceptable tissue reactions.
    c. Treatment set up that limits dose to non-affected area, hence limits integral dose i.e. the overall dose and volume of body tissue irradiated.

4.1 Positioning

The patient can be positioned supine with both arms elevated above the head and resting on a breast board (Figure 4). Or in a semi-recumbent position with the ipsilateral arm elevated above the head and a back support to stabilize the patient (Figure 5).

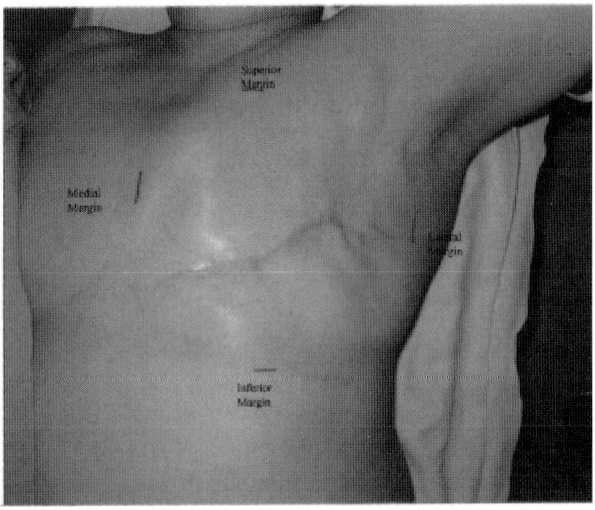

Figure 4. Chest wall post mastectomy.

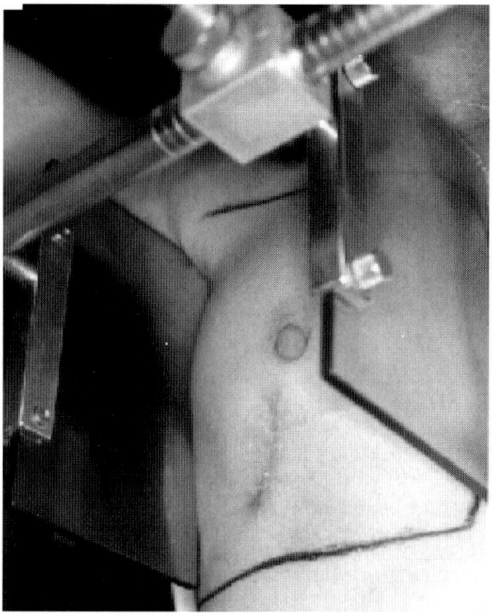

Figure 5. A semi-recumbent position for treatment of the intact breast following segmental resection.

The advantages of the supine positioning are: the patient is more comfortable in this position, it is better reproduced on a day to day basis, this position could be used to treat areas related to breast cancer treatment i.e. axilla, supraclavicular, internal mammary nodes as well as the chest wall. It also allows isocentric breast set up.

In the supine position, CT-Simulation can be used for virtual simulation of the breast; this type of simulation allows better definition of dose to the regional fields, better visualization of the tissues being irradiated.

4.2 Immobilization

This is not required for all cases, but it is very useful in patients with large pendulous breast, it permits easier and daily reproducible set up. For this a plastic shell is made for each patient, vacuum formed from a Plexiglas to the patient's shape. This keeps the breast in a well-defined position and lifted off the chest wall (Figure 6).

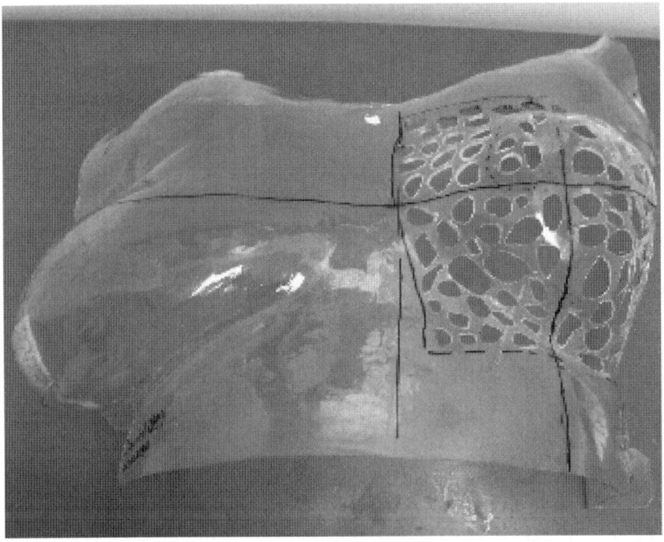

Figure 6. A plastic shell to keep large breast in reproducible position.

5. TREATMENT OF THE INTACT BREAST AND CHEST WALL

5.1 Margins

The medial margin is located medial to the medial margin of the breast, not beyond the middle of the sternum. The medial margin may be extended if the internal mammary nodes are being included in the tangential fields of radiation treatment used to treat the breast.

The lateral margin is along the mid-axillary line, or 1 cm beyond the lateral border of the palpable breast. For patients who have had mastectomy, the lateral margin should extend beyond the mastectomy scar (Figures 4&5).

The inferior margin is defined by a line drawn 2 cm below the inferior curvature of the breast and the superior margin is at the level of the base of head of the clavicle.

It is important in treatment of the breast to limit the volume of lung tissue and heart tissue (when treating the left breast) in the radiation field. The acceptable depth of lung tissue to be included in the tangential field should be limited to below 2 cm to 3 cm (Figure 7). This can be achieved by modifying the angles of the beam, or by the use of half beam either by collimation or beam block. In treating the left breast, attempts should be made to limit the cardiac tissue.

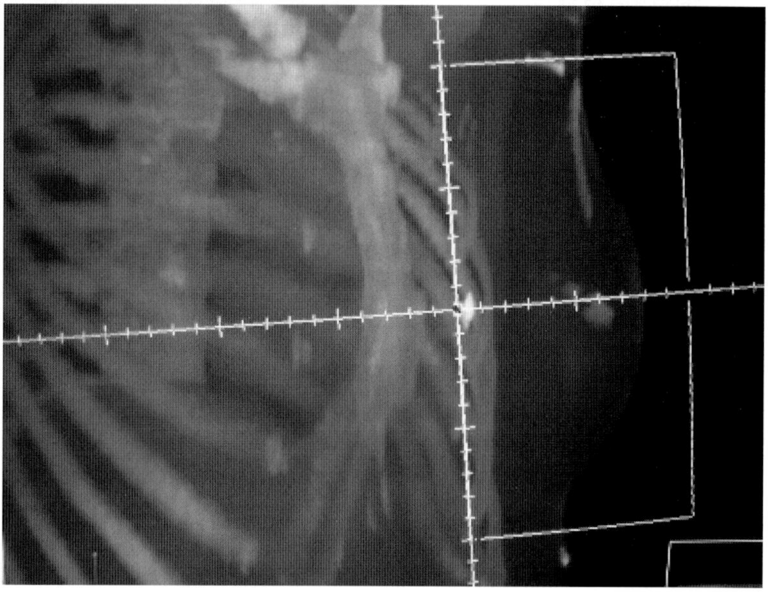

Figure 7. A radiograph of irradiated breast.

When there is need to treat the chest wall, the technique is the same as for the intact breast, except that a build up (bolus) is needed to bring the radiation dose given to the skin of the chest wall to an acceptable level. Applying a bolus material of tissue equivalent density to the chest wall suffices.

5.2 Dose

The standard dose for the treatment of the breast or chest wall is 50Gy in 25 fractions over five weeks at 2Gy per fraction or 45Gy at 1.8Gy per fraction over 25 fractions in 5weeks. Other dose schedules have been tried successfully. This is supported by the results of a Canadian Trial in which node negative breast cancer patients who underwent breast-conserving surgery (BCS) were randomly assigned to either the standard 50Gy in 25 fractions at 2Gy per fractions over 35 days or 42.5Gy in 16 fractions over 22 days. The median follow up is 5.8 years; no difference is noted in the local recurrence at 5 years: 3.2% and 2.8% respectively. The cosmetic outcome was not different, good cosmesis was achieved in 70% in both arms (2).

Other dose schedules are hyper-fractionation and accelerated fractionation: in a study involving cases with inflammatory breast cancer, a dose of 1.5Gy was given twice daily for a total of 66Gy successfully. Loco-regional control was better when compared with historical control, 84% and 58% respectively. The overall 5-year survival was also improved: 77% and 58% respectively (3). The cosmetic result is less optimal with higher total dose or increased fraction dose.

6. AXILLARY / SUPRACLAVICULAR NODES

The positioning of the patient for irradiating the axillary nodes is the same as for treatment of the breast. The lower axillary nodes are radiated in conjunction with the irradiated breast.

Following adequate axillary dissection, only the apex of the axilla need to be irradiated even when the nodes removed are positive for metastatic disease. Treatment of the axilla is required if the dissection is inadequate or if it is not done, or when the nodes are matted, or if there is extranodal spread.

6.1 The Field Margins

The superior margin is at the level of the thyrocricoid groove, and not beyond the superior margin of the supraclavicular fossa. The inferior margin is made to coincide with the superior margin to the breast radiation field with no overlap, along the head of the clavicle. The medial margin is medial to the head of the clavicle, along the pedicle of the cervical spine. The lateral margin is at the insertion of pectoralis minor at the coracoid process, if the apex of the axilla only is treated. For full axillary treatment i.e. levels 2 and 3 axillary nodes, the lateral border is extended laterally to include the medial third of the humerus to encompass the lateral border of the axillary wall, with the head of the humerus shielded (Figure 8).

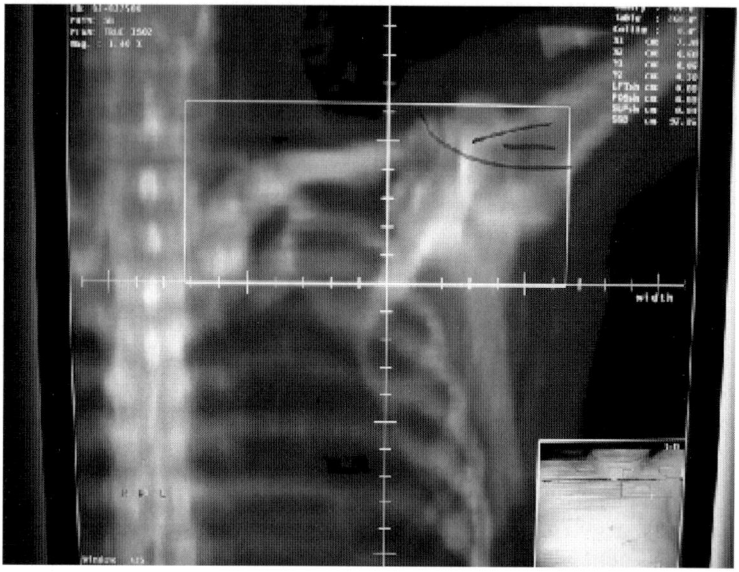

Figure 8. The supraclavicular and axillary fields margins.

6.2 Beam Arrangement

For treatment of the apex of the axilla, a single anterior field is used. For the full axilla a posterior field is needed to supplement the mid plane axillary dose from the anterior field. The posterior field is smaller in dimension, the superior border is at the level of the clavicle, and the inferior border matches the superior border of the breast field. The lateral border is at the medial border of the humeral head and medially it takes in about 2 cm of the lateral border of the lung (Figure 9).

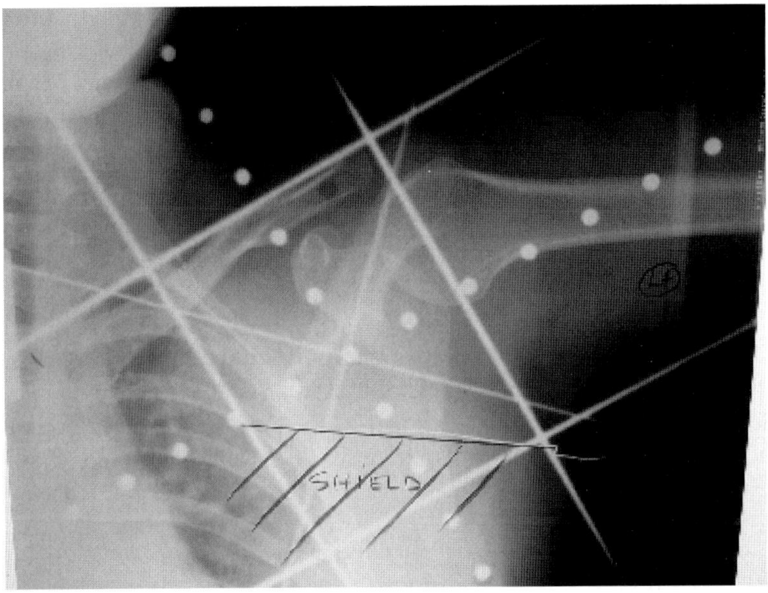

Figure 9. The posterior axillary field.

6.3 Dose

For the apex of the axilla and the supraclavicular the prescribed dose is 50Gy at 3cm in 25 fractions over 5 weeks. For full axilla, the anterior field provides about 70 to 80% of the axilla dose, the supplementary dose from the posterior field should bring the dose in the axilla to 50Gy in 25 fractions over 5 weeks.

7. INTERNAL MAMMARY NODES

The intramammary nodes draining the breast are located in the 2nd, 3rd and 4th ipsilateral intercostal spaces. These are the lymph nodes likely to be involved pathologically (4).

7.1 Technique

The internal mammary nodes are irradiated either by increasing the width of the fields used to treat the breast or chest wall, to include these nodes. The drawback with this technique is the increased volume of lung tissue in the field of radiation (5). The other technique is a direct single anterior field. If a

Co^{60} or linear energy is used, high doses will be delivered to the heart muscle, which is unacceptable, hence electron beam therapy of 15 Mev with 80% depth at 3 cm is preferred, because the location of the nodes is within this distance from the skin of the chest wall. The drawback of this technique is difficulty with junctioning the direct intramammary field with the tangential fields used in treating the chest wall or breast. Areas of overlap tend to occur which could result in overdosing areas of the skin.

7.2 The Field Margins

The medial border is 1cm across the midline, the lateral border is 4cm into the medial border of the ipsilateral breast, or matching the medial border of the tangential field used in treatment of the breast. The upper border is at the head of the clavicle, matching the inferior margin of the supraclavicular field.

8. BOOST TREATMENT

This is the extra dose of radiation treatment given to the surgical scar or tumor bed either in intact breast or following mastectomy, when the surgical resection margin is positive, or when the tumour tissue is very close to the surgical resection margin.

8.1 Technique

The tumour bed is radiated with a field size sufficient to cover the tumour bed. In the intact breast the treatment could be given by external photon beam, electron beam and brachytherapy. Post mastectomy, the surgical scar is treated with either photon or electron beam with a field covering the whole length of the surgical scar with a minimum of 1 cm on either side of the scar.

8.2 Dose

In our Institute, boost treatment is given with either external beam or interstitial brachytherapy when the recurrence margin is positive and no further surgical resection is not possible or when the surgical resection margin is close i.e. less than 1 cm. A dose of 10Gy is prescribed after the initial 50Gy to bring to the total tumour dose to 60Gy.

There have been studies in which boost treatment was given in the absence of close or involved margins. These studies were conducted to determine if the administration of boost treatment to breast cancer cases post-surgery will improve the local recurrence rate or have significant influence on overall survival. There are two positive studies, the EORTC and the French study. In the EORTC study (6), 5,318 cases after initial treatment to 50Gy in 25fractions were randomised to receive boost treatment of either 16Gy with external beam or 15Gy by interstitial brachytherapy; this was compared with cases with no boost treatment. The 5-year recurrence rates were 4.3% and 6.8% in favour of the boost arm. There was higher number of distant metastatic cases in the no-boost arm, 8.3 times more.

In the French study, (7), 1,024 cases were randomised after initial treatment to 50Gy in 25 fractions to receive a dose of 10Gy as boost treatment or no boost treatment. After 3.3 years of follow-up, the 5-year recurrence rate was 3.6% and 4.5% in the boost and no boost arms respectively. From the above studies it appears that boost treatment decreases the local recurrence rate by a small percentage, this has to be weighed against the extra dose and the decreased cosmesis associated with this treatment.

9. POTENTIAL COMPLICATIONS OF RADIATION TREATMENT

The potential complications of radiation treatment are classified into acute and late complications. The acute complications are the complications or reactions noted while the patient is undergoing radiation treatment and the late complications occur at varying times after the treatment has been given.

10. ACUTE REACTIONS

10.1 Skin Changes

The skin shows changes secondary to the radiation treatment. The severity of the skin reaction increases as the radiation treatment progresses and as more dose is delivered to the breast tissue and the skin. The initial change noted is erythema, which progresses to dry desquamation and then moist desquamation. In erythema, there is inflammation of the skin with increasing redness and at times tenderness; this colour change is not often noted in the dark skin patients.

The dry desquamation is more severe, the superficial layer of the skin peels but the skin is intact. This reaction is limited to the superficial dermal layer. Patient may experience a moderate degree of discomfort at this level; this skin change is noted about the 3rd week of treatment onward.

Moist desquamation is when there is break down of the skin, with secretion of serous fluid. The dermis is involved in this reaction. This is associated with considerable discomfort. These changes are more common when there is overlap of irradiated fields, e.g. supraclavicular and chest wall or chest wall and internal mammary fields and in cases where the dose gradient over the treated region is higher than 20-25%.

If severe, the treatment should be withheld until resolution occurs. In most cases the area involved is not large and superficial treatment with Flamazine or similar agent may allow the treatment to be completed without stoppage. Skin reaction varies with individuals; more skin reaction is noted in the Scandinavian type skin, which easily burns when exposed to excessive sunlight. Skin reaction in the dark skin patients is often mild comparatively. The initial skin reaction noted is increase darkening of the irradiated area, however, excessive doses do result in most desquamation and loss of skin pigmentation. The incidence and severity of skin changes are higher in women with large breasts.

10.2 Fatigue

Patients undergoing treatment experience fatigue and do complain of being tired, more so immediately after each dose of radiation treatment. This is independent of the patient's blood count. The complaint of fatigue is not restricted to breast cancer cases alone, it is universal complaint expressed by patients undergoing radiation treatment. It is more common in patients undergoing both chemotherapy and radiation treatment. The severity increases as the treatment progresses (8). The tiredness dissipates within four weeks of completion of treatment.

11. LATE COMPLICATIONS

11.1 Late Skin Changes

Late skin changes are noted from about three months onwards after radiation treatment and they are fully established by one year. These are telangiectasia of the dermal blood vessels, loss of skin elasticity and thinning of the skin more common following chest wall irradiation. Telangiectasia of the blood vessels is less prominent in the dark skin patients. In the Caucasians, there is permanent increased pigmentation of the skin of the chest wall and of the breast. These changes are also noted in the dark skin patients. In the latter, vitiligo type skin change is noted in areas of excessive skin doses. The irradiated breast is firmer due to increase fibrosis, and there is loss of axilla hairs and sweating from the armpit following irradiation of the axilla. Necrosis of the skin of the chest wall can occur resulting in non-healing ulceration of the skin, which could be very painful to touch and very resistant to healing. This is secondary to destruction of the blood supply of the skin when the dermal layer of the skin has been heavily irradiated. Treatment involves daily dressing and in very severe cases, full thickness skin graph is needed.

11.2 Lymphedema

Lymphedema results from interruption of the axillary lymphatic system after surgery and or radiation treatment. Swelling of the ipsilateral arm as a result of lymphedema becomes noticeable within three months of completion of treatment. The extent of axillary dissection is an important determinant of post-treatment lymphedema. The incidence is higher with full axillary dissection compared with levels 1/II dissection (9). The incidence increases with radiation of the whole axilla especially after extensive axillary dissection (10-11). There is no significant increase in its occurrence when the intact breast or the chest wall is irradiated or when radiation treatment is limited to the apex of the axilla along with chest wall or breast irradiation compared with surgery with axillary dissection; about 3% risk (12).

11.3 Rib Fracture

Fracture of the ribs is uncommon. The administration of radiation treatment devitalises the structure of the bone and if excessive pressure or trauma is applied to the ribs at a later date, fracture of the ribs can occur. This complication is a related to the dose of radiation treatment. The incidence is less than 3% if radiation dose is limited to 50Gy; doses higher than 50Gy and concurrent chemotherapy increase the incidence (13).

11.4 Brachial Plexopathy

The incidence of permanent plexopathy is less than 1% in patients treated to a dose of 50Gy at 2Gy per fraction (13-14). This complication often occurs when excessive dose greater than 50Gy or higher fraction dose is delivered to the supraclavicular and axillary regions with concomitant chemotherapy. It is also found when no compensation is made for the decreased thickness of the tissue in the superior margin of the supraclavicular region in calculating the dose of radiation given. The symptoms are shoulder discomfort and progressive paraesthesia of the ipsilateral arm noticed within four to six following completion of treatment. Some patients may develop an early and transient brachial plexopathy within weeks of completion of treatment even when radiation dose is within 50Gy, this often resolves within weeks or a couple of months.

11.5 Radiation Pneumonitis

About 5% of the lung volume is irradiated when tangential fields are used to treat the breast or chest wall. Most patients are not symptomatic from the treatment; the incidence of symptomatic pneumonitis is about 1-5%, (15-16) starting about 2 to 6 months following radiation treatment. The patient presents with dry cough, low-grade fever and shortness of breath on exertion. X-ray of the chest shows fluffy infiltrate corresponding with the area of lung irradiated. In some cases it may be more extensive. It is a self-limiting illness; however in patients that have compromised lung function before radiation treatment permanent changes can occur leading to respiratory insufficiency. The role of steroid in limiting radiation pneumonitis is controversial.

Higher incidence of radiation pneumonitis about 3.3% is reported when a third field is added to treat the supraclavicular and axillary apex or when the medial aspect of the tangential field used in treating the breast is extended to cover the internal mammary nodes. The incidence increases significantly (8.8%) when RT and chemotherapy are given concurrently (17).

11.6 Cardiac Injury

This is a long-term complication associated with older methods of treating the internal mammary nodes (18). It is as a result of damage to cardiac vasculature from radiation treatment resulting in myocardial infarction. The cardiac muscle is more resistant to radiation treatment except when high doses of anthracyline-containing chemotherapy are given concurrently with radiation treatment, that the cardiac musculature is also prone to damage (19).

A significant proportion of the cardiac muscle can be in the field of treatment when treating the intact breast or chest wall, especially on the left side. This can be avoided to a significant degree by modification of the beam direction. Virtual simulation of patients on the CT-Sim has also allowed visualization of the extent of the cardiac volume included in the radiation field and more accurate treatment planning. A recent Canadian study noted an increase in myocardial infarction of 1% at 10years in patients with left side breast cancer treated with irradiation (20).

11.7 Secondary Malignancies

Secondary malignancies have been reported following irradiation treatment. The time lag between treatment and development of secondary malignancy is about 10yrs. Sarcoma of the chest wall and angiosarcoma of the skin of the breast are the commonest occurring radiation induced malignancy (13). In one series from Gustave-Roussy Institute in France, the cumulative frequency of occurrence at 10, 20 and 30 years was 0.2, 0.43, and 0.78 percent respectively (21). Other reported malignancies are very rare; these are lung cancer, leukaemia, and contralateral breast cancer. The latter is seen when patient are treated at a younger age (22).

12. THE ROLE OF RADIATION TREATMENT IN THE MANAGEMENT OF DIFFERENT STAGES OF BREAST CANCER

The following factors: the tumour, the surgery performed and the patient parameters determine the use of radiation in the management of breast cancer.

a) The tumour: size, pathology, and stage.
b) The surgery: mastectomy, partial mastectomy (segmentectomy) and axillary dissection.
c) The patient: breast size, expected cosmesis, clinical status.

13. CARCINOMA IN SITU

T1s (Intraductal Carcinoma in Situ and Paget's Disease of the nipple with no associated tumour).

This is a non-invasive breast cancer in which only about 10% of women have palpable breast mass. Most are detected on screening mammography. The risk of progression to invasive cancer is depended on a combination of factors i.e. the site of the mass, the histological type and the status of the resection margin.

13.1 Treatment

In the past, mastectomy was the choice of treatment; this is curative in about 98% of cases (23). Modern day practice is segmental resection of the tumour with breast preservation if the volume of tumour involvement is not greater than 3 cm in diameter, with low or intermediate nuclear grade histology (24). Larger tumour volume leaves the breast severely deformed following segmental resection. Such cases could benefit from prosthetic replacement of lost breast tissue.

Following breast conserving surgery, with negative surgical resection margin, the local failure rates were 11% and 26% with and without post-operative radiation treatment respectively after an average of 46 months follow up as per NSAPB B-17 study (25). Half of the recurrences had the same histology as the primary lesion. There were no significant differences in survival in those who recurred and those that did not recur, 94 and 98% 5-year survival respectively (26). Radiation treatment is recommended after breast conserving surgery in these cases.

13.2 Technique

The breast is treated as outlined above by using either of the two set up techniques. The radiation dose is 50Gy fractions in 25 fractions over 5 weeks at 2Gy per fraction. The axillary lymph region does not need to be irradiated, because the possibility of involvement is about 3.6% (27). It is significantly lower if the DCIS detected by mammography (27-30).

13.3 Positive Resection Margin

If the margin of surgical resection is positive, the probability of local recurrence is higher (31). This is best managed by re-resection until a clear margin is obtained or mastectomy should be considered if a clear margin couldn't be achieved. If this is not clinically possible, extra radiation treatment should be given to the tumour by bed using photon or electron beam RT to a dose of 10Gy to 15Gy given in 5 fractions or interstitial brachytherapy.

14. LOBULAR CARCINOMA IN SITU

Lobular carcinoma in situ tends to be multicentric in the breast and bilateral in 25 to 35% of cases (32). Radiation treatment dose does not play a role in its management.

15. MICRO INVASIVE CARCINOMA

These are tumours, which have extended beyond the basement membrane of the cells into adjacent tissue no more than 1mm in greatest depth (T1mic). The post-surgical radiation treatment management is as in situ carcinoma.

16. EARLY STAGE INVASIVE BREAST CANCERS

Stages 1 to III (node negative, T1, T2, T3; N0).

Breast conserving surgery (BCS) and ipsilateral axillary node dissection followed by radiation treatment is the preferred surgery as this achieves the same result as mastectomy and axillary node dissection. The review of randomised trials on radiotherapy and early breast cancer found no difference in survival in both groups of patients (33-34).

16.1 Treatment

Ipsilateral breast radiation is recommended in all sub groups of cases following BCS irrespective of the size of tumour, the histology, the menopausal status of the patient or other adjuvant treatments. Previous NSABP studies did show significant decrease in the rate of local recurrence with whole breast RT following BSC. The recurrence rates after 20 year follow up were 14 vs. 39% for RT and no RT cases (34). No significant survival benefit was reported with addition of RT. Following Mastectomy in patients not suitable for BCS, chest wall irradiation is not recommended when the axillary nodes are negative. However when the size of the primary tumour is 4 cm and above chest wall irradiation is recommended to decrease the probability of local recurrence.

16.2 Technique

Whole ipsilateral breast or chest wall is treated to a dose of 5000cGy in 25 fractions over 5 weeks. The benefit of extra treatment to the tumour site when the resection margin is negative is debatable. This should however be considered when the surgical resection margin is close, i.e. less than 10mm.

16.3 Timing of Radiation Treatment

In patients not undergoing chemotherapy, radiation treatment can commence after the surgical site is healed and swelling of the breast has subsided. If patient is scheduled for chemotherapy, RT can be delayed until completion of chemotherapy. A delay up to 6 months to permit the administration of adjuvant chemotherapy has not been found to compromise local control or survival (35).

Concurrent administration of radiation treatment and chemotherapy increases significantly the skin reaction to radiation treatment as well as other reactions associated with treatment. The probability of cardiac complications increases especially with anthracylines-containing chemotherapy e.g. Doxorubicin (also known as Adriamycin) (19).

No advantage is noted in concurrent chemotherapy and RT compared with consecutive treatment with both modalities (36).

Axillary nodal radiation is not suggested in patients with negative nodal involvement following adequate dissection of the axilla. The risk of nodal recurrence is less than 1% (37). In patients with clinically negative axilla nodal involvement, who have not undergone axillary dissection, irradiating the axilla decrease the possibility of recurrence. The incidence of axillary recurrence in this situation is 3% Vs 1.4% if radiation treatment was given as reported in NSABP B 04 study (38).

17. EARLY DISEASE NODE POSITIVE CASES

(T0N1, T1N1, T2 N1).

The status of the axillary node is an important prognostic factor in patients with early breast cancer. The probability of lymph node involvement increases with increasing tumour size (39-41). The decision whether to radiate the axilla or not in the presence of axillary node involvement is dependent on the number of nodes involved, the presence or absence of extra-nodal spread and the number of nodes examined.

17.1 Treatment

The number of lymph nodes involved is an independent prognostic factor (42), as well as the grade of the tumour and the presence or absence of lymphatic vessel invasion. In patients with less than 4 lymph nodes involved following adequate axillary node dissection and in the absence of extra nodal disease, the benefit of irradiation treatment is uncertain, hence treatment of the axilla is not recommended. Such patients should receive ipsilateral breast RT following BCS, or no RT following mastectomy. However if axillary nodal dissection is limited to level 1 of the axilla, the probability of presence of involved node at higher level is up to 67% (43), such cases should benefit from axilla irradiation.

The number of nodes retrieved from the surgical specimen is another factor. When less than five nodes are retrieved, the risk of axillary recurrence is high. The risk of axillary recurrence has been noted to be inversely related to the number of removed nodes. Failure in the axilla was about 5-21% when less than 5 nodes were removed and 3-5% when greater than 5 nodes were removed (41-46). In this situation axillary nodal irradiation is advised along with breast irradiation following BCS, and axilla alone following mastectomy.

In the presence of four or more positive axillary nodes, the risk of axillary recurrence and the probability of loco-regional recurrence are high. Loco-regional recurrence is associated with high rate of distant metastases and decrease overall survival by about 10% (47-49). The apex of the axilla and the supraclavicular region should be irradiated as well as ipsilateral breast following BCS or chest wall following mastectomy.

17.2 TECHNIQUE

The technique of treatment of the breast and the axilla is described above.

17.3 Dose

The dose to is 50 Gy in 25 fractions over five weeks at daily fraction dose of 2Gy for both the breast and axilla.

18. INTERNAL MAMMARY NODES

The internal mammary (i.m.) nodes that are most important in breast lymphatic drainage are the nodes in the upper three interspaces; the 2nd, 3rd and 4th interspaces. The nodes in these interspaces are most likely to contain metastases if it occurs. In the absence of axillary nodal involvement, the probability of isolated i.m. nodal involvement is very low. With laterally located tumour the incidence is about 2-9%, but could go as high as 7% in medially located tumour (50).

As a rule, the i.m. nodes are not irradiated when the axillary nodes are negative in laterally situated tumours. There is enough justification to radiate the i.m. nodes in medially situated tumours when the axillary nodes are positive especially if the number of nodes involved is four or greater. The probability of involvement could be as high as 22% (50-51).

With the advent of adjuvant chemotherapy, the thinking is that the chemotherapy should be able to sterilize microscopic disease in the i.m. nodes if present. Hence there is increasing tendency to not irradiate the i.m.

nodes even in the presence of medially situated tumour. This could be justified because of the infrequent cases of isolated i.m. nodal recurrence in the absence of recurrence at other sites (52-54).

Also the morbidity and long-term complications associated with radiation of these nodes may not justify the gains realized from irradiation.

There have been three randomised trials in which irradiating the i.m. nodes following mastectomy have shown survival benefit (55-57). There are several randomised studies that have failed to show a benefit for i.m. nodal irradiation (58-60). A definitive trial of the necessity of i.m. nodal irradiation is being conducted by NCI Canada and EORTC. There are also on-going isotope studies to identify the presence or absence of involvement of these nodes prior to instituting treatment, the studies are in early stages and the results are not conclusive at present. Before a decision is made as to irradiating the i.m. nodes or not, each case should be studied as to the benefit of treatment by combining the physical and biological characteristics of the tumour as well as other adjuvant treatments given. The technique for irradiating the i.m. nodes is outlined above. The dose is 50Gy in 25 fractions over 5 weeks, at 2Gy per fraction. If a direct electron beam is used, 15Mev electron energy, to a dose of 45Gy at 80% in 15 fractions over 3 weeks is an alternative method of treatment.

19. LOCALLY ADVANCED CANCER OF THE BREAST

(T3, T4; N1, N2, N3).

These are large tumours with either extensive nodal involvement, skin and or chest wall infiltration. Inoperable supraclavicular nodal involvement (N3) is included with this group. The advance cancer of the breast constitutes up to 50% of incident breast cancers in the developing world as opposed to 5% incidence where screening mammography is practiced (61-63). The management of advance cancer of the breast involves multimodality treatment. Although surgery is still central in its management, chemotherapy treatment plays a major role in the management of these cases. There is a probability of 30 to 50% loco-regional recurrence if surgery or radiation treatment alone is given without chemotherapy (64). The point of controversy is whether surgery should be preceded by chemotherapy or not. Neo-adjuvant chemotherapy has been very successful in downgrading the tumour and making borderline-resectable cases suitable for surgery. It also allows breast-conserving surgery to be performed in cases, which would have been treated with mastectomy. Clinical complete response rate of 8 to 63% has been reported and average overall response of 75% (65-68). The pathological complete response rate is about half of the clinical response identified. The argument against neo-adjuvant chemotherapy is the progression of tumour seen in non-responders to chemotherapy; this percentage is low (5%). A randomised control trial of neoadjuvant chemotherapy against postoperative chemotherapy found an increase survival and disease free survival benefit in patients given neoadjuvant chemotherapy (69). A 10% increase in the number of breast conserving surgery was also noted in these cases (68).

Even when there is a favourable response to neo-adjuvant chemotherapy and surgery, radiation treatment is still required in these cases to decrease the probability of loco- regional recurrence (70-71). ECOG study (72) randomised 352 cases following six months of chemotherapy and hormonal therapy to surgery vs. surgery plus RT. After nine years of follow-up, the disease free survival and overall survival showed no statistically significant difference, but local relapse rate was 60% and 46% in surgery arm and surgery plus RT arm respectively. The median time to recurrence was 4.7 and 5.2 years respectively. Isolated loco regional recurrence was higher in the surgery alone arm, 27% vs. 4%.

19.1 Treatment

Patients who have had pathologic complete remission should have breast irradiation and supraclavicular treatment if the surgical treatment was BCS and axillary node dissection. Treatment of the internal mammary node in this group of patients will depend on the location of the tumour and other physical and biological parameters as espoused above. For patients with residual disease in the surgical specimen, all the potential areas of recurrence should be irradiated, i.e. chest wall following mastectomy or breast following BCS, the axilla and the supraclavicular regions. The same criteria as above should be used to decide if the internal mammary nodes need to be treated. The technique is as above and the dose of radiation treatment for each of the site is 50Gy in 25 fractions over 5 weeks at 2Gy per fraction.

Boost treatment to the tumour bed is required if the surgical resection is positive or if the surgical resection is close. This could be done with either by brachytherapy, photon or electron beam therapy. The patients treated with the trimodality approach have about 50% long-term survival.

20. INOPERABLE BREAST CANCER

These are cases that even after adequate number of courses of chemotherapy are not suitable for surgical resection. The dose to the intact breast is increased to 60Gy in 6weeks, in 30 fractions at 2Gy per fraction. In cases of involved supraclavicular node involvement, extra dose of 5Gy in 2 fractions need to be given to the nodal region using a reduced field size.

21. INFLAMMATORY BREAST CANCER: (T4D)

This is a locally aggressive and relatively rare breast cancer. It is responsible for 0.5 to 5% of all breast cancers. The incidence tends to be higher in the black population when compared with the white population (73-74) (Please, see also the contributions of Paul Levine and Carmela Veneroso – Chapter 6 –, and of Hussein Khaled – Chapter 5 - in this book). It is often of the ductal type breast cancer with at times no palpable mass. Diagnosis is by its clinical presentation. There is diffuse brawny induration of the skin of the breast due to dermal lymphatic infiltration by tumour cells. About one third of the cases present with distant metastases at the time of diagnosis (75), (77).

Management is primarily with chemotherapy, followed by surgery and radiation treatment as per advanced cancer of the breast in cases that respond to the chemotherapy with non-metastatic disease at diagnosis. The internal mammary node should be treated in these cases due to extensive lymphatic involvement. Response to chemotherapy is fairly high, about 60-80%. The loco-regional failure rate is very high about 70% when surgery is not a component of the management. About 39% are alive and disease free following 20-year follow-up. Five year survival with surgical management alone is 10% and with RT alone 17% (70), (76). Combined surgery and RT improves loco-regional control without significant impact on long-term survival (77).

22. BREAST CANCER DURING PREGNANCY

The incidence is about 0.2 to 3.8% of breast cancer in patients above 30 years of age, but this can get as high as 10-25 % in patients less than 30 years old. The gestational age and expected date of delivery are important factors in the management of breast cancer occurring during pregnancy. Surgery is well tolerated by both the fetus and the mother without increase incidence of complications or fetal loss (78). Radiation treatment during pregnancy is avoided because of probable fetal malfunction from the treatment. There is no safe dose of radiation. The fetus is more sensitive to radiation treatment during the first trimester. Also at third trimester, the uterus rises superiorly toward the breast thereby increasing the fetal dose from secondary scatter radiation.

The treatment strategies used for pregnant and non-pregnant breast cancer cases are similar. If a patient requires treatment during the 1st trimester, abortion of the fetus is advised. In the third trimester, treatment could be withheld until the fetus matures to be delivered. It is not known if delay in giving radiation or chemotherapy treatment has adverse effect on the outcome of the cancer. Considering however that up to six month delay occurs before radiation treatment is given when patients are on chemotherapy, it may sound to reason that delay will not be harmful. The difference in this situation is that chemotherapy is also not given.

23. PALLIATIVE TREATMENT

Radiation treatment plays an important role in relieving of symptoms in patients with metastatic disease. The commonest sites of metastases are liver, bone and brain.

The liver is not amenable to high doses of radiation treatment due to the sensitivity of the liver parenchyma cells to radiation. The dose that the liver can tolerate is not sufficient to control the tumour within it. Hence liver metastases are best managed by chemotherapy or other modalities that are suited for the liver.

24. BONE AND BRAIN

These sites respond favourably to radiation treatment.

24.1 Bone

Symptomatic bone metastases respond in more than 60% of cases to radiation treatment. Subjective changes are noted prior to objective changes in the bone. Patients start to experience pain relieve within a week of completion of treatment. The early relieve of symptom may be due to the turning off of the prostaglandin produced by the growth of the metastatic cancer in the bone.

The dose of radiation given varies from 30Gy in 10 fractions over 2 weeks to a single dose of 8 to 10Gy. The duration of relieve of symptom tend to be longer with the longer duration and higher dose of radiation treatment. There is no difference in the proportion of cases that respond to the various dose regimens. Hence in patients in whom the expected life span is more than 6 months, the longer duration of treatment with higher dose may be preferred. The commonest dose regimen is 20Gy in 5 fractions both to axillary and appendicular skeleton. For appendicular skeleton a single high dose is not advised especially if the spinal cord is in the field of treatment due to a possibility of damage to the spinal cord.

In most cases treatment is limited to symptomatic site, when there is a presenting symptom. However non-symptomatic bone metastases identified on bone X-ray or on bone scan with significant bony erosion should be treated if there is possibility of fracture of the bone or spinal cord compression in the nearest future.

24.2 Brain Metastases

Breast cancer is responsible for 15 to 20% of all brain metastases (79-81). The management is depended on the following factors: the number of metastases, the location of the metastases, the performances status of the patient, the expected life span, the presence or absence of other sites of metastases or local recurrence and availability of neurosurgical facilities.

The initial management is medical before definitive treatment is given, this involves the control of symptom of raised intracranial pressure with steroid, and administration of anticonvulsant if the patient presents with convulsive symptoms.

Single metastatic lesion suitably located is best managed by surgical resection followed by whole brain radiation treatment to a dose of 30Gy in 10 fractions over 2 weeks, in good performance patient in which expected life span is 6 months or longer. Stereotactic radiation treatment followed by whole brain irradiation treatment gives similar result.

A patient with a single brain lesion but with other metastatic sites and with poor prognosis may not benefit from surgical resection of the tumour. Such patients are best managed by whole brain radiation to a dose of 30Gy in 10 fractions over 2 weeks or 20Gy in 5 fractions in 1 week. The patients with multiple brain metastases are also managed in the same fashion.

The advantage of surgical resection followed by whole brain RT is decrease recurrence in the brain, improved quality of survival as opposed to whole brain RT alone. The overall survival of the patient is determined by the state of the primary cancer rather the mode of management of the brain metastases. About 50 to 85% of cases treated respond with symptomatic relieve and positive changes on CAT scan of the brain following radiation treatment (82-83). Improved response is noted in the cases managed by combined modality, i.e. surgery and RT.

25. PROBLEMS CONFRONTING WOMEN OF AFRICAN DESCENT WITH REGARD TO ACCESS TO RADIOTHERAPY TREATMENT

The problem with access to medical treatment is accentuated when dealing with radiotherapy treatment due to its unique nature. There are fewer facilities in the developing world. As of 1998 the total number of megavoltage teletherapy machines in the whole of Africa was one hundred and fifty five. To appreciate the magnitude of the problem, the developing world represent 85% of the world population, the developed world (North America, Western Europe, Australia and Japan) have two thirds of all radiotherapy facilities, 85% of all Linac Accelerators and 30% of all Cobalt ^{60} machines. Most machines in the developing world are of the latter type, 2100 in all. To appreciate the magnitude of shortage of facilities, the IAEA figures show that there are 30 radiotherapy centres in Middle Africa consisting of 28 countries of which 8 countries do not have any radiotherapy facility. These countries have a total of 28 Cobalt ^{60} and 4 Linac Accelerators. In comparison, Canada, with a population of 36 million, has 34 centres, 108 Linac Accelerators and 36 Cobalt units. By International Atomic Energy Agency (IAEA) estimate, 10,000 machines will be needed in the developing world by year 2015 to provide the needed treatment for the population estimating that 50% of all cancer patients will need radiotherapy treatment (84).

The centres available are often less equipped with regard to personnel and equipment. Due to the high technology involved in delivering radiation treatment, the personnel required to run the centres need training which often are not available within the country i.e. radiation physicists, dosimetrists, radiation technologists, electronic personnel and radiation oncologists. Hence the centres operate at a sub-optimal level compared with similar centres in the developed world. The shortage of maintenance personnel leads to greater downtime of the machines due poor servicing and unavailable spare parts. The infrastructure of the countries are often less capable of supporting sophisticated machinery, hence the increased frequency of breakdown. One of the most important infrastructure often lacking is uninterrupted and steady power supply without significant fluctuation in electrical voltages. In addition, the high humidity confronted in the tropical countries is not conducive to the proper running of electronic equipment. On account of these, less sophisticated machines are found in the developing world i.e. Cobalt ^{60} machines, which are more robust in terms of day-to-day use. Radiation is produced from an isotope resident within the machine, which does not need electrical power for its production. The drawback with this machine is the fact that the energy of radiation produced is low and it is not suitable for the various situations encountered in the management of malignant lesions.

In the developed world, the problem confronting women of African Descent is not the availability of radiotherapy centres but accessibility. Studies carried out in the US found that, this sector of the population tend to be more in the low socio-economic stratum. In this stratum are 26.3% of African Americans, 22.3% of Latin Americans, 11.6% of Asia-Pacific Americans females compared with 7% of white females (85). This sector of the population is depended on public funded health care. Access to state of the art in surgical, radiotherapeutic and systemic treatment are limited on account of limited resources available in publicly funded hospitals compared with teaching hospitals (86). It was noted that African American women who were diagnosed with early stage breast disease and who received BCS as initial treatment are less likely to receive postoperative radiotherapy (87).

There is a common trend in both the developed and developing worlds amongst women of African descent, higher percentage of the women present with advanced disease when compared with their white counterpart. In African-Americans, this was explained on the basis of limited access to and less use of baseline prevalence screening i.e. clinical breast examination and mammography and delay in subsequent follow up of abnormal screening results. These are due to patients and clinician's behaviour, health care system barrier and inequities in the system. In the developing world, the limited access to health care due to decreased availability, cultural patterns and most importantly poor health education are responsible for delay in reporting illnesses. Significant proportion of women with low education are not aware of breast cancer as a disease that cannot be treated successfully with known native measures. It is not unusual for these patients to have received treatment from native herbalists (Also, please, see the contribution of Christopher Williams on "Barriers to successful management of breast cancer" – Chapter 14 – in this book).

Irrespective of increase in health care funding towards more radiotherapy centres in the developing world; if health education is not improved, patients will continue to present with late disease resulting in poor overall survival from treatment.

26. PRACTICE OF RADIOTHERAPY IN THE DEVELOPING COUNTRIES

The practice of radiotherapy in the developing world in general and in breast cancer in particular presents a set of unique situations not confronted in the developed world. A high proportion of cases up to 60% present with late disease and radiotherapy facilities are few and distant from the homes of the patients. The late presentation may have a cultural undertone coupled with the paucity of available radiotherapy centres and physicians.

To practice in this situation, innovative ways of managing these cases have to be developed. One such innovative approach is the use of hypofractionation, in which single large doses at interval of about seven days apart are given to patients. A dose of 8 to 10 Gy is given with simple set up for one to three treatment to cases with locally advanced disease with or without metastases, in which surgery is not an option, and staying away from home for long duration of treatment is not feasible. The treatment helps alleviate the presenting symptom and provides good palliation of symptoms, which in most cases is pain. It also decreases the size of the breast tumour and the fungation that may be present for better cosmesis. This type of technique has been successfully used in the treatment of advanced non-surgical lung cancer in which the prospect of cure is low (88).

This practice was found to be very useful at the University College Hospital, Ibadan, Nigeria, when the number of functioning radiotherapy centres in the country of 100 million population was not more than two and the surrounding West African countries had no such facility. The equipment consisted of a Co^{60} and an Ortho-Voltage machine.

The unstable power supply especially voltage fluctuation often causes frequent machine break down, far more than what is experienced in the developed world. We found the Co^{60} to be best suited for treatment of our cases because of its γ-rays are generated from in dwelling isotope that requires no power supply. The electrical power needed for running of other aspects of the machine was easily supplied from a stand-by electrical generator.

ACKNOWLEDGMENTS

I would like to acknowledge the help of Ms. Betty McAndrew for the secretarial work, Dr. Laval Grimard for Figures 3 & 8 and Mr. Steve Andrusyk for Figures 5, 7-8.

REFERENCES

(1) Vicini FA, Baglan KL, Kestin LL, Mitchell C, Chen PY, Frazier RC, et al: Accelerated treatment of breast cancer. J Clin Oncol 2001;19:1993-2001.

(2) Whelan T, MacKenzie R, Julian J, Levine M, Shelley W, Grimard L, et al: Randomized trial of breast irradiation schedules after lumpectomy for women with lymph node-negative breast cancer. J Natl Cancer Inst 2002;94:1143-50.

(3) Liao Z, Strom EA, Buzdar AU, Singletary SE, Hunt K, Allen PK, et al: Locoregional irradiation for inflammatory breast cancer: effectiveness of dose escalation in decreasing recurrence. Int J Radiat Oncol Biol Phys 2000;47:1191-200.

(4) Stibbe: The internal mammary lymphatic glands. anatomy 1918:52-258.

(5) Lichter AS, Fraass BA, Yanke B: Treatment Techniques in the Conservative Management of Breast Cancer. Semin Radiat Oncol 1992;2:94-106.

(6) Vrieling C, Collette L, Fourquet A, Hoogenraad WJ, Horiot JH, Jager JJ, et al: The influence of patient, tumor and treatment factors on the cosmetic results after breast-conserving therapy in the EORTC 'boost vs. no boost' trial. EORTC Radiotherapy and Breast Cancer Cooperative Groups. Radiother Oncol 2000;55:219-32.

(7) Romestaing P, Lehingue Y, Carrie C, Coquard R, Montbarbon X, Ardiet JM, et al: Role of a 10-Gy boost in the conservative treatment of early breast cancer: results of a randomized clinical trial in Lyon, France. J Clin Oncol 1997;15:963-8.

(8) Irvine DM, Vincent L, Graydon JE, Bubela N: Fatigue in women with breast cancer receiving radiation therapy. Cancer Nurs 1998;21:127-35.

(9) Larson D, Weinstein M, Goldberg I, Silver B, Recht A, Cady B, et al: Edema of the arm as a function of the extent of axillary surgery in patients with stage I-II carcinoma of the breast treated with primary radiotherapy. Int J Radiat Oncol Biol Phys 1986;12:1575-82.

(10) Liljegren G, Holmberg L: Arm morbidity after sector resection and axillary dissection with or without postoperative radiotherapy in breast cancer stage I. Results from a randomised trial. Uppsala-Orebro Breast Cancer Study Group. Eur J Cancer 1997;33:193-9.

(11) Segerstrom K, Bjerle P, Graffman S, Nystrom A: Factors that influence the incidence of brachial oedema after treatment of breast cancer. Scand J Plast Reconstr Surg Hand Surg 1992;26:223-7.

(12) Pierce LJ, Oberman HA, Strawderman MH, Lichter AS: Microscopic extracapsular extension in the axilla: is this an indication for axillary radiotherapy? Int J Radiat Oncol Biol Phys 1995;33:253-9.

(13) Pierce SM, Recht A, Lingos TI, Abner A, Vicini F, Silver B, et al: Long-term radiation complications following conservative surgery (CS) and radiation therapy (RT) in patients with early stage breast cancer. Int J Radiat Oncol Biol Phys 1992;23:915-23.

(14) Olsen NK, Pfeiffer P, Johannsen L, Schroder H, Rose C: Radiation-induced brachial plexopathy: neurological follow-up in 161 recurrence-free breast cancer patients. Int J Radiat Oncol Biol Phys 1993;26:43-9.

(15) Lingos TI, Recht A, Vicini F, Abner A, Silver B, Harris JR: Radiation pneumonitis in breast cancer patients treated with conservative surgery and radiation therapy. Int J Radiat Oncol Biol Phys 1991;21:355-60.

(16) Marks LB, Clough R, Fan M, Hollis MG, Bowen D, Lucas A, et al: Radiation-induced pneumonitis following tangential breast/chest wall irradiation. Int J Radiat Oncol Biol Phys 2000;48 (Suppl):294 (abstract).

(17) Wong JS, Recht A, Beard CJ, Busse PM, Cady B, Chaffey JT, et al: Treatment outcome after tangential radiation therapy without axillary dissection in patients with early-stage breast cancer and clinically negative axillary nodes. Int J Radiat Oncol Biol Phys 1997;39:915-20.

(18) Freedman GM, Fowble BL, Nicolaou N, Sigurdson ER, Torosian MH, Boraas MC, et al: Should internal mammary lymph nodes in breast cancer be a target for the radiation oncologist? Int J Radiat Oncol Biol Phys 2000;46:805-14.

(19) Shapiro CL, Hardenbergh PH, Gelman R, Blanks D, Hauptman P, Recht A, et al: Cardiac effects of adjuvant doxorubicin and radiation therapy in breast cancer patients. J Clin Oncol 1998;16:3493-501.

(20) Paszat LF, Mackillop WJ, Groome PA, Schulze K, Holowaty E: Mortality from myocardial infarction following postlumpectomy radiotherapy for breast cancer: a population-based study in Ontario, Canada. Int J Radiat Oncol Biol Phys 1999;43:755-62.

(21) Taghian A, de Vathaire F, Terrier P, Le M, Auquier A, Mouriesse H, et al: Long-term risk of sarcoma following radiation treatment for breast cancer. Int J Radiat Oncol Biol Phys 1991;21:361-7.

(22) Vallis K, Pintilie M, Chong N, al e: Assessment of coronary heart disease morbidity and mortality following radiation therapy for early breast cancer. Int J Radiat Oncol Biol Phys 2000;66:144 (abstract).

(23) Schuh ME, Nemoto T, Penetrante RB, Rosner D, Dao TL: Intraductal carcinoma. Analysis of presentation, pathologic findings, and outcome of disease. Arch Surg 1986;121:1303-7.

(24) Schwartz GF, Solin LJ, Olivotto IA, Ernster VL, Pressman PI: Consensus Conference on the Treatment of In Situ Ductal Carcinoma of the Breast, April 22-25, 1999. Cancer 2000;88:946-54.

(25) Fisher B, Costantino J, Redmond C, Fisher E, Margolese R, Dimitrov N, et al: Lumpectomy compared with lumpectomy and radiation therapy for the treatment of intraductal breast cancer. N Engl J Med 1993;328:1581-6.

(26) Fisher B, Dignam J, Wolmark N, Mamounas E, Costantino J, Poller W, et al: Lumpectomy and radiation therapy for the treatment of intraductal breast cancer: findings from National Surgical Adjuvant Breast and Bowel Project B-17. J Clin Oncol 1998;16:441-52.

(27) Winchester DP, Menck HR, Osteen RT, Kraybill W: Treatment trends for ductal carcinoma in situ of the breast. Ann Surg Oncol 1995;2:207-13.

(28) Pendas S, Dauway E, Giuliano R, Ku N, Cox CE, Reintgen DS: Sentinel node biopsy in ductal carcinoma in situ patients. Ann Surg Oncol 2000;7:15-20.

(29) Silverstein MJ, Gierson ED, Waisman JR, Senofsky GM, Colburn WJ, Gamagami P: Axillary lymph node dissection for T1a breast carcinoma. Is it indicated? Cancer 1994;73:664-7.

(30) Patchefsky AS, Schwartz GF, Finkelstein SD, Prestipino A, Sohn SE, Singer JS, et al: Heterogeneity of intraductal carcinoma of the breast. Cancer 1989;63:731-41.

(31) Fisher ER, Costantino J, Fisher B, Palekar AS, Redmond C, Mamounas E: Pathologic findings from the National Surgical Adjuvant Breast Project (NSABP) Protocol B-17. Intraductal carcinoma (ductal carcinoma in situ). The National Surgical Adjuvant Breast and Bowel Project Collaborating Investigators. Cancer 1995;75:1310-9.

(32) Haagensen C, Bodian, C, Haagensen, DE: Lobular neoplasia (lobular carinoma in situ) breast carcinoma: risk and detection. Philadelphia: WB Saunders; 1981.

(33) Group EBCTC: Effects of radiotherapy and surgery in early breast cancer. An overview of the randomized trials. Early Breast Cancer Trialists' Collaborative Group. N ngl J Med 1995;333:1444-55.

(34) Fisher B, Anderson S, Bryant J, Margolese RG, Deutsch M, Fisher ER, et al: Twenty-year follow-up of a randomized trial comparing total mastectomy, lumpectomy, and lumpectomy plus irradiation for the treatment of invasive breast cancer. N Engl J Med 2002;347:1233-41.

(35) Wallgren A, Bernier J, Gelber RD, Goldhirsch A, Roncadin M, Joseph D, et al: Timing of radiotherapy and chemotherapy following breast-conserving surgery for patients

with node-positive breast cancer. International Breast Cancer Study Group. Int J Radiat Oncol Biol Phys 1996;35:649-59.
(36) Valagussa P, Zambetti M, Biasi S, Moliterni A, Zucali R, Bonadonna G: Cardiac effects following adjuvant chemotherapy and breast irradiation in operable breast cancer. Ann Oncol 1994;5:209-16.
(37) Recht A, Houlihan MJ: Axillary lymph nodes and breast cancer: a review. Cancer 1995;76:1491-512.
(38) Fisher B, Redmond C, Fisher ER, Bauer M, Wolmark N, Wickerham DL, et al: Ten-year results of a randomized clinical trial comparing radical mastectomy and total mastectomy with or without radiation. N Engl J Med 1985;312:674-81.
(39) Nemoto T, Vana J, Bedwani RN, Baker HW, McGregor FH, Murphy GP: Management and survival of female breast cancer: results of a national survey by the American College of Surgeons. Cancer 1980;45:2917-24.
(40) Carter CL, Allen C, Henson DE: Relation of tumor size, lymph node status, and survival in 24,740 breast cancer cases. Cancer 1989;63:181-7.
(41) Silverstein MJ, Skinner KA, Lomis TJ: Predicting axillary nodal positivity in 2282 patients with breast carcinoma. World J Surg 2001;25:767-72.
(42) Ravdin P: Can patient and tumor characteristics allow prediction of axillary lymph node status? Breast Discovery, 1998, p. 141.
(43) Turner RR, Chu KU, Qi K, Botnick LE, Hansen NM, Glass EC, et al: Pathologic features associated with nonsentinel lymph node metastases in patients with metastatic breast carcinoma in a sentinel lymph node. Cancer 2000;89:574-81.
(44) Fowble B, Solin LJ, Schultz DJ, Goodman RL: Frequency, sites of relapse, and outcome of regional node failures following conservative surgery and radiation for early breast cancer. Int J Radiat Oncol Biol Phys 1989;17:703-10.
(45) Graversen HP, Blichert-Toft M, Andersen JA, Zedeler K: Breast cancer: risk of axillary recurrence in node-negative patients following partial dissection of the axilla. Eur J Surg Oncol 1988;14:407-12.
(46) Fisher B, Wolmark N, Bauer M, Redmond C, Gebhardt M: The accuracy of clinical nodal staging and of limited axillary dissection as a determinant of histologic nodal status in carcinoma of the breast. Surg Gynecol Obstet 1981;152:765-72.
(47) Arriagada R, Rutqvist LE, Mattsson A, Kramar A, Rotstein S: Adequate locoregional treatment for early breast cancer may prevent secondary dissemination. J Clin Oncol 1995;13:2869-78.
(48) Hayward J, Caleffi M: The significance of local control in the primary treatment of breast cancer. Lucy Wortham James clinical research award. Arch Surg 1987;122:1244-7.
(49) Schmoor C, Sauerbrei W, Bastert G, Schumacher M: Role of isolated locoregional ecurrence of breast cancer: results of four prospective studies. J Clin Oncol 2000;18:1696-708.
(50) Morrow M, Foster RS, Jr.: Staging of breast cancer: a new rationale for internal mammary node biopsy. Arch Surg 1981;116:748-51.
(51) Harlow S, Krag D, Weaver D, Ashikaga T: Extra-Axillary Sentinel Lymph Nodes in Breast Cancer. Breast Cancer 1999;6:159-65.
(52) Katz A, Strom EA, Buchholz TA, Thames HD, Smith CD, Jhingran A, et al: Locoregional recurrence patterns after mastectomy and doxorubicin-based chemotherapy: implications for postoperative irradiation. J Clin Oncol 2000;18:2817-27.

(53) Halverson KJ, Perez CA, Kuske RR, Garcia DM, Simpson JR, Fineberg B: Isolated local-regional recurrence of breast cancer following mastectomy: radiotherapeutic management. Int J Radiat Oncol Biol Phys 1990;19:851-8.
(54) Schwaibold F, Fowble BL, Solin LJ, Schultz DJ, Goodman RL: The results of radiation therapy for isolated local regional recurrence after mastectomy. Int J Radiat Oncol Biol Phys 1991;21:299-310.
(55) Overgaard M, Hansen PS, Overgaard J, Rose C, Andersson M, Bach F, et al: Postoperative radiotherapy in high-risk premenopausal women with breast cancer who receive adjuvant chemotherapy. Danish Breast Cancer Cooperative Group 82b Trial. N Engl J Med 1997;337:949-55.
(56) Ragaz J, Jackson SM, Le N, Plenderleith IH, Spinelli JJ, Basco VE, et al: Adjuvant radiotherapy and chemotherapy in node-positive premenopausal women with breast cancer. N Engl J Med 1997;337:956-62.
(57) Overgaard M, Jensen MB, Overgaard J, Hansen PS, Rose C, Andersson M, et al: Postoperative radiotherapy in high-risk postmenopausal breast-cancer patients given adjuvant tamoxifen: Danish Breast Cancer Cooperative Group DBCG 82c randomised trial. Lancet 1999;353:1641-8.
(58) Obedian E, Haffty BG: Internal mammary nodal irradiation in conservatively-managed breast cancer patients: is there a benefit? Int J Radiat Oncol Biol Phys 1999;44:997-1003.
(59) Fowble B, Hanlon A, Freedman G, Nicolaou N, Hoffman J, Sigurdson E, et al: Internal mammary node irradiation neither decreases distant metastases nor improves survival in stage I and II breast cancer. Int J Radiat Oncol Biol Phys 2000;47:883-94.
(60) Yamashita TS, K et al: Efficacy of Loco-regional lymph nodes irradiation after mastectomy for breast with biopsy proven lymph nodes metastasis. Int J Radiat Oncol Biol Phys 1996;36:277.
(61) Smart CR, Byrne C, Smith RA, Garfinkel L, Letton AH, Dodd GD, et al: Twenty-year follow-up of the breast cancers diagnosed during the Breast Cancer Detection Demonstration Project. CA Cancer J Clin 1997;47:134-49.
(62) Hortobagi GS, SE Strom, EA: Treatment of locally advanced and inflammatory breast cancer. In Harris J, Lippman, ME, Morrow, M, Osborne, CK, editor. Diseases of the Breast. Philadelphia: Lippincott Williams and Wilkins, 2000.
(63) Ezzat AA, Ibrahim EM, Raja MA, Al-Sobhi S, Rostom A, Stuart RK: Locally advanced breast cancer in Saudi Arabia: high frequency of stage III in a young population. Med Oncol 1999;16:95-103.
(64) Perloff M, Lesnick GJ, Korzun A, Chu F, Holland JF, Thirlwell MP, et al: Combination chemotherapy with mastectomy or radiotherapy for stage III breast carcinoma: a Cancer and Leukemia Group B study. J Clin Oncol 1988;6:261-9.
(65) Hortobagyi GN, Ames FC, Buzdar AU, Kau SW, McNeese MD, Paulus D, et al: Management of stage III primary breast cancer with primary chemotherapy, surgery, and radiation therapy. Cancer 1988;62:2507-16.
(66) Merajver SD, Weber BL, Cody R, Zhang D, Strawderman M, Calzone KA, et al: Breast conservation and prolonged chemotherapy for locally advanced breast cancer: the University of Michigan experience. J Clin Oncol 1997;15:2873-81.
(67) Graham MV, Perez CA, Kuske RR, Garcia DM, Fineberg B: Locally advanced (noninflammatory) carcinoma of the breast: results and comparison of various treatment modalities. Int J Radiat Oncol Biol Phys 1991;21:311-8.
(68) Makris A, Powles TJ, Ashley SE, Chang J, Hickish T, Tidy VA, et al: A reduction in the requirements for mastectomy in a randomized trial of neoadjuvant chemoendocrine therapy in primary breast cancer. Ann Oncol 1998;9:1179-84.

(69) Recht A, Come SE, Henderson IC, Gelman RS, Silver B, Hayes DF, et al: The sequencing of chemotherapy and radiation therapy after conservative surgery for early-stage breast cancer. N Engl J Med 1996;334:1356-61.

(70) Brito RA, Valero V, Buzdar AU, Booser DJ, Ames F, Strom E, et al: Long-term results of combined-modality therapy for locally advanced breast cancer with ipsilateral supraclavicular metastases: The University of Texas M.D. Anderson Cancer Center experience. J Clin Oncol 2001;19:628-33.

(71) De Lena M, Zucali R, Viganotti G, Valagussa P, Bonadonna G: Combined chemotherapy-radiotherapy approach in locally advanced (T3b- T4) breast cancer. Cancer Chemother Pharmacol 1978;1:53-9.

(72) Olson JE, Neuberg D, Pandya KJ, Richter MP, Solin LJ, Gilchrist KW, et al: The role of radiotherapy in the management of operable locally advanced breast carcinoma: results of a randomized trial by the Eastern Cooperative Oncology Group. Cancer 1997;79:1138-49.

(73) Berg JW, Hutter RV: Breast cancer. Cancer 1995;75:257-69.

(74) Chang S, Parker SL, Pham T, Buzdar AU, Hursting SD: Inflammatory breast arcinoma incidence and survival: the surveillance, epidemiology, and end results program of the National Cancer Institute, 1975-1992. Cancer 1998;82:2366-72.

(75) Kleer CG, van Golen KL, Merajver SD: Molecular biology of breast cancer metastasis. Inflammatory breast cancer: clinical syndrome and molecular determinants. Breast Cancer Res 2000;2:423-9.

(76) Lamb CC, Eberlein TJ, Parker LM, Silver B, Harris JR: Results of radical radiotherapy for inflammatory breast cancer. Am J Surg 1991;162:236-42.

(77) Jaiyesimi IA, Buzdar AU, Hortobagyi G: Inflammatory breast cancer: a review. J Clin Oncol 1992;10:1014-24.

(78) Mazze RI, Kallen B: Reproductive outcome after anesthesia and operation during pregnancy: a registry study of 5405 cases. Am J Obstet Gynecol 1989;161:1178-85.

(79) Patchell R: Brain Metastases. Handbook of Neurology, 1997, p. 135.

(80) Posner JB: Management of brain metastases. Rev Neurol 1992;148:477-87.

(81) Sawaya R, Bindal RK: Metastic brain tumors. In Kaye AL, ER, editor. Brain Tumors. Edinburgh: Churchill Livingstone, 1995, p. 923.

(82) Sneed PK, Larson DA, Wara WM: Radiotherapy for cerebral metastases. Neurosurg Clin N Am 1996;7:505-15.

(83) Cairncross JG, Kim JH, Posner JB: Radiation therapy for brain metastases. Ann Neurol 1980;7:529-41.

(84) Levin V, Meghzifene, A, Izewska, J, Tatsuzaki, H: Improving Cancer Care: Increased need for radiotherapy in the developing countries, 2001, p. 25-32.

(85) Census UBot: 1990 Census of population, social and economic characteristics. Washington, 1990.

(86) Ayanian JZ, Guadagnoli E: Variations in breast cancer treatment by patient and provider characteristics. Breast Cancer Res Treat 1996;40:65-74.

(87) Joslyn SA: Racial differences in treatment and survival from early-stage breast rcinoma. Cancer 2002;95:1759-66.

(88) Macbeth FR, Bolger JJ, Hopwood P, Bleehen NM, Cartmell J, Girling DJ, et al: Randomized trial of palliative two-fraction versus more intensive 13- fraction radiotherapy for patients with inoperable non-small cell lung cancer and good performance status. Medical Research Council Lung Cancer Working Party. Clin Oncol 1996;8:167-75.

Chapter 8

ADJUVANT THERAPY OF BREAST CANCER

[1]Christopher K .O. Williams, MD, FRCPC and [2]Carla Falkson, MD
[1]B.C. Cancer Agency, Nanaiamo, British Columbia, and University of British Columbia, Vancouver, Canada; [2]University of Alabama at Birmingham Comprehensive Cancer Center, Birmingham, AL, USA

1. INTRODUCTION

One of the most important factors that have led to the modern management of primary breast cancer has been the acceptance of the fact that the natural history of the disease is associated with early systemic dissemination, treatment failure and death (*1*). This led in the 1950's and 1960's to the realisation of the need for systemic therapy in conjunction with the local therapy of the cancer either by surgical resection with or without radiation therapy. This type of chemotherapy, which is administered in the absence of overt manifestation of disseminated cancer, is referred to as adjuvant chemotherapy. The rationale for the development of this type of treatment modality was based on observations that disseminated tumour cells in experimental animals could be destroyed with systemically administered chemotherapeutic agents (*2-4*).

The global annual incidence of breast cancer is approximately one million cases (*5*). In North America and Europe, breast cancer accounts for 30% of all cases of malignant disease in women, and is the second most frequent cause of cancer death (after lung cancer) (*6*). The incidence of breast cancer is over 180,000 cases per year in the United States alone. It is estimated that more than 100,000 out of the newly diagnosed cases are candidates for adjuvant chemotherapy. The disease therefore represents a significant workload of the practising oncologist. The benefit from chemotherapy, however, is modest, being only 2% at 10 years. In other words, of the 100,000 women who are candidates for adjuvant chemotherapy annually, about 2,000 of them can be saved from disease recurrence or

C. Williams et al. (eds.), Breast Cancer in Women of African Descent, 153-219.

progression. On the other hand, however, the modest improvement is achieved at the cost of significant morbidity from the treatment modalities, including systemic chemotherapy, hormonal therapy and perhaps, radiation therapy. On the whole, breast cancer constitutes a significant public health challenge in much of the developed world. It is becoming increasingly so in the developing parts of the world as well. While there can be no question that major advances have been made, breast cancer continues to be a major public health problem. The 1990 National Institutes of Health (NIH) Consensus Development Conference on early-stage breast cancer has emphasised that from 1990 to 2000, 1.5 million women in the United States would be newly diagnosed with breast cancer and that up to 30% of them will ultimately succumb to the disease (*7*). The plethora of information that has been generated over the years has contributed to management confusion in terms of description of disease prognosis and designing appropriate treatment for the patients who are increasingly becoming knowledgeable about their disease as a result of pervasive on-line information. Furthermore, the traditional role of the physician as the authority of how his or her patients should be treated is rapidly yielding grounds, at least in the developed world, to one of joint participation in decision making between the physician, the patient and the family.

1.1 Factors that are used to Identify Patients who should Receive Adjuvant Chemotherapy

The process of identifying which women with breast cancer should be recommended for adjuvant chemotherapy involves determining the presence in such women of prognostic and predictive factors. Prognostic factors are measurements available at the time of diagnosis or at the time of surgery, that, in the absence of adjuvant chemotherapy, are associated with recurrence rate, death rate, or other clinical outcome. Such factors address the question: "How bad is my cancer?" Predictive factors on the other hand are measurements associated with the degree of response to a specific therapy. Predictive factors address the question: "How should my cancer be treated?" Any factor has the potential of being both prognostic and predictive. Thus, a breast mass that is 5cm or more in any woman carries a worse prognosis than one that is 1.0cm or less. The size of the former tumour is predictive of the need for adjuvant intervention after local therapy, while the later would tend to predict a more favourable outcome and a more conservative management approach.

Prognostic and predictive factors fall into three categories: patient's characteristics (such as age and race), disease characteristics (tumour size and number of involved regional lymph nodes), and biomarkers (measurable parameters in tissue, cells and body fluids). Accepted prognostic and predictive factors include age (*8*), tumour size (*9*), axillary node status (*9*), histological tumour type, standardised pathologic grade and hormonal status (*10*). Race appears to be a prognostic but not predictive factor. Women of African descent with breast cancer are generally younger (*11-17*) often have larger tumours at diagnosis (*13-14*), (*17-18*) and a higher percentage of hormone receptors negative tumours (*14*),(*16*),(*18-20*). These factors contribute to a poorer prognosis (*12-13*), (*17-18*), (*20-24*). In cases of similar clinical presentation, however, adjuvant treatment confers similar benefit to women of African descent and their Caucasian counterpart (*16*), (*25*). The value of newer biomarkers and the new technologies for identifying them remains unclear. Overexpression of HER2/neu is associated with poor prognosis in node positive patients (*26-27*) and may predict response to therapy (*28-29*). However, laboratory methods and the reporting of results would need to be standardised before the predictive performance of this biomarker, like several other newer markers, can be determined (*30*). The NIH Consensus Conference panel of 2000 listed the currently accepted prognostic and predictive factors to include: age, race (prognostic factor only), tumour size, histological tumour type, axillary nodal status and standardised pathologic grade. The only two biomarkers in the list were hormone receptor status and mitotic rate. These and other prognostic/predictive factors as they occur in African American as compared to Caucasian American women are listed in Table 1.

The conference observed that more benefit accrues from chemotherapy in which antimetabolites, such as 5-Fluorouracil and Methotrexate, are given intravenously rather than orally. It also recommended that, given the numerous unanswered questions in the adjuvant systemic treatment of node negative breast cancer, all suitable patients should be encouraged to enrol in appropriate clinical trials. Those who are not candidates for clinical trials or who refuse to participate in such trials should be made aware of the benefit and the risk of adjuvant systemic therapy. The estimated risk of recurrence without chemotherapy should be discussed with the patients. The degree of benefit should be discussed in comparison to the potential for toxicities and impact on quality of life. In some cases of node negative breast cancer, the degree of improvement may be outweighed by the disadvantage of therapy. Prognostic factors are used to provide an estimate of risk of recurrence in women with early-stage breast cancer. In no case of node-negative breast cancer can the risk be regarded as zero, although most patients are probably cured by local/regional therapy. The 1990 conference reviewed at least 6 prognostic factors of node negative breast cancer, including tumour size, oestrogen and progesterone receptor status, nuclear grade, histological grade, proliferative grade, and other factors.

Table 1: Prognosis of Early-Stage Breast Cancer in African American Compared to Caucasian American Women.

Author's Name	Study Type	AA vs.. CA study population	AA/CA age<50 ratio	AA/CA age<35 ratio	AA/CA Adverse Prognostic features	Prognosis in AA	DFS	OS
Boyer-Chammard et al, 1999	Population-based	185 vs. 10,937 (vs. 875 HSP)	increased	unknown	unknown	unfavorable	unknown	reduced
Bradley et al, 2002	Detroit, MI SEER and Medicaid ED	593 Medicaid out of 5719	unknown	unknown	unknown	unfavorable	unknown	reduced
Dignam, 2001	NSABP	543 vs. 7582	increased	?increased	increased	unfavorable	equivalent for n-ve and n+ve	reduced for n-ve and n+ve
Dignan et al, 1997	NSABP	? vs.?	unknown	unknown	increased	unfavorable	equival	equival
El-Tamer et al, 1999	Tumor Registry data	1297 vs. 448	reduced	unknown	?	unfavorable	unknown	reduced
English et al, 2002	Review of Tumor Registry data	Total = 585	increased	?increased	increased	unfavorable	unknown	equivalent
Franzini et al, 1997	Cancer Centre MDACC	163 vs. 964(vs.. 205 Hispanics)	unknown	unknown	unknown	normal	N/a	normal
Joslyn	NCI Surveillance	10,073 vs. 123,127	unknown	unknown	unknown	N/a.	unknown	reduced
Joslyn and West, 2000	SEER	11,159 vs. 124,265	increased	unknown	increased	unfavorable	unknown	reduced
Joslyn, 2002	SEER	12,303 vs.142,038 (incl. 993 men)	unknown	unknown	unknown	unfavorable by race	unknown	reduced
Muss, 2001	Literature review	n/a	n/a	n/a	increased	unfavourable	n/a	reduced
Newman et al, 2002	SEER	507 vs. 1378	unknown	unknown	increased	unfavorable of <40	unknown	reduced
Newman et al, 2002	Meta-analysis	10,001 vs. 42,473	unknown	unknown	unknown	unfavorable	unknown	reduced
Newman et al, 2003	Literature review	n/a	n/a	n/a	n/a	unfav. -incr. Rate of local relapse	reduced	unknown
Shavers et al, 2003	SEER	n/a	increased	increased	increased	unfavorable	unknown	reduced
Simon and Severson, 1997	Population-based	Total = 10,502 (82% AA, 18% CA)	unknown	unknown	increased	unfavorable	unknown	reduced
Trock, 1996	Literature review	n/a	increased?	?increased	increased	unfavorable due to dx at late stage	unknown	unknown
Winchester et al, 1998	Population-based National Cancer Data Base	Total = 508,724	unknown	unknown	increased	unfavorable	unknown	unknown
Wojcik et al, 1998	Military Health System	698 vs.5879	unknown	unknown	unknown	unfavorable	unknown	reduced

N/a = not available; SEER= Surveillance, Epidemiology, and End Results; AA= African American; CA= Caucasian American; HSP=Hispanic; +ve = node positive; -ve = node negative; NSABBP = National Surgical Adjuvant Breast and Bowel Project; MDACC = M.D. Anderson Cancer Centre; ED = Enrolment Date.

Among the other factors reviewed were protease cathepsin D, HER2/neu, epidermal growth factor receptor and stress-response (heat shock) protein. There is a strong correlation between tumour size and risk of recurrence (*9*). Even within the category of T_1 (i.e less than 2.0cm) the risk of relapse varies significantly, being as low as <10% in 10 years for tumours less than 1.0cm in maximum dimension, as compared to 1.1 to 2.0cm. Patients with receptor positive tumours have a better prognosis than those with receptor negative tumours, although the difference in recurrence rate is of the order of only 8% to 10% at 5 years (*26*). Nuclear grade is a well-documented prognostic factor, if it can be reliably determined. High nuclear grade is associated with a higher risk of recurrence. Histological subtypes that carry favourable prognosis include tubular, colloid (mucinous), and papillary subtypes. Proliferative rate as determined by DNA flow cytometry is used to determine the S-phase fraction and ploidy status. Unfortunately, methodological problems limit the usefulness of this test in up to 25% of specimens. S-phase fraction has been observed as an independent prognostic factors in some studies (*26*). High levels of protease cathepsin D are associated with unfavourable prognosis (*26*). At the time of the 1990 conference, there was not enough information on HER2/neu, epidermal growth factor receptor, and stress-response protein to allow for conclusive comments. In the process of evaluating individual risk in node negative breast cancer, it should be recognised that prognostic factors are associated with a wide range of risk of recurrence. Outside clinical trials, it is reasonable not to treat a node negative individual with a tumour of <1.0cm because of the risk of recurrence of <10% in 10 years. However, with increasing tumour size, other prognostic factors should be considered in addition to the tumour size. Risk categories for patients with node-negative breast cancer, as recommended at the 6th International Conference on adjuvant therapy of primary breast cancer in St. Gallen, Switzerland in 1998 as summarised by Goldhirsch et al (*31*) is presented in Table 2.

Table 2. Risk Categories for Patients with Node-Negative Breast Cancer.

Factors*	Minimal/low risk (has all listed factors)	Intermediate risk (risk classified between the other two categories)	High Risk (has at least one listed factor)
Tumour size†	≼1 cm	> 1-2 cm	>2 cm
Oestrogen receptor (ER and/or progesterone receptor (PgR) status‡	Positive	Positive	Negative
Grade§	Grade 1 (uncertain relevance for tumours ≼ 1 cm)	Grade 1 – 2	Grade 2-3
Age in years	≽35		<35

Some panel members also recognize lymphatic and/or vascular invasion as an important feature that indicates an increased risk.

† Pathologic tumour size (i.e. size of the invasive component) the most important prognostic factor for defining the additional risk of relapse.

‡ ER status and PgR status are important biologic characteristics that identify responsiveness to endocrine therapies.

§ Histological and/or nuclear grade.

‖ Young age is high risk: Exact age threshold undefined.

Adapted from Goldhirsch et al. Journal of the National Cancer Institute: 1998;90 (21); 1601 - 1609

2. AFRICAN AMERICAN ETHNICITY AS A PROGNOSTIC FACTOR IN EARLY-STAGE BREAST CANCER

The subject of the prognosis of early-stage breast cancer in African American women has received a considerable attention in the medical scientific community over the last decade. Several studies indicate that the disease carries a more unfavourable prognosis among African American women as compared to Caucasian American women (*8*), (*10*), (*12-13*), (*15*), (*19*), (*21-24*), (*32*). The unfavourable prognosis appears to be related to the higher frequency among the African American patients on presentation of prognostic factors that are universally accepted as indicative for poor outcome (Table 1). These include: age ≤ 50 year (*12-14*), (*16-17*), age < 35 year (*17*), large primary tumours (*17-18*), (*20*) and involvement of regional lymph nodes (*8*), (*12*), (*14*), (*19*), (*25*). Other adverse prognostic features at presentation include a higher stage , hormone receptor negative disease (*10*), (*13*), (*21*), (*24*), and other adverse histological features including poor differentiation, high-grade nuclear atypia, and more aggressive histology (e.g. more medullary and less lobular histological variants) (*11*). Trock reviewed published data on the epidemiology, pathology and molecular biology of breast cancer in African American women in order to identify how the aetiology and presentation of the disease differ from those in Caucasian American women. He indicated that the survival difference could be primarily attributed to diagnosis at a later stage among African American women. He however cited evidence from a number of studies suggesting that the tumour in African American women might have a more aggressive phenotype, which could also contribute to the survival disparity. He pointed out that some studies have reported higher S-phase fraction in the African American breast cancer. However, overexpression of p53 and erbB-2 occurs with similar frequency in African American and Caucasian women, although limited data suggest the African Americans might exhibit different p53 mutation spectra, including the suggestion of a specific polymorphism in the African Americans women (also see the contribution of David F. Chhieng et al., - Chapter 3 - and O.F. Ikpatt and O. Olopade – Chapter 2 - in this book).

At least one group of investigators has suggested that the reported difference in prognosis may not be on the basis of racial disparity (*33*). In their study of 163 African American, 205 Hispanic, and 964 Caucasian women treated at the MD Anderson Cancer Centre between 1987 and 1991, Franzini et al. observed that race was not a significant predictor of survival after adjusting for socio-economic status, and "other confounding factors such as demographic and disease characteristics". They concluded that other factors associated with low socio-economic status, such as "life-style and behaviour" might affect survival. Simon and Severson (*21*) reported that after controlling various factors that are known to be associated with adverse prognosis, as well as census-derived socio-economic status, and the presence of a residency training program at the treatment hospital, the relative risk for dying of breast cancer for African American women compared to Caucasian American women was 1.68 for women less than 50 years old, and 1.33 for women older than 50 years. They concluded by emphasising the need to focus more attention on public health efforts towards younger women. Haybittle et al, (*34*) used the data from the Cancer Research Campaign trial of the Medical Research Council of the United Kingdom to study the effect of social class and weight on prognosis after treatment either a simple mastectomy plus postoperative radiotherapy or by a simple mastectomy followed by watch-and-wait policy. When the survival curves of patients in manual classes were compared with those of non-manual classes, there was a tendency for the latter to do better. However, the difference was not statistically significant ($p=0.12$). By contrast, there was a statistical highly significant difference ($p=0.002$) in survival favouring patients weighing less than or equal to 60 kg compared to those weighing greater than 60 kg. Thus, while the effect of weight is to increase the mortality due to breast cancer, the same was not clearly so for social class.

Dignam and colleagues (*25*) examined the survival and related endpoints among African American and Caucasian American women with lymph node negative breast cancer who participated in two randomised clinical trials of the National Surgical Adjuvant Breast and Bowel Project (NSABP). One of the studies involved oestrogen-receptor positive, while the other involved oestrogen receptor negative patients. The goals of the study included the determination of whether African American and Caucasian American had comparable outcomes, accounting for ER status and differences in patient characteristics at diagnosis. The investigators also sought to establish whether treatment response was similar for African American and Caucasian American women with these variants of breast cancer. They observed that adjusted relative risk estimates indicated similar prognosis for African American women compared with Caucasian American women for mortality and disease-free survival, regardless of the oestrogen receptor status. They concluded that African American and Caucasian American women with localised breast cancer had similar outcomes and benefited equally from systemic therapy. They suggested that early detection and appropriate therapy among African American women could result in a reduction in the current disparity in breast cancer mortality between African American and Caucasian American women.

3. AFRICAN AMERICAN ETHNICITY AS A PREDICTOR OF BREAST CANCER MANAGEMENT

The NIH Consensus Conference of 2000 recognised African American ethnicity as a prognostic factor, that is, a marker for aggressive behaviour of early-stage breast cancer as compared to the Caucasian American population. However, this does not imply that the African American woman with early-stage breast cancer needs to be treated differently compared to the Caucasian American woman. A number of publications suggest that optimal treatment of early-stage breast cancer in African American woman leads to a similar outcome as in Caucasian American woman stage for stage. Thus, Dignam (*14*) reported on the National Surgical Adjuvant Breast and Bowel Project (NSABP) clinical trials experience in which the outcomes of treatment for African-American and Caucasian-American patients participating in selected randomised studies were analysed. The objective of the analysis was to determine whether prognosis or efficacy of systemic adjuvant therapy differed between the groups, since randomised clinical trials offer the advantage of a similar disease stage and a uniform treatment plan for all participants. The patients in the analysis included those with axillary node positive and axillary node negative diseases. Among patients with lymph node-negative disease, African-Americans had disease-free survival rates similar to Caucasians. However, the African Americans experienced a modestly greater mortality rate than their Caucasian-American counterparts. Among lymph node-positive patients, African-American patients experienced disease-free survival rates that were similar to those of Caucasian Americans. However, survival was again less favourable among African Americans than Caucasian Americans. Survival (excluding deaths most likely attributable to causes other than cancer) was similar between African-American and Caucasian-American patients. In summary, African-American women and Caucasian-American women who were diagnosed at a comparable disease stage and were similarly treated appeared to experience similar breast cancer prognosis. However, an excess of mortality rate persisted among African American women. The excess may be due to greater mortality from non-cancer causes among African-American women. Dignam concluded from the study findings that African-American women and Caucasian-American women with breast cancer derive a similar benefit from systemic adjuvant therapy when it is administered in accordance with their clinical and pathologic disease presentation. Newman and her colleagues (*32*) observed from a literature review that breast-conservation therapy is under-used among African-American women. They also observed higher rates of locoregional recurrence among them compared to Caucasian-American women whether they receive breast-conserving surgery or undergo mastectomy. They attribute this to the more aggressive nature of the disease in African Americans. They noted, however, that the response rates to appropriately delivered systemic therapy were similar for African-American and Caucasian-American women.

4. IDENTIFYING PATIENTS FOR WHOM CHEMOTHERAPY SHOULD BE RECOMMENDED

Clinical research performed in various parts of the world over the last two decades have made it possible to identify women who would benefit from adjuvant chemotherapy. The question was the subject of discussion at the NIH Consensus Conferences of 1990 and 2000. The 1990 conference limited its deliberations to the role of adjuvant therapy for patients with node negative breast cancer. The conference observed from the data of 10 randomised trials reviewed showed that adjuvant systemic therapy reduces the rate of recurrence by approximately one-third. This would imply that among a group of patients with a risk of breast cancer relapse of about 30%, adjuvant chemotherapy would decrease the risk to 20%. However, the benefit in terms of improvement in survival or quality of life could not be defined. This was probably because of the small sizes of the studies, and because the risk of death from node negative breast cancer is low. Thus, the impact of chemotherapy in further reducing the low-risk of mortality in node negative breast cancer would probably require a much longer period of time of observation to achieve statistical significance. Since the NIH Consensus Statement on Treatment of Early Stage Breast Cancer was issued in 1990, additional information has become available that supplements the original statement. This includes the observation of Fisher and his colleagues (*35*) that adjuvant chemotherapy does provide a survival benefit in clinical trials in selected women with node-negative, hormone receptor negative breast cancer status. The B-20 study of the National Surgical Adjuvant Breast and Bowel Project (NSABP) was conducted to determine whether chemotherapy plus Tamoxifen would be of greater benefit than Tamoxifen alone in the treatment of patients with axillary lymph node-negative, oestrogen receptor-positive breast cancer. Eligible patients (n = 2306) were randomly assigned to one of three treatment groups following surgery. A total of 771 patients with follow-up data received Tamoxifen alone; 767 received Methotrexate, Fluorouracil, and Tamoxifen (MFT); and 768 received Cyclophosphamide, Methotrexate, Fluorouracil, and Tamoxifen (CMFT). Through 5 years of follow-up, chemotherapy plus Tamoxifen resulted in significantly better disease-free survival than Tamoxifen alone (90% for MFT versus 85% for Tamoxifen [P = .01]; 89% for CMFT versus 85% for Tamoxifen [P = .001]). A similar benefit was observed in both distant disease-free survival (92% for MFT versus 87% for Tamoxifen [P = .008]; 91% for CMFT versus 87% for Tamoxifen [P = .006]) and survival (97% for MFT versus 94% for Tamoxifen [P = .05]; 96% for CMFT versus 94% for Tamoxifen [P = .03]). Compared with Tamoxifen alone, MFT and CMFT reduced the risk of ipsilateral breast tumour recurrence after lumpectomy and the risk of recurrence at other local, regional, and distant sites. Risk of treatment failure was reduced after both types of chemotherapy, regardless of tumour size, tumour oestrogen or progesterone receptor level, or patient age; however, the reduction was greatest in patients aged 49 years or less. No subgroup of patients evaluated in this study failed to benefit from chemotherapy. The authors concluded that this and other NSABP studies indicate that patients with breast cancer who meet NSABP protocol criteria, regardless of age, lymph node status, tumour size, or oestrogen receptor status, are candidates for chemotherapy.

The most relevant factors for the estimation of risk in node-positive early-stage breast cancer are the nodal status and the number of involved lymph nodes. For patients with node-negative breast cancer, tumour size, tumour histological and nuclear grade, steroid hormone receptor status, lymphatic and/or vascular, and age are among factors that should be considered. This is according to a report on the deliberations of a panel at the 6th International Conference on Adjuvant Therapy of Primary Breast Cancer in St. Gallen, Switzerland in 1998. Additional considerations in respect of decision-making about treatment should include patient's preference. Table 2 list risk categories that were discussed by the St. Gallen 1998 panel (*31*).

5. EVOLUTION OF ADJUVANT CHEMOTHERAPY OF BREAST CANCER

The development of adjuvant chemotherapy has been a worldwide process geared towards the definition of chemotherapy regimens that would be most effective in eliminating micrometastatic disease.

The concepts of cell kinetics of Skipper, Schabel and Cox as well as that of Norton and Simon (*36-37*) led to the designs of treatment regimens based on cell kinetic principles. The Norton-Simon model, however, likens the cellular growth kinetics to Gompertzian growth pattern. According to this model, the growth rate of the unperturbed tumour is always changing. Thus, the proportion of cells killed is higher when cell number is low. Conversely, the re-growth of a population of cells is more rapid when the cell number is decreased. This probably explains why although varying adjuvant chemotherapeutic regimens could result in greatly different log kill, only a slight impact on eventual clinical results measured in years is observed. It is clear that such moderate impact could be missed easily as a result of other experimental fluctuations, especially when the clinical experimental sample is small (*37*).

Principles of chemotherapy that have been explored both in experimental and clinical research include dose escalation, dose intensity and dose density. Dose escalation involves simple increase in the dose of individual treatment agent. Dose intensity is the amount of drug in relation to the body surface area and the period of therapy, e.g. the amount in milligram per m^2 (body surface area) per week. Dose density is the term applied when the period between treatment doses is reduced. These three characteristics of cancer chemotherapy have been evaluated with reference to their role in tumour control both in animal experiments (*38-39*) and in retrospective (*40*) as well as in prospective randomised clinical trials (*40-46*). An extreme example of dose escalation is the modality that involves the use of consolidation of conventional adjuvant chemotherapy with autologous bone marrow support (ABMS). It is clear that the higher the dose or dose intensity of treatment, the higher the response rate and the longer the disease-free survival period. However, there is no consensus that dose escalation or higher treatment dose intensity in adjuvant therapy leads to augmentation of cure rate in early-stage breast cancer (*36*). This is because the major impediment to cure is the Gompertzian type re-growth phenomenon. Norton has suggested that for this reason one should realise that it is unlikely that dose escalation alone will make a dramatic difference to the outcome of breast cancer. He also suggested that the combination of dose escalation and dose-intensity might be an extremely efficient method of increasing cell kill and improving therapeutic results. The mathematical models of tumour cytoreduction and re-growth following conventional dose-escalated and dose dense treatment regimens are illustrated in Figure 1.

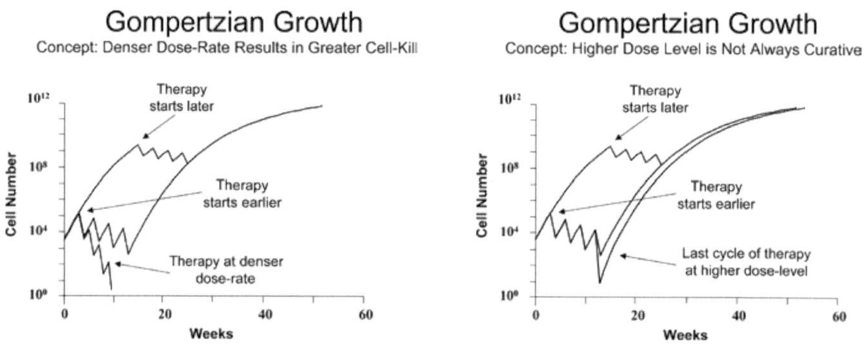

Figure 1. Gompertzian mathematical models of tumour cytoreduction and regrowth following conventional, dose-escalated, and dose-dense treatment regimens. Provided with the courtesy and reproduced with the permission of Larry Norton, MD. For relevant text, please, see - Larry Norton: Evolving Concepts in the Systemic Drug Therapy of Breast Cancers (Seminars in Oncology, 1197; 24 (No. 4, Suppl) S10-3-S10-10.

Combination chemotherapy, which is one of the products of modern approaches to pharmacological control of cancer, has been shown to be superior to single agent chemotherapy in some tumours. Its use, however, involves a reduction in the dose of each agent compared to its dose as a single agent. It is possible that the reduced dose may be less effective in controlling the tumour as compared to the full dose of the drug, thus leading to treatment failure. Norton *(36)* has argued, as illustrated in Figure 2, that sequential use of agents (or regimens) is potentially superior to using them in alternating fashion. In support of this concept is the outcome of a randomised clinical trial study of adjuvant chemotherapy of high-risk breast cancer, in which patients who received cycles of Doxorubicin followed by CMF sequentially had a better outcome as those who received Doxorubicin cycles alternatingly with CMF (*44*). Sequential use of drug agents or regimens assumes a greater importance when the agents or regimens used are known to be non-cross resistant. Thus, the taxanes (Docetaxel and Paclitaxel) are known to be non-cross resistant to the Anthracylines. In a pilot study by Hudis and his colleagues (*47*), high-risk breast cancer patients were treated with sequential single-agent schedule involving Doxorubicin, followed by Paclitaxel and high dose Cyclophosphamide. All agents were given every 2 weeks for three cycles each and supported with G-CSF. More than 80% of the 42 patients, who presented initially with stage II/IIIA disease, were disease free at 42 months. The concept of sequential use of non-cross resistant agents forms the background of clinical trial studies of adjuvant chemotherapy of early-stage breast cancer by various co-operative groups (e.g. Intergroup Trial C9741/Cancer and Leukemia Group B Trial 9741, please, see later).

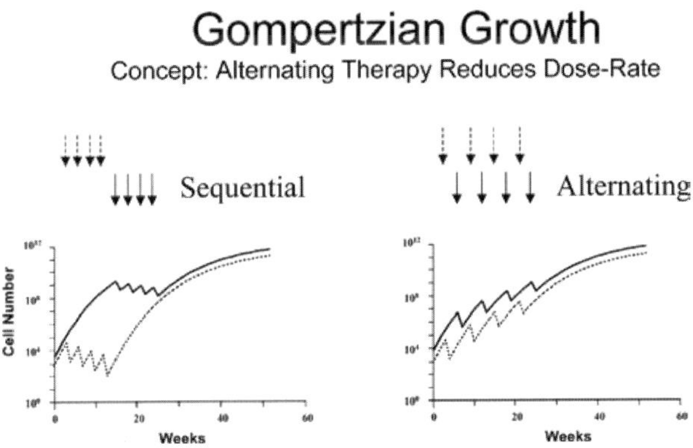

Figure 2. Gompertzian mathematical models of tumour cytoreduction and regrowth following alternating and sequential dose-dense cytotoxic treatment regimens. Provided with the courtesy and reproduced with the permission of Larry Norton, MD. For relevant text, please, see - Larry Norton: Evolving Concepts in the Systemic Drug Therapy of Breast Cancers (Seminars in Oncology, 197, 24 (No. 4, Suppl) S10-3-S10-10.

In the last three decades, more than 200 clinical trials have been conducted on adjuvant chemotherapy of breast cancer, thus making it the most studied clinical problem in oncology. The trials have addressed various aspects of treatment design, including drug combinations, types of agents, their dose, dose intensity, schedules, duration of treatment and associated side effects to name a few of the subjects examined. The complexities of the clinical presentations of breast cancer and the availability of a wide variety of treatment regimens have made consensus building a significant requirement for the practising oncologist. This has been achieved in recent years as a result of two unrelated efforts. Between 1985 and 2000, the United States of America's National Institutes of Health organised three consensus conferences on the management of early-stage breast cancer (*7*), (*48-50*). Leading investigators in the field were invited to these conferences to share their experience, and formulate consensus statements to guide physicians, their patients and the public in general on the appropriate management measures for the disease. The other measure was undertaken by a group of investigators operating under the name "Early Breast Cancer Trialists' Collaborative Group". The group used a novel method of statistical analysis called "meta-analysis," with which they analysed the massive data gathered on studies performed in various parts of the world, and involving in total tens of thousands of women. The objective has been the discovery of small changes which otherwise would be lost due to the reduced sizes of study populations involved in small studies. The reports of the group are often referred to as "Oxford Overviews", in reference to their operation base of Oxford, United Kingdom. Recommendations emanating from the NIH Consensus Conference Program, especially the most recent one of 2000, and the Oxford Overviews have greatly clarified the confusion in the literature on the management of early-stage breast cancer. These recommendations would be referred to from time to time in the course of this chapter.

6. SINGLE-AGENT ADJUVANT CHEMOTHERAPY OF BREAST CANCER

Shortly after the concept of micro-metastatic disease and the need to control it with adjuvant systemic treatment was accepted, a group of investigators at the National Institutes of Health Cancer Chemotherapy National Service Centre devised a protocol to investigate its validity. The group later became known as the National Surgical Adjuvant Breast and Bowel Project (NSABP). The group's first clinical trial was initiated in 1958. In the trial, 826 women receiving care at 23 institutions and who had undergone the conventional Halstedian radical mastectomy procedure were treated subsequently with triethylene thiophosphoramide (Thiotepa) or placebo. The drug was administered at the time of operation and on each of the first two post-operative days. 99.3% of the patients were available for 5-year follow-up. In 1968, the investigators reported a significant increase in the 5-year survival of premenopausal women who were treated with Thiotepa (*51*). The difference was seen mainly in patients with more than four or more positive axillary lymph nodes, and it persisted after ten-year follow-up, at which time 95.3% of the patients had been observed (*52*). 76% of all patients, including 65% of those with one to three positive axillary lymph nodes, and 86% of those with four or more positive axillary lymph nodes, demonstrated treatment failure, while 24.9% survived. The survival rate of those with 1-3 positive axillary lymph nodes was 37.5%, while the survival rate of those with $\geq$ 4 positive axillary lymph nodes was 13.4%. One of 4 node-negative patients demonstrated treatment failure by 10 years. The investigators expressed considerable disappointment from their findings, which were the first of its kind in oncologic literature. In a randomised trial of perioperative adjuvant chemotherapy carried out by a Scandinavian group of investigators (*53*), 507 patients received one single course of chemotherapy with Cyclophosphamide 5mg/kg/day for six days, the first dose being given immediately after mastectomy. 519 control patients received no adjuvant chemotherapy. Relapse free survival was better among the chemotherapy treated women by 13.5% and the difference was statistically significant. The benefit was observed in node-positive and node-negative patients, as well as over and under 50-year age cut-off. The immediate side effects of Cyclophosphamide were moderate, and after median follow-up time of 17.1 years, no late side effects of Cyclophosphamide chemotherapy had been observed. The investigators also remarked that in a parallel-randomised study with 110 patients, to whom the same course of treatment had been given 2-4 weeks after mastectomy, no benefit could be observed.

The first full-scale trial of prolonged adjuvant therapy to be performed in the United States was undertaken jointly by NSABP and the Eastern Cooperative Oncology Group (ECOG). It was a prospective randomised study of 1-phenylalanine mustard (L-PAM), which was administered for 24 months to women with node positive breast cancer. Treatment failure occurred in 22% of 108 patients receiving placebo, and 9.7% of 103 women given L-PAM. A statistically significant difference (p=0.02) was observed in favour of L-PAM relative to disease-free survival. Treatment failure occurred in 30% of premenopausal patients on the placebo arm, and in 3% of the L-PAM arm (p=0.008). A similar trend was observed among the postmenopausal patients although this was not statistically significant. Thus, single-agent L-PAM chemotherapy was found to be effective in the treatment of women with early-stage breast cancer, especially in the premenopausal state.

Single-agent L-PAM chemotherapy has been compared in a number of randomised trials against combination chemotherapy. The Southwest Oncology Group randomised 441 women with operable breast cancer to receive either combination of Cyclophosphamide (60mg/m^2 IV orally every day for 1 year), Fluorouracil (300mg/m^2 IV weekly for 1 year), Methotrexate (15mg/m^2 IV weekly for 1 year), Vincritine (0.625mg/m^2 IV for 10 weeks), Prednisone (30mg/m^2 orally days 1 to 14, 20mg/m^2 days 15-28, and 10mg/m^2 days 29 to 42) (CMFVP) or single-agent Melphalan (L-PAM) (5mg/m^2 orally every day for 5 days, and then every 6 weeks for 2 years) chemotherapy after modified or radical mastectomy between January 1975 and February 1978. Disease free survival and overall survival were superior with CMFVP (p = 0.002 and 0.005, respectively). At 10 years, 48% of patients treated with CMFVP remained alive and disease-free and 56% remained alive, compared to 35% alive and disease free and 43% alive on the L-PAM arm. Disease free survival (DFS) and overall survival (OS) was better with CMFVP compared with L-PAM only in premenopausal patients and patients with four or more positive nodes. Both regimens were well tolerated although toxicity was more severe and more frequent with CMFVP.

7. COMBINATION ADJUVANT CHEMOTHERAPY OF BREAST CANCER

The development of cell kinetic principles in the 1960's (*37*), (*54-56*) (discussed elsewhere in this chapter) led to new approaches in the design of clinical trial studies of adjuvant chemotherapy. The fractional cell-kill hypothesis of Skipper and Schabel of the 1960's proposed that cancer should be curable by chemotherapy, if enough drugs at adequate dose could be applied against a tumour of sufficiently small size. This is because after a number of treatment cycles of the drugs, a tumour cell population of less than one could be achieved. In other words, the tumour would be eradicated. The concept became the underlying principle of combination chemotherapy. It was first validated in childhood leukaemia (*57*). The concept was subsequently transposed to solid tumour "with considerable enthusiasm" (*58*) since micrometastases could be regarded as small collections of rapidly dividing neoplastic cells not unlike those of childhood leukaemia. It was postulated that the growth fraction of a tumour cell population is inversely related to its size. Therefore, micrometastases should be more sensitive to antimetabolites (cell-cycle-specific) anticancer agents than the larger, grossly apparent, primary tumours from which they derive. The NSABP investigators focused their research effort on L-PAM single-agent chemotherapy for early-stage breast cancer because it had limited toxicity. This consideration was particularly important in patients who had undergone the radical Halstedian mastectomy for curative intent. In spite of the attractiveness of the curative concept of combination chemotherapy, the therapeutic community resisted the idea. The first polychemothepeutic regimen, a combination of Cyclophosphamide, Methotrexate and 5-fluorouracil (CMF), was first studied under the auspices of a Milan-based team of investigators led by Bonadonna in 1973. The study was a randomised trial of CMF administered at 28-day cycles for 6 cycles after mastectomy as compared to a similar surgical intervention followed by no chemotherapy. In their first report of the study outcome in 1976 after the first 27 months the investigators observed statistically significant reduction of treatment failure among all subgroups of treated women (*59*). These early positive results have remained demonstrable after 20 years of the study subjects, with continued benefit being demonstrated in premenopausal but not in postmenopausal patients (*44*).

The clinical trials of the last 30 years have consistently shown that adjuvant chemotherapy of early-stage breast cancer is associated with reduced probability of recurrence and the development of grossly demonstrable (stage IV) disease. Unfortunately, however, the optimistic expectations engendered by the cell kinetic theories have not been borne out by clinical trials. It has been suggested that the concept of cure failed because acquisition of drug-resistance by the cells had not been factored into the models. This poorly understood drug resistance mechanism of cancer cells represents a formidable barrier to progress in the development of a curative management of the micrometastatic tumour burden of early-stage breast cancer. The benefits that accrue from adjuvant chemotherapy of early-stage breast cancer depend on various factors that have been extensively studied in clinical trials, and are the subjects of numerous publications, some of which have recently been summarised by Hortobagyi (*60*).

8. THE CONCEPT OF REDUCTION OF RELATIVE AND ABSOLUTE RISK OF RECURRENCE AND MORTALITY

Every individual case of early-stage breast cancer is associated *ab initio* with a roughly quantifiable degree of risk of recurrence and mortality. This degree of risk is related to the biology of the disease. The reduction of this risk by a systemic intervention occurs in all subgroups in which the hypothesis has been adequately tested. Thus, the relative reduction in the odds of cancer recurrence and death is similar in patients with lymph node-negative and lymph node-positive disease. The absolute clinical benefit depends on the initial risk of cancer recurrence and death. Table 3, expresses this as "baseline risk of recurrence of 10% per year or 5% per year". The understanding of the concept of the odds of recurrence of cancer as well as the occurrence of death from cancer and the impact of therapeutic intervention in influencing these odds is often quite challenging. The data in Table 3 could be interpreted as follows: at a baseline risk of 10%/year, 10% of patients depicted in panel A of the table would experience disease recurrence by the end of the first year following surgery alone. By the end of the second year, 10% of those who are free of relapse by the end of the first year would be expected to have developed a recurrence. Therefore, the proportion remaining free of relapse is 90% of 90% = 81% of the original number. By the end of the third year, the proportion remaining free of relapse is 90% of 81%, i.e. 72.9%. By the end of the fifth year, 90% of the 65.61% of those who were free of disease at the end of the 4th year are expected to be disease by the end of the 5th year. The benefit of adjuvant chemotherapy is to reduce the risk of relapse by 35%. Thus, the residual risk is at the rate of 10% minus 35% of 10% (3.5%), i.e. 6.5%. Thus, a year after appropriate surgery and adjuvant chemotherapy, the percentage of women remaining free of recurrence will be 100% minus 6.5%, i.e. 93.5%. At the end of the second year, the proportion of women remaining free of recurrence would be 93.5% of 93.5%, i.e. 87.42%. At the end of the 3rd year, the proportion of those remaining free of recurrence would be 93.5% of 87.42, i.e. 81.74, and so on. The absolute reduction derived from chemotherapy is computed from the relative risk of recurrence after surgery and chemotherapy minus the relative risk after surgery alone. Panel B of Table 3 indicates that although the relative reduction of risk of relapse derived from chemotherapy is the same (35%), the absolute benefit from chemotherapy at the lower baseline risk of recurrence is lower at all stages of the 5-year observation.

*Table 3**. Relative and absolute reductions in annual odds of recurrence after adjuvant chemotherapy.

Disease-free survival after	1 yr	2 y	3 y	4y	5 y
A) Benefit over a 5–y period, assuming a relative reduction in odds of recurrences of 35% and a baseline risk of recurrence of 10%/y					
Surgery	90%	81%	72.9%	65.61%	59.05%
Relative reduction with chemotherapy	35%	35%	35%	35%	35%
Surgery + chemotherapy	93.5%	87.42%	81.74%	76.43%	71.46%
Absolute reduction	3.5%	6.42%	8.84%	10.82%	12.41%
B) Benefit over a 5-y period, assuming a relative reduction in odds of recurrence of 35% and a baseline risk of recurrence of 5%/y					
Surgery	95%	90.25%	85.74%	81.45%	77.38%
Relative reduction with chemotherapy	35%	35%	35%	35%	35%
Surgery + chemotherapy	96.74%	93.61%	90.57%	87.63%	84.78%
Absolute reduction	1.75%	3.36%	4.83%	6.18%	7.40%

* From Hortobagyi, GN: Journal of the National Institute. Monographs 2001 (30): 72-79.

There is now overwhelming evidence, largely from the Oxford Overviews, that adjuvant chemotherapy reduces the annual odds of recurrence and death in a highly statistically significant fashion. The overview of 1988 reported on the meta-analysis of 61 randomised trials among 28,896 women, including 40 chemotherapy trials among 13,442 women. There were also 28 trials of Tamoxifen among 16,513 women. At the time of the analysis, nearly 4000 of the Tamoxifen-treated women, and slightly more than 4000 of the chemotherapy trial patients had died. The 8106 deaths were approximately evenly distributed over years 1, 2, 3, 4 and 5+ years of follow-up, with little useful information beyond 5 years. Clinical trials in which women were treated with Tamoxifen revealed reductions in mortality when compared with the results of other trials in which no Tamoxifen was given to patients. The difference in the rates of reduction of mortality in these two sets of patients was highly statistically significant (p less than 0.0001). Similarly, clinical trials in which any chemotherapy was given revealed statistically significant reductions in mortality as compared to studies in which no chemotherapy was given (p = 0.03). Clinical trials in which polychemotherapy was used as compared to those in which single-agent chemotherapy was used revealed statistically reduced mortality (p = 0.001). In the Tamoxifen trials, there was a clear reduction in mortality only in women 50 years and older, for whom assignment to Tamoxifen reduced the annual odds of death during the first five years by about one fifth. In chemotherapy trials, there was a clear reduction only among women under 50, for whom the assignment to polychemotherapy reduced the odds of death during the first five years by about one quarter. Direct comparison of polychemotherapy with single-agent chemotherapy showed the former to be more effective. The overview also revealed that administration of chemotherapy for 8 to 24 months might offer no survival advantage over administration of the same chemotherapy for 4 to 6 months.

9. ANTHRACYCLINE-BASED ADJUVANT CHEMOTHERAPY

The anthracyclines, including Doxorubicin and Epirubicin, have proven to be significantly effective agents in the management of metastatic breast cancer (*61-62*). The drugs have therefore been incorporated into several clinical trial regimens, which have been studied in comparison with non-anthracycline-containing ones (*63-75*). Hortobagyi (*60*) has outlined the role of the anthracylines in the adjuvant chemotherapy of early breast cancer. Table 4 is an update of the information in his publication on the subject. It is generally agreed that anthracycline-containing adjuvant chemotherapeutic regimens for early-stage breast cancer have a small but statistically significant benefit compared to those that contain no anthracylines (*64), (66), (71*). The findings have been confirmed in the Oxford Overviews (*76-78*).

Table 4. Randomized trials comparing anthracycline-containing with non-anthracycline-containing combination chemotherapy regimens as adjuvant therapy for operable breast cancer*.

Author (study)	No. of patients: A/non-A	Regimens†	Year	Lymph Node-positive	Median Follow-up, mo.	Disease-free survival rate		Overall survival rate		Notes
						A	Non-A	A	Non-A	
FAC/CAF/FEC/PAF vs. CMF±VP or PF(T)										
Hutchins et al	1340/1351	CAF/CMF	1998	None	84	85	82‡	NA	NA	
Mouridsen et al	1195	CEF/CMF	1999	None	61	NA	NA	93	83	Premenopausal
Fisher (B12) et al	548/558	PAF/PFT	1989	All	64	64	63	77	78	Postmenopausal or ER positive
Fisher (B11) et al	347/360	PAF/PF	1989	All	51	51	44‡	65	59‡	Premenopausal or ER negative
Budd et al	??	FAC-M /CMFVP	1995	All	60	NA	NA	64	61	Premenopausal Postmenopausal All ER negative
Levine (MA5 et al)	351/359	CEF/CMF	1998	All	60	63	53‡	77	70‡	Premenopausal
Carpenter et al	260/268	CAF/CMF	1991	All	42	NA	NA	71	70	Premenopausal
Coombes et al	380/379	FEC/CMF	1996	All	54	NA§	NA§	NA§	NA §	Premenopausal
AC or EC vs. CMF										
Fisher (B23) et al	1003/1005	AC/CMF	2000	None	67	82	82	88	90	
Fisher (B15) et al	781/776	AC/CMF	1990	All	26	62	63	83	82	
Di Leo et al	562/255	EC/CMF	1999	All	50	64-74	71	78-86	85	
Galligioni et al	103/104	EC/CMF	1997	All	36	72	63	91	89	Premenopausal
Piccart et al	522/255	EC/CMF	2001	All	52-58	72-80**[a]	78[a]	89-92**[b]	95[b]	Age <70

*A = Anthracycline-containing regimen; ER = estrogen receptor.
†FAC = 5-Fluorouracil, Doxorubicin (i.e. Adriamycin), and Cyclophosphamide; CAF = Cyclophosphamide, Doxorubicin and 5-Fluorouracil; FEC = 5 Fluorouracil, Epirubicin, and Cyclophosphamide; PAF = L-Phenylalanine Mustard, Doxorubicin, and 5-Flourouracil; T= Tamoxifen; CMF = Cyclophosphamide, Methotrexate, and 5 Fluorouracil; NA = not available; CMFVP = CMF+ Vincristine and Prednisone.
‡ Statistically significant difference (two-sided $P<0.05$).
§ Separate sets of numbers given for different schedules of FEC and CMF. FEC1 (the dose schedule used in the first half of the study by Coombes) and CMF1 (the dose schedule used in the first half of the study by Coombes) appear equivalent. FEC2 (the dose schedule used in the second half of the study by Coombes) appears to be superior to CMF2 (the dose schedule used in the second half of the study by Coombes) in disease-free survival rate and overall survival rate.
** = High-dose EC arm.
[a] = 3 - year event-free survival.
[b] = 3 - year overall survival.
Updated from and based on Hortobagyi , GN (2001).

According to the 1998 Overview, adjuvant anthracycline containing regimens are associated with higher disease-free and overall survival compared to the non-anthracycline containing ones. The benefit observed in most studies are, however, only marginal.

Anthracycline-containing regimens that have been studied against non-anthracycline containing ones (e.g. CMF, CMFVP or PF±T – see below for details), include two-agent combinations, such as Adriamycin plus Cyclophosphamide (AC) (*65), (74*), and Epirubicin plus Cyclophosphamide (*70), (72), (79*). Others are three-drug regimens combining an alkylating agent such as Cyclophosphamide or L-PAM, and 5-fluorouracil with either Doxorubicin or Epidoxorubicin (Epirubicin). Both Doxorubicin and Epidoxorubicin (Epirubicin) are cardiotoxic. However, Epidoxorubicin is less so, thus enabling its use in high-dose treatment regimens as has been done in a number of European studies (*70), (72-73), (79*). The marginal difference in improvement in survival reported in association with anthracycline-containing regimens reaches statistical significance only in two studies. These include the outcome of B-11 (*64*), in which the combination of L-PAM, Adriamycin and 5-fluorouracil (PAF) was compared against L-PAM plus 5-fluorouracil (PF). In the study, which involved 707 Tamoxifen unresponsive women, those randomised to PAF had a significantly better disease-free survival (p=0.003) and overall survival (p=0.05) than those who received PF. This was in contrast to another NSABP study (B-12), in which 1,106 women considered responsive to Tamoxifen were randomised to receive L-PAM, Adriamycin, 5-fluorouracil and Tamoxifen (PAFT) or L-PAM, 5-fluorouracil and Tamoxifen (PFT). There was no significant difference in disease-free survival (p=0.6) or overall survival (0.7) among these women. There was no disparity in the amount of drug or the dose intensity of drugs received in the two studies to account for the difference in findings in the two studies. Hutchins and colleagues (*71*) reported the outcome of the Intergroup trial INT 0102 comparing the efficacy of triple combination regimen of Cyclophosphamide, Adriamycin and 5-fluorouracil (CAF) against CMF in node negative breast cancer patients with high-risk features. The 5-year disease-free survival and overall survival among the CAF treated women were with 85% and 92% marginally superior to 82% and 90% observed in CMF treated patients respectively. CAF was somewhat more toxic than CMF in that study. Another study in which a three-drug anthracycline containing regimen has been found to be superior to a non-anthracycline containing regimen in the adjuvant setting is the National Cancer Institute of Canada Clinical Trials Group (NCIC-CTG) MA.5 (*75*). In that study, 351 women were randomised to receive Cyclophosphamide, Epirubicin, and 5-fluorouracil (CEF) (together with oral antibiotic prophylaxis) or Cyclophosphamide,

Methotrexate, and 5-fluorouracil (CMF). After a median follow-up period of 59 months, relapse and overall survival rates were much improved in association with CEF as compared to CMF (p=0.009 and p=0.03 respectively). The rate of hospitalisation for febrile neutropaenia was 8.9% in the CEF treated patients, compared to 1.1% among the CMF patients. There was a case of congestive cardiac failure in a CMF treated woman, but none among the CEF treated patients. Anthracycline-based two drug regimens for early-stage breast cancer include the combination of Cyclophosphamide and Doxorubicin (Adriamycin), (commonly referred to as the AC regimen), and the combination of Cyclophosphamide and Epidoxorubicin (Epirubicin). Both regimens have been compared with the CMF regimen. The AC regimen was studied by NSABP in their B-15 (*65*) three-arm randomised study in which patients were assigned to one of three arms. These included: Adriamycin plus Cyclophosphamide every 21 days for four cycles (regimen I), Adriamycin plus Cyclophosphamide every 21 days for three cycles followed by Cyclophosphamide, Methotrexate, 5-Fluorouracil days 1, 8 and 28 for three cycles (regimen II), and standard CMF (regimen III). After three years of follow-up, findings from 2,194 patients indicated no significant difference in disease-free survival ($p = 0.5$), distant disease-free survival ($p = 0.5$) and overall survival ($p = 0.8$). The AC regimen was completed in 63 days compared to 154 days for the standard CMF regimen. Patients visited health professionals three times as often for CMF as for AC. Women on AC received treatment on 4 days as compared to 84 days of treatment for women on the CMF regimen. AC-regimen treated women required antiemetic agents for about 12 days as compared to 84 days in those treated with the CMF regimen. The convenience of administration and apparently better quality of life on AC has made regimen more popular than CMF among medical oncologists, especially in the US. Another NSABP study, B-23 *(74)*, which involved 2008 women also revealed no significant difference in the outcome of patients who received AC or CMF.

Studies of the EC regimen versus CMF have been reported by Di Leo and colleagues (*72*) and Galligioni and colleagues (*70*), and Piccart and others (*79*) recently provided an update to the report of Di Leo of 1999. In the more recent report of the EC vs CMF study, which was carried out at the Jules Bordet Institute in Belgium, node-positive women who were aged 70 years or younger were randomly allocated to one of the following regimens: CMF for six cycles (oral Cyclophosphamide) (i.e. 24 week duration), EC regimen at the dose of Epirubicin of 60mg/m^2 IV and Cyclophosphamide 500mg/m^2 IV, every three weeks for 8 cycles (24-week duration), and HEC (high-dose EC), Epirubicin 100mg/m^2 IV, and Cyclophosphamide 830mg/m^2 IV every three weeks for 8 cycles (24 weeks). 255, 267 and 255 women were assigned to the three study arms respectively. After 4 years of median follow-up time, no statistically significant differences have been observed between CMF and HEC in terms of event-free survival (EFS) ($p = 0.7$), distant EFS ($p = 0.87$), and overall survival ($p = 0.87$). HEC was however superior to EC. The authors concluded that there was no advantage in favour of an adequately dosed Epirubicin-based (two-drug) regimen over classical CMF in the adjuvant therapy of node-positive pre- and postmenopausal women with breast cancer. The study, however, confirmed the dose-response curve for Epirubicin in breast cancer adjuvant therapy.

Neither the AC nor the EC regimen has ever been compared to any of the clinically more active three-drug anthracycline containing regimens, including CAF/FAC, CEF/FEC, or PAF. However, given the fact that at least two studies (*65), (74*) have indicated equivalence in efficacy between AC and CMF, while the anthracycline containing triple regimens have indicated at least marginal superiority to CMF, it could be inferred that these triple regimens are superior to AC. As Hortobagyi (*60*) has suggested, it is not clear whether this apparent superiority of the triple combination should be attributed to the presence of a third drug in the combination or a longer period of treatment (4 months with the doublets as compared to 6-8 months with the triplets). He also suggests that prudence should dictate that the triple combination of FAC (or CAF, FEC, or CEF) should be considered the standard treatment rather than AC. In Canada, CEF is a popular regimen and would seem to be the regimen of choice in high-risk node positive premenopausal women with early stage breast cancer. The treatment regimen is less commonly used in the U.S. where Epirubicin has not been available until relatively recently. Other standard regimens in Canada and Europe include 6 cycles of FEC or FAC, 6 cycles of CEF-120, 4 cycles of Doxorubicin followed by 3-4 cycles of CMF, and 4 cycles of AC followed by 3 cycles of CMF (see Table 5 for regional variation in the use of treatment regimens for high-risk breast cancer).

Table 5. Regional Variation of Standard Adjuvant Chemotherapy for breast cancer.

Regional Variation of Standard Adjuvant Chemotherapy		
USA*	CANADA	EUROPE*
AC x 4	CEF‡	FAC or FEC
ACT→T	NCIC-CTG MA.21§	?
CAF/FAC	CAF/FAC	A x 4 → CMF x 3 – 4
		AC x 4 → CMF x 3

*Based on Highlights of the 2000 NIH Consensus Conference on Adjuvant Therapy for Breast Cancer [Sledge G.W. (Editor)].
‡ Levine et al, 1998.
§ NCIC-CTG MA.21: AC →T vs. CEF‡ vs. EC [+G-CSF + EPOIETIN – ALFA] + T (in progress).
A = Adriamycin
AC = Adriamycin + Cyclophosphamide
CEF = Cyclophosphamide, Epirubicin, 5-fluorouracil
FAC = 5-fluorouracil, Adriamycin, Cyclophosphamide
FEC = 5-fluorouracil, Epirubicin, Cyclophosphamide
CAF = Cyclophosphamide, Adriamycin, 5-fluorouracil
CMF = Cyclophosphamide, Methotrexate, 5-fluorouracil

10. TAXANES IN ADJUVANT CHEMOTHERAPY OF BREAST CANCER

The taxanes, including Paclitaxel and Docetaxel, are recent additions to the armamentarium in the management of advanced breast cancer. The success of their use in this situation *(80,81)* has engendered the optimism of their role in the adjuvant setting. Studies currently evaluating the role of the taxanes in the adjuvant management of breast cancer include CALGB 9344, NSABP B 28, and NCIC CTG MA.21. Henderson et al *(82)* reported on CALGB 9344, a 3x2 factorial designed trial that involved 3,170 women who were randomly assigned to receive Cyclophosphamide 600 mg/m^2 plus Adriamycin at 60 mg/m2, 75 mg/m^2 or 90 mg/m^2 (+G-CSF) every 4 weeks for 4 cycles, followed either by no Paclitaxel, or followed by Paclitaxel at 175 mg/m^2. At the first pre-planned interim analysis (after 450 events), there was no difference related to the dose of Adriamycin seen. However, the use of Paclitaxel reduced recurrence rate by 22% ($p = 0.0390$) and the death rate by 26% ($p = 0.0077$). A 52-month follow-up analysis, however, revealed that the initial improvements in disease-free and overall survival observed with AC followed by Paclitaxel compared with AC alone no longer reached statistical significance ($p = 0.0745$). Furthermore, the benefit in the risk of recurrence and death from the addition of Paclitaxel to AC had decreased to 13% and 14%, respectively. There were no differences in outcome between treatment groups in patient subsets defined by number of positive axillary lymph nodes, tumour size, and menopausal status (*49*). The original protocol for the study called for a final analysis 4 years after the end of patient accrual. That analysis showed that there continued to be no Doxorubicin dose-effect. At 5 years, DFS was 69%, 66%, and 67% for patients randomly assigned to 60, 75, and 90 mg/m^2 respectively. The hazard reductions from adding Paclitaxel to AC were 17% for recurrence (adjusted Wald χ^2 p = 0.0023; unadjusted Wilcoxon p = 0.0011) and 18% of death (adjusted p = 0.0064; unadjusted p = 0.0098). At 5 years, the DFS (± SE) was 65% (± 1) and 80% (± 1) after AC alone, or AC plus Paclitaxel, respectively. Thus, the addition of four cycles of Paclitaxel after the completion of a standard course of AC improves the disease-free and overall survival of patients with early breast cancer (Henderson, Berry et al. 2003). The role of sequencing four cycles of Paclitaxel after four cycles of AC is not clear in explaining the superiority of the Paclitaxel-containing arm. It is possible that eight cycles of AC could have compared more favourably with the Paclitaxel-containing arm. However, eight cycles of AC using conventional-dose schedules would be considerably more toxic, with unacceptable levels of cardiomyopathy (*82*).

Manounas outlined the outcome of NSABP B 28 at the NIH Consensus Conference on the Adjuvant Therapy for Breast Cancer (*50*). The study, which involved 3060 patients was a prospective randomisation of patients to receive either the standard AC regimen alone, or followed by Paclitaxel 225 mg/m^2 every 21 days for 4 cycles. With a median follow-up of 34 months, patients receiving Paclitaxel following AC did not benefit in terms of disease-free and overall survival compared to those who received AC alone.

In the first investigation of Docetaxel-containing triplet regimen, Nabholtz and his colleagues (*83*) from the Breast Cancer International Research Group (BCIRG) reported the results of the first interim analysis of the BCIRG 01 trial. The phase II study, which involved 1491 patients, compared six cycles of FAC (5-fluorouracil 500 mg/m^2, Adriamycin 50 mg/m^2, and Cyclophosphamide 500 mg/m^2) every 3 weeks against 6 cycles of TAC (Docetaxel 75 mg/m^2, Adriamycin 50 mg/m^2, and Cyclophosphamide 500 mg/m^2) every 3 weeks. Using a Cox analysis adjusted for prognostic factors, TAC was found to be associated with a significant improvement in disease-free survival (hazard ratio for recurrence: 0.64; p = 0.0002), and overall survival (hazard ratio for death: 0.71; p = 0.049). In subgroup analysis, benefits in disease-free survival were seen in patients with oestrogen receptor and/or progesterone receptor positive disease, in patients with 1 to 3, but not in patients with at least 4 positive nodes.

Among a number of ongoing studies aiming at defining the role of taxanes in adjuvant management of breast cancer is a phase III study of the

National Cancer Institute of Canada Clinical Trials Group, the MA.21 study. The study compares an experimental regimen of Epirubicin and Cyclophosphamide (+G-CSF, + Epoietin-alfa) plus Paclitaxel, against both the U.S. taxane-containing regimen AC+Paclitaxel and the Canadian anthracycline-containing standard regimen CEF. The study has the potential of clarifying the role of Paclitaxel in the adjuvant setting, as well as the relative efficacy of the two North American standard adjuvant treatment regimens, namely the Canadian CEF and the U.S. AC+T regimens.

11. OPTIMAL ADJUVANT SYSTEMIC THERAPY

The optimal adjuvant systemic therapy of operable breast cancer is unknown. There is consensus that all patients who are at significant risk for recurrence and death from operable breast cancer should receive multi-disciplinary management to reduce the risk of systemic and local recurrence as well as the risk of death from breast cancer. Such patients are those with a calculated risk of recurrence of over 10% over a 10-year period. A judicious combination of good and appropriate surgical management, systemic and loco-regional radiation therapy would be the benchmark of optimal treatment especially in the presence of high-risk features. All patients with hormone receptor positive tumour, and features of intermediate or high-risk disease, should benefit from endocrine therapy with or without chemotherapy. Currently, this translates to 5-years of Tamoxifen, with or without polychemotherapy. These recommendations are consistent with the outcome of the 2000 NIH Consensus Conference (*49-50*). Patients with hormone receptor negative tumour do not benefit appreciably from endocrine therapy and should not be managed with this treatment modality. Such patients, like the high-risk hormone receptor positive patients should receive polychemotherapy (i.e. chemotherapy with more than one agent). From available evidence some of which have been outlined earlier in this chapter, three-drug regimens (FAC, CAF, FEC or CEF) would seem to be preferred to AC or EC (also, see (*60*). The role of the taxanes in adjuvant treatment of breast cancer remains to be defined. Hortobagyi (*60*) recommends that their use should be considered for high-risk patients with lymph node-positive breast cancer, especially if the cancer is hormone receptor negative. The process of making treatment choices and consensus building between the health care provider, the patient and family members is usually difficult for all concerned. Ravdin and his colleagues (*84*) have recently developed a computer program called Adjuvant! to assist in making such decisions. It was developed from actuarian analysis, which was used to project outcomes of patients with, and without adjuvant therapy based on estimates of prognosis largely derived from Surveillance, Epidemiology, and End-Results data. To these are added estimates of the efficacy of adjuvant therapy based on the 1998 overviews of randomised trials of adjuvant therapy. From the entries of patient's information (age, menopausal status, co-morbidity estimates) and tumour staging as well as characteristics (tumour size, number of positive lymph nodes, hormone receptor status) baseline prognostic estimates is made. Estimates of the efficacy of planned systemic therapy (endocrine ± polychemotherapy), including definition of chemotherapy type (e.g. anthracycline-containing or non-anthracycline containing) can then be used to project outcomes. These projected outcomes are presented in both numerical and graphical formats. Thus, the computer programs Adjuvant! can play practical and educational roles in clinical settings. While information from sources such as individual clinical trials and the Oxford Overviews could provide estimates of benefit for the average patients who participate in them, the tool Adjuvant! can provide estimates for the individual patients. The designers of the tool believe that the output format is useful for the clinician and "easily understood by the patient." A randomised clinical trial is, however, in progress to validate the use of the program. Institutions around the world have their own guidelines that may mimic the facilities that are available in Adjuvant!. Tables 6 and 7 outline the guidelines of the British Columbia Cancer Agency (based in Vancouver, Canada) for adjuvant breast cancer for the period 2001-2002. The policy guidelines are based on recurrence risk which is expressed as "low", "intermediate", "high - node negative", "high – node positive", "extreme". The guidelines reflect the practice in the cancer centres of British Columbia, Canada and the associated community oncology centres.

Guidelines For The Adjuvant Therapy Of Breast Cancer

Table 6. Categories of Risk.

Extreme Risk	High risk	Intermediate Risk	Low Risk
Pathological T4, or N2 or clinically locally advanced/inflammatory (T3N1or T4 and any N or any T and N2)	Positive axillary nodes or Node-negative – Lymphatic or Vascular Invasion (LVI) or -T size ≥ 2 cm, any grade or -T size 1-1.9 cm and Grade 3	-T size 1-1.9 cm and Grade 1 or 2 and No LVI and Node negative	-T size <1 cm and No LVI and Node negative or DCIS

Premenopausal /Peri-menopausal or < Age 50 (Note 5,6)

	Extreme Risk *	High Risk **	Intermediate and Low Risk
Estrogen Positive	CEF followed by Tamoxifen (Note 1,2)	Node Positive: CEF or AC (if risks of CEF unacceptable) followed by Tamoxifen Node Negative: AC followed by Tamoxifen (Note 1, 2)	Tamoxifen (Note 3,4)
Estrogen Receptor Negative	CEF (undesignated ‡ use of AC+T) (Note 1)	Node Positive: CEF or AC (BRAJAC) (if risks of CEF unacceptable) or (undesignated ‡ use of AC+T) Node Negative: AC (Note 1)	(Note 4)

Post Menopausal Patients or Age 50-59 (Note 5,6)

Estrogen Positive	FAC or CEF followed by Tamoxifen (Note 1)	Tamoxifen +/_ AC or CEF Node Positive: CEF or AC (if risks of CEF unacceptable) Node Negative: AC (Note 1)	Tamoxifen (Note 3,4)
Estrogen Receptor Negative	FAC or CEF (undesignated ‡ use of AC+T (Note 1)	Node Positive: CEF or AC (if the risks of CEF unacceptable) or (undesignated ‡ use of AC+T) Node Negative: AC (Note 1)	(Note 4)

Guidelines For The Adjuvant Therapy Of Breast Cancer

Table 6. Categories of Risk (*cont*).

Post Menopausal Patients or Age ≥ 60 (Note 5,6)

	Extreme Risk* ≥ 8 positive nodes	High Risk < 8 positive nodes	Intermediate Risk and Low Risk
Estrogen Postive	FAC and Tamoxifen (Note 1,3)	Tamoxifen ±AC (Note 1,3)	Tamoxifen (Note 3,4)
Estrogen Receptor Negative	FAC (Note 1,3)	AC (Note 1,3)	No Treatment (Note 4)

Note 1: Anthracycline-based regimens are superior to CMF regimens. CMFpo (BRAJCMFpo***) is preferred for patients at significant risk of cardiac toxicity.

Note 2: Premenopausal women with estrogen receptor positive tumors who decline chemotherapy should be offered ovarian ablation or Tamoxifen (BRAJTAM).

Note 3:Except where risk outweighs benefit.

Note 4: Some patients in the intermediate risk category may elect to pursue chemotherapy only after a discussion of the risks and benefits.

* British Columbia Cancer Agency (BCCA) locally advanced study; ** NCIC CTG MA 21; ***: "BRAJ" is the BCCA prefix for breast adjuvant regimens.

‡ Application for authorization for use is required.

Explanation of abbreviations of treatment regimens:

CEF: Cyclophosphamide: 75 mg/m^2/day p.o. x 14 days (day 1-14); Epirubicin: 60 mg/m^2 I.V. on days 1 and 8; Fluorouracil: 500 mg/m^2 I.V. on days 1 and 8 – repeat on day 29 for a total of six cycles. AC: Doxorubicin (Adriamycin): 60 mg/m^2 I.V. on day 1; Cyclophosphamide 600 mg/m^2 I.V. on day 1; repeat every 21 days x 4 cycles; AC+T: Doxorubicin (Adriamycin): 60 mg/m^2 I.V. on day 1; Cyclophosphamide 600 mg/m^2 I.V. on day 1; (repeat A and C every 21 days x 4 cycles), to be then followed 21 days after the final A+C cycle by 4 consecutive cycles of Paclitaxel (Taxol): 175 mg/m^2 I.V. at 21 day intervals; CAF: Cyclophosphamide: 500 mg/m^2 I.V. or 100 mg/m^2/day p.o. x 14 days; Doxorubicin (Adriamycin) 30 mg/m^2 I.V. on days 1 and 8; Fluorouracil: 500 mg/m^2 I.V., - repeat every on day 29 for 6 cycles; CMF: Cyclophosphamide: 600 mg/m^2 I.V. on day 1 or 100 mg/m^2/day x 14 days (days 1-14); Methotrexate 40 mg/m^2 I.V. on day 1 and Fluorouracil: 600 mg/m^2 I.V. on day 1 – repeat every 28 days x 6 cycles.

Reproduced (with modification) from BCCA sources, with permission.

Guidelines For The Adjuvant Therapy Of Breast Cancer

Table 7. Proportional Risk Reduction (PRR) for Recurrence/ Death for Polychemotherapy Alone by Age-Groups 1998 Overview[#]

Age	PRR Recurrence	PRR Death
< 40	37% (SD 7)	27% (SD 8)
40 – 49	34% (SD 5)	27% (SD 5)
50- 59	22% (SD 4)	14% (SD 4)
60-69	18% (SD 4)	8% (SD 4)

Proportional Risk Reduction (PRR) for Recurrence/ Death for 5 years of Tamoxifen Alone for ER+ Tumors, by Age-Groups 1998 Overview[§]

Age	PRR Recurrence	PRR Death
< 50	45% (SD 8)	32% (SD 10)
50 - 59	37% (SD 6)	11% (SD 8)
60 - 69	54% (SD 5)	33% (SD 6)
≥ 70	54% (SD 13)	34% (SD 13)

Note 5: Proportional risk reduction of recurrence and death for chemotherapy by age (see table).
Note 6: Proportional risk reduction of recurrence and death for tamoxifen for estrogen receptor positive tumors by age(see table).
#: Early Breast Cancer Trialists' Collaborative Group. Polychemotherapy for early breast cancer: an overview of the randomised trials. Lancet 1998;352: 930-942.
§: Early Breast Cancer Trialists' Collaborative Group. Tamoxifen for early breast cancer: an overview of the randomised trials. Lancet 1998;351: 1451-1467.
SD: standard deviation.

To roughly estimate the absolute difference in recurrence or death between no adjuvant treatment and adjuvant treatment, multiply the PRR times the estimated risk of recurrence or death at 10years without adjuvant treatment. Always assess competing causes of death (other than breast cancer) over the next the 10years, particularly for those over 70 years of age.

Reproduced (with modification) from BCCA sources, with permission.

12. DURATION OF TREATMENT

The duration of administration of adjuvant chemotherapy has varied between peri-operative to protracted delivery. In spite of the brevity of peri-operative administration of Thiotepa or 5-fluorouracil, the drug was effective enough for a slight delay in recurrence to be demonstrable compared to the experience in placebo-treated control patients (*51*). 1-phenylalanine mustard (L-PAM) was subsequently studied in 24-month course of administration, with resultant significant reduction of treatment failure as compared to placebo-treated control patients (*85*). Long-term chemotherapy apparently produced acceptable toxicity, thus allowing the administration of a high percentage of drug doses. The practice of protracted adjuvant polychemotherapy became the standard of care in the 1970's (*59), (86*). Several clinical trials were subsequently designed to address the question of optimal treatment duration. The effect of the duration of administration of CMF-like regimens on the outcome of the management of operable breast cancer has been the subject of a number of clinical trials carried out by investigators at various centres. Henderson (*87*) reviewed the experience of the Dana Farber Cancer Institute, Milan Cancer Institute, the Southeastern Cancer Group, the Swiss Group for Clinical Cancer Research and the Southwest Oncology Group. These clinical trials compared short treatment courses ranging from 15 weeks to 6 months against those ranging from 12 to 24 months. None of the studies showed a clear advantage for longer treatment duration. The reason for this observation appears to be that longer treatment duration did not correlate with longer disease free or overall survival status while being associated with additional toxicity. The Oxford Overview of 1998 also did not reveal any survival advantage in polychemotherapy regimens comparing longer against shorter (3-6 months) treatment durations. Thus, the use of a single-combination chemotherapy regimen for longer than 6 months is not more beneficial than 6 months of treatment (*78*). It would therefore seem that with the currently available regimens (FAC, CAF, CEF, FEC, CMF, AC and EC), the three-drug single combination regimens that are usually give over 4-6 months duration tend to be superior to the two-drug single combination regimens. Hortobagyi, in fact, questions the adequacy of the AC, which seems to be attractive for its brief, 12-week delivery course (*50, 60*). The views about 6 months being the optimal treatment duration may not apply to the newer treatment regimens that involve two different non-cross resistant regimens requiring more than 6 months to deliver.

13. TREATMENT DOSE AND DOSE INTENSITY

Golding was the first to propose the principle of dose-response of cancer to chemotherapeutic agents (*38*). It has since become clear that the property varies by tumour types. This has been well illustrated in the pre-clinical experiments of Bruce and Lau (*39*), who elegantly showed the marked variation of the steep dose-response curve to Cyclophosphamide in transplantable and spontaneous leukaemia, as compared to the shallow dose-response curve of CH3 spontaneous mammary tumour. The dose-response curve in the latter tumour is even shallower than that of the normal bone marrow CFU, thus indicating that an increase in the dose of Cyclophosphamide is unlikely to effect a cure of the experimental animals of their CH3 mammary tumour. Given the complexity of dose-response patterns observed in pre-clinical experiments, testing the concept in clinical investigations is bound to be more challenging, because of the multiplicity of drugs, tumour types and individual human variations (Henderson, Hayes et al. 1988). It is therefore not surprising that observations of clinical investigators in respect of increased dose of CMF or CMF-like regimens have ranged from being advantageous for disease-free and/or overall survival (*88-91*), to non-beneficial (*41) (63), (89), (92) (93*). Increased dose of chemotherapeutic agents has also been observed to influence outcome negatively (*8,90,93*).

The concept of dose-intensity, which was introduced by Hryniuk and his colleagues (*40,94*) differs from that of Bonadonna and his group (*44,59*) on the basis of the time element that enables the expression of the treatment characteristic as dose-rate. Hryniuk and Levine (*40*) analysed retrospectively the data of a number of clinical trials with a view to determining the projected relative dose-intensity of the treatment regimens as compared to the dose-intensity of a standard treatment regime, the CMFVP of Cooper (*95*). They showed that the projected relative dose-intensity in the investigated treatment regimens correlated significantly with disease free survival. These results, which involved a number of controversial assumptions, have, however, been questioned by Gelman and others (*96*) who have reanalysed the data. Gelman and Henderson (after doing away with a number of the controversial assumptions, but otherwise keeping to the methodology of dose rate calculations of Hryniuk and Levine (*40*), did not only fail to confirm a correlation between dose intensity and disease free survival. They also suggested a negative impact of dose intensity in the survival of premenopausal women. As pointed out by Henderson and others (*97*) a higher dose intensity of CMF (± VP) would actually result in a decreased survival. A number of clinical trials have been designed to evaluate the role of dose intensity of CMF-like regimens in adjuvant chemotherapy of operable breast cancer. Thus, CALGB 8082 (*98*) and Ludwig Breast Cancer Group (*99*), which studied such treatment regimens at significant differences in dose intensities observed no differences in disease-free and overall survival. The dose intensity studies of NSABP in B22 and B25 (*46*), (*100*) also failed to demonstrate an advantage of increasing the dose intensity of Cyclophosphamide, which is the most amenable to this strategy since it has the steepest dose-response curve of all the agents of CMF in pre-clinical models. Thus, no prospective clinical trial study of CMF-based regimen has revealed evidence of advantage of dose intensity strategy in the adjuvant setting. The study of Tannock and his colleagues (*42*) that showed advantage of dose intensity applied to palliative doses of the regimen in metastatic breast cancer. The effects of dose and dose intensity of anthracycline containing treatment regimens are somewhat different from those of non-anthracycline containing regimens in the management of early breast cancer. The French Adjuvant Study Group studied 595 premenopausal women with operable breast cancer between 1986 and 1990 (FASG 1). They were randomly assigned to one of three groups. Group A patients received six cycles with FEC 50 (Fluorouracil 500 mg/m^2, Epirubicin 50 mg/m^2, and Cyclophosphamide 500 mg/m^2) every 21 days. Group B patients received FEC 50 every 21 days for three cycles and Group C received three cycles of FEC 75 (Epirubicin 75 mg/m^2 and same dosages of Cyclophosphamide and Fluorouracil) every 21 days. The 5-year disease-free survival was 64.2% for group A, 55.6% for group B and 55.2% for group C. The differences were not statistically significant. The five-year distant disease-free survival was 72.3% for group A, 64.1% for group B, 65.8% for group C. The differences were not statistically significant. The overall survival was 82.6% for group A, 74.9% for group B and 79.5% for group seen. The differences were not statistically significant. There was a trend towards a higher disease-free survival with FEC 50 administered for six cycles and the best overall survival was in group A in which patients received six cycles of FEC 50 (*101*). Thus, the treatment regimen with the highest cumulative dose of Epirubicin (50 mg/m^2 x 6 versus 50 mg/m^2 x 3 versus 75 mg/m^2 x 3) would seem to be marginally superior. The data could also be interpreted to reflect the superiority of longer treatment duration (6 x 21 days) over shorter treatment duration (21 x 3 days). Fumoleau and his colleagues (*102*) reported on an updated analysis of FASG 1, which revealed that after a long-term follow-up (103 months), the benefit for 6 cycles of FEC 50 compared to 3 cycles, whatever the doses, was highly significant in terms of disease-free survival. As regards overall survival, the results were better in the group receiving 6 cycles of FEC 50 and significantly better compared to 3 cycles of FEC 50. Piccart and her colleagues (*79*) in a three-arm study compared two dose levels of an Epirubicin containing 2-drug regimen (Epirubicin plus Cyclophosphamide) against standard CMF. The study did not show an advantage in favour of an adequately dosed Epirubicin-based regimen over classical CMF in the adjuvant therapy of node-positive pre- and postmenopausal women with breast cancer. It, however, confirmed a dose-response curve for Epirubicin in the adjuvant setting. Thus, although Cyclophosphamide is known to show no benefit when given in higher dose intensity per cycle or at higher cumulative doses of conventional treatment program (*46,100*), the situation is different with anthracycline containing regimens. For anthracycline, increased number of cycles of treatment, dose intensity and cumulative doses are beneficial at least up to a certain threshold beyond which they do not enhance benefit (*79,82,97,(101-102*). In general, it would seem that suboptimal doses of chemotherapy are associated with reduced survival (*43,103*). The observation by Bonadonna et al (*44*) that the delivery of >85% of planned dose of all three drugs of CMF manifests in superior disease-free survival and overall survival is consistent with the benefit of treatment of early breast cancer with optimal treatment doses in the adjuvant setting. The stated benefits of anthracycline-based regimens compared to non-anthracycline-based ones may account for the superiority of the former in the Overview of 1998 of outcome of polychemothepy of early breast cancer *(78)*. In that report, anthracycline-based regimens yielded 12% additional proportional reduction in the risk of recurrence (2p = 0.006), and marginally significant additional 11% proportional reduction in the risk of mortality (2p = 0.02) *(78)*.

14. TIMING AND NEOADJUVANT CHEMOTHERAPY

The subject of timing of chemotherapy of early-stage breast cancer has been extensively studied in the last 30 years. Perioperative treatment, i.e. administration of chemotherapy within a few hours of surgical management of the primary tumour, was the earliest management model tested, and the concept was based on evidence from preclinical experiments (*55*), with a view, apparently, to minimise the risk of dissemination of cancer cells through operation associated manipulation. Perioperative chemotherapy was done historically with single agents, including Thiotepa and L-PAM. As discussed earlier in this chapter, evidence has been adduced for the significant but limited effectiveness of this management model (*51-52*), especially in view of the greater efficacy of adjuvant poly-chemotherapy (Early Breast Cancer Trialists' Collaborative Group, (*91*)). The appropriate timing of initiation of adjuvant chemotherapy is however not clearly defined in the literature, although a few publications have suggested some benefit for early initiation of chemotherapy in high-risk cases of breast cancer (*104-105*). Most clinical trials are designed with eligibility for enrolment being within 60 days of primary surgical procedure (Hortobagyi 2001). The term "neoadjuvant chemotherapy" is synonymous with "primary medical therapy" and "induction chemotherapy" (*106*). Its use originated from the conventional management of locally advanced and inoperable breast cancer. In such situations, neoadjuvant chemotherapy would be used to "downstage" the tumours, thus rendering them surgical resectable. In more recent years, however, the use of the model has become less clear. Jacquillat and his colleagues (*107-108*) were the first to report the use of multiple drug regimens as primary therapy in women with tumours ranging from stage I to stage IIIB. Most of the experience in the treatment of locally advanced breast cancer has been reported as small trials or case series from single-institutions, with limited data available from multi-institution, randomised trials. Thus, treatment guidelines for the management of this type of breast cancer have been derived from level III types of evidence (*109*). The National Adjuvant Breast and Bowel (NSABP) B-18 study is a striking exception in this regard with its 1523 cases (*110*). The result of this study and 5 other ones reported by others (*111-117*) that have studied the value of neoadjuvant chemotherapy as compared to adjuvant chemotherapy in randomised clinical trials have recently been summarised by Cleator et al (*106*) (see Table 8). The analysis of the trials suggests that neoadjuvant chemotherapy results in modest reduction of mastectomy rates. Otherwise, the trials shown in the table suggest that there is no difference between preoperative and postoperative chemotherapy in terns of impact on disease free and overall survival. In B-18, for example, more patients treated preoperatively than postoperatively underwent lumpectomy (67.8% versus 59.8% respectively). B-18 also showed that outcome was better in women whose tumours showed pathologic complete remission (pCR) than in those with a pathologic residual invasive cells (pINV), pathologic partial response (pPR) and pathologic non-response (pNR) (85.7%, 76.9%, 68.1%, 63.9% - $p<0.0001$). It has also been suggested that pCR in both the breast and the axillary lymph nodes might be the biologic marker of success of neoadjuvant chemotherapy. In this connection, a non-randomised study of the University of Texas M.D Anderson Cancer Center (*118*) reported 5-year overall and disease free survival rates that were significantly higher in a group that had pCR in the breast and axillary lymph nodes (89% and 87% respectively) than in a group that had less than a pCR (64% and 58% - $p<0.01$). None of the regimens in Table 8 can be considered the ideal or universally accepted standard for postoperative (adjuvant) chemotherapy model, with the possible exception of FEC (see earlier discussion in this chapter). Thus, the lack of impact in the neoadjuvant setting of these regimens on survival should not be of undue concern. One would like to see a repeat of these type of experiment (preoperative versus postoperative chemotherapy) with the use of effective >2 drug combinations (e.g. the Canadian CEF (*75*) or sequential therapy (e.g. AC+T *(82)*). NSABP B-27 is another study of interest in this category. It is a phase III randomised trial designed to evaluate whether sequencing Docetaxel (Taxotere) in postoperative phase to neoadjuvant Doxorubicin/Cyclophosphamide prolongs disease-free and overall survival in patients with operable breast cancer (*119*).

Preoperative chemotherapy provides a number of opportunities, including downstaging of the breast primary so that inoperable tumours become operable, while those that are otherwise candidates for mastectomy could be downgraded to a size manageable by breast conserving operation. Neoadjuvant chemotherapy also provides a veritable opportunity for *in vivo* chemotherapy sensitivity testing. What is not known is how well the biology of the disease of the breast primary site mirrors the disease in the axillary lymph nodes or that in the micro-metastatic environment. If one can assume that the correlation of the diseases in these sites is close, then neoadjuvant chemotherapy may provide an opportunity to develop a more effective treatment regimen for the control of micrometastatic disease, with resultant impact on survival outcome in the management of early breast cancer. Cleator and colleagues (*106*) have suggested that neoadjuvant chemotherapy of breast cancer provides a unique opportunity to derive biological information related to tumour response given the fact that pCR is an independent predictor of improved disease-free survival. Accessibility of the breast tumour mass for tissue sampling could provide the opportunity of performing critical studies on biological markers, such as p53, bcl-2, and HER-2/neu and other markers that can be identified by emerging technologies (e.g. cDNA microarray) for their role in chemoresistance/response (*106*).

Table 8. Randomised Clinical Trials of Neoadjuvant Compared with Adjuvant Chemotherapy in operable Breast Cancer.

Study	Author	Clinical Stage (all MO)	Year	No. of Patients	Preop. arm treatment	Postop. arm treatment	DFS preop. (%)	DFS postop. (%)	P	OS preop. (%)	OS postop. (%)	P	Notes
NSABP-B18	Fisher et al.	T1-3 N0-1	1997	1523	AC	AC	67	67	NS	80	80	NS	
	Fisher et al.		1998										
EORTC	VanDer Hage et al	T1c-T4b N0-1	2001	698	FEC	FEC	65	70	NS	82	84	NS	
RMH	Powles et al	T0-T4 N0-1	1995	309	2M/3M+ tam.	2M/3M+ tam.	N/A	N/A	NS	N/A	N/A	NS	
	Makras et al		1998										
Institute Bergonie	Mauriac et al.	T2-T3 N0-1	1991	272	EVMx3 MTV x 3	EVMx3 MTV x 3 if high risk	50	50	NS	82	75	NS	Preop. Arm underwent RT or Sx
	Mauriac et al.		1999										
Institute Curie	Scholl et al.	T2-T3 N0-N1	1994	414	CAF	(RT ± Sx) CAF	N/A	N/A	NS	65	60	NS	Sx if incomplete response to chemo/RT
	Broet et al		1999										
St. Petersburg	Semiglazov et al	T1-2N2 T2N1 T3N0-1	1994	271	1-2 x TFM RT→Sx →MF x 4-5	RT→Sx→ TMF x 6	81	72	0	86	78	NS	

DFS: disease free survival; OS, overall survival; AC: Adriamycin, Cyclophosphamide; CAF, Adriamycin, Fluorouracil, Cyclophosphamide; EVM: Epirubicin, Vincritine, Methotrexate; FEC: 5-fluorouracil, Epirubicin, Cyclophosphamide; 2M/3M: Mitoxantrone, Methotrexate/Mitoxantrone, Methotrexate, Mitomycin, C: MTV: Mitomycin, Thiotepa, Vindesine; TFM: Thiotepa, Fluorouracil, Methotrexate; N/A: not available, RT: radiotherapy; Sx surgery; tam: tamoxifen.
NS. $p>0.05$
Adapted from: Cleator, S. Parton, M, Dowsett, M. The Biology of Neoadjuvant Chemotherapy for Breast Cancer. Endocrine - Related Cancer (2002) 9: 183-195.

15. COMBINATION OF CHEMOTHERAPY AND ENDOCRINE THERAPY IN EARLY BREAST CANCER

The subject of endocrine therapy of breast cancer is covered in Chapter 9 in this book. Hormone responsive subtypes of breast cancer are the ones that are associated with hormone receptors, namely oestrogen and/or progesterone receptors. The relationship between the degree of risk reduction of early breast cancer by chemotherapy and endocrine therapy has been controversial for a long time. The subgroups of patients that the controversy applies to are the premenopausal and perimenopausal patients among whom chemotherapy could invoke a double action: a direct anti-neoplastic cytotoxicity and indirect anti-neoplastic endocrine withdrawal effect from ovarian ablation (*48,120,121*). The Overviews of 1988, 1992, 1996 and 1998 (*76,77,91,122,123*) have reviewed the clinical trials and the large numbers of patients involved in relevant trials in order to define the benefit of endocrine therapy, chemotherapy and, indirectly, combined effects of chemotherapy and endocrine therapy. The Overview of 1992 is an update of the one of 1988, and the data reviewed included those of 30,000 women in Tamoxifen trials, 3000 in ovarian ablation trials, 11,000 in polychemotherapy trials. In the 1992 publication of the first 10-year follow-up, only Tamoxifen had been extensively tested among women over 70 years. In this age group, the drug produced a risk reduction of 28% (SD 5) ($p<0.00001$) in recurrence, and 21% (SD 6) in mortality (Table 9). For women 50-69, two modalities, namely Tamoxifen alone and chemotherapy alone produced risk reduction of 30% (SD 2) ($p<0.00001$) and 22% (SD 4) ($p<0.0001$) in recurrence, respectively. These two modalities can interact only in one way to predict the outcome of their combined use. Given the independent impact of the two modalities on treatment outcome, the overview estimated the effects of combined chemo-endocrine therapy by combining the results from trials of chemotherapy with the results from trials of Tamoxifen using three different methods of combination. All three led to virtually identical results. The Overview estimated that treatment with polychemotherapy and Tamoxifen as prescribed in the clinical trials examined would produce risk reduction of about 45% (SD 3) in recurrence, and 30% (SD 4) in mortality (Table 9). These figures could go up to 50% and 33% respectively if the analysis were restricted to trials of over 2 years of Tamoxifen.

Table 9. Age ≥ 50: Indirect estimation of the effects of concurrent prolonged polychemotherapy and about 2 years of tamoxifen.

Types of systemic adjuvant therapy being compared	No. aged ≥ 50 yr.	Typical reduction % (SD)[a] in annual odds of: Recurrence or prior death	Death from any cause
1.Effects of adding Tamoxifen (TAM) [b]			
(1a) Tam. alone vs Nil	13114	30(2)	
(1b) CTX + Tam vs CTX alone	8148	28(3)	
(1c) Tam vs same but without Tam (overview of la and 1b)	21262	29(2)	
2.Effects of adding polychemotherapy (CTX)			
(2a) CTX alone vs. Nil	3745	22(4)	14(5)
(2b) CTX + Tam vs. Tam alone	3932	26(5)	10(7)
(2c) CTX vs. same but without CTX (overview of 2a and 2b)	7677	23(3)	12(4)
3. Indirectly estimated effects of C+ Tam vs. Nil [b]			
(3a) Tam plus adding CTX top Tam (1a & 2b)		48(4)	27(6)
(3b) CTX plus adding Tam to CTX (1 b & 2a)		44(4)	31(5)
(3c) Ignoring any interactions (1c & 2c)		45(3)	30(4)

Few in the chemotherapy comparisons were aged over 70, so all but the first of these estimates are of directly demonstrated relevance only at 50-69 years of age.

[a] Mean scheduled Tamoxifen durations for those dying in the trials of Tamoxifen: (1a) 2-1 (1b) 1-8, and (1c) 2-0 yr. [b] The process of estimation simply involves letting one percentage reduction follow another; for example, in (3a) a 30% reduction in recurrence (from 1.0 down to 0.7) followed by a 25% reduction (from 0.7 down to 0.52) combine to yield a 48% reduction (from 1.0 down to 0.52). Likewise, prevention of ¼ and then prevention of 1/3 would combine to prevent ½.

Adapted with modification from Early Breast Cancer Trialists' Cooperative Group. Lancet 1992; 339: 71-85.

Below the age of 50 years, three treatment modalities each produced a highly significant (p<0.001) percentage reduction in recurrence: Tamoxifen alone 27% (SD 7), chemotherapy alone 37% (SD 5) and ovarian ablation alone 30% (SD 9). With the three significantly effective treatment modalities, there are several "interactions" to be studied, for which the overview did not have enough data on women below the age of 50 to estimate possible outcomes of the possible interactions. Focusing on mortality alone among women less than 50 years, only two of the three treatment modalities had clearly significant effects (Table 10). Chemotherapy alone effected a percentage risk reduction of mortality of 27% (SD 6), and ovarian ablation alone had a corresponding effect of 28% (SD 9). An interaction of these two modalities would probably result in 30%-40% proportional reduction in annual mortality and 12-16 per 100 women treated (Tables 10 and 11), although the investigators felt that the estimates might not be reliable (*76-77*).

Table 10. Proportional and absolute Improvements in 10-year survival from treatment of 100 Middle-Aged Women with Stage 11 Breast Cancer.

Types of systemic adjuvant therapy being compared	No. aged <50 yr	Typical reduction %(SD)[a] in annual odds of: Recurrence or prior death	Death from any cause
1.Effects of adding Tamoxifen (TAM) [b]			
(1a) Tam. alone vs. Nil	2216	27(7)	[17 (10)]
(1b) CTX + Tam vs CTX alone	6362	7(4)	3(5)
(1c) Tam vs same but without Tam (overview of la and 1b)	8578	12(4)	6(5)
2. Effects of adding polychemotherapy (CTX)			
(2a) CTX alone vs. Nil	2976	37(5)	27(6)
(2b) CTX + Tam vs. Tam alone	386	[32(16)]	[-6(23)]
(2c) CTX vs. same but without CTX (overview of 2a and 2b)	3363	36(5)	25(5)
3. Effects of adding ovarian ablation.			
(3a) Ablation alone vs. Nil	878	[30 (9)]	[28 (9)]
(3b) CTX + Ablation vs. CTX alone	939	[21 (9)]	[19 (11)]
(3c) Ablation vs, same but without ablation (overview of 3a and 3b)	1817	26(6)	25(7)
4. Indirectly estimated effects of ovarian ablation + CTX vs. Nil			
(4a) CTX plus adding ablation to CTX (2a & 3b)		Not Reliably known 40-50%?	30-40%?

[a] Square parentheses [ ] denote statistically unstable results with SD > 9.
[b] Mean scheduled Tamoxifen duration's for those dying in the trials of Tamoxifen: (1a) 2-6, (1b) 1-6, and (1c) 1-8 yr.
Adapted with modification from Early Breast Cancer Trialists' Cooperative Group. Lancet 1992; 339: 71-85.

Table 11. Proportional and absolute Improvements in 10-year survival from treatment of 100 Middle-Aged Women with Stage 11 Breast Cancer.

Allocated Treatment	Proportional Reduction in annual mortality % (SD), from tables V & V1	Absolute reduction in 10-year mortality per 100 women treated
Age ≥ 50 yr.		
(i) Polychemotherapy alone (eg, ≥ 6 mo CMF)	12% (4)	5(2)
(ii) Tamoxifen alone (for a median of 2 yr)	20% (2)[a]	8(1) [a]
(iii) Polychemotherapy + Tamoxifen	30%(5) [a]	12(2) [a]
Age < 50 yr.[b]		
(i) Polychemotherapy Alone	25%(5)	10 (3)
(ii) Ovarian ablation Alone	28%(9)	11(4)
(iii) Polychemotherapy + Ovarian ablation	30-40%??	12-16??

Absolute benefit might be about half as great for stage 1.
[a] Results cited are for analysis (by allocated treatment) of women in all Tamoxifen trials, so the median Tamoxifen duration is only 2 year. But more prolonged Tamoxifen appears better.

[e] Estimates of the effects of one or other treatment alone for age <50 yr. Are somewhat unreliable (1a, 2b, or 3c in table 10), owing to small numbers and data-dependent emphasis on this particular age range.
Adapted with modification from Early Breast Cancer Trialists' Cooperative Group. Lancet 1992; 339: 71-85.

The 1996 publication of the overview (*122*) reported on updated results of 12 randomised trials that studied ovarian ablation by either surgery or radiation. These trials, some of which began before 1980 (especially the ovarian ablation only trials), while others began much later (especially the ovarian ablation plus chemotherapy trials), enrolled 2102 premenopausal and 1354 postmenopausal women. Five of the trials included the use of standard adjuvant chemotherapy, while the others did not. In the trials in which chemotherapy was given and ovarian ablation was performed, chemotherapy preceded ovarian ablation. Tamoxifen was not given in any of the trials. Oestrogen receptor status was assessed in only 4 of the 12 trials, all of which included chemotherapy. Surgical or radiation-induced oophorectomy as sole treatment modalities resulted in a reduction of 25% (SD 7) in the odds of recurrence and 24% (SD 7) in the odds of death, regardless of nodal status. The impact of ovarian ablation was significantly less for women < 50 years of age treated with chemotherapy followed by ovarian ablation. The observed rate of reduction in the annual odds of recurrence and death was 10% (SD 9) and 8% (SD 10) respectively. This degree of benefit is much less than what was projected earlier (*76,77*) when the percentage 10-year reduction in the annual odds of recurrence from this treatment modality was 30%-40%. The lack of impact of ovarian ablation following chemotherapy has been attributed to chemotherapy-induced amenorrhoea. The discrepancy between the results published 1992 and 1996 overview publications is attributed to the fact that the trials of ovarian ablation in the presence of routine cytotoxic chemotherapy started much later than those in its absence, and therefore made a greater proportional contribution to the overall 15-year analyses of in the latter publication. Furthermore, Emens and Davidson (*124*) have suggested that the poor results of ovarian ablation and chemotherapy combination in the 15-year analyses could be attributed to the use of less efficacious chemotherapy regimens, and that the results may be different with the use of modern treatment regimens. More information on ovarian ablation plus chemotherapy trials should be forthcoming in near future. For example, a Chinese trial of ovarian ablation that began in 1991 is reported to have randomised more than 3000 women aged < 50years between adjuvant chemotherapy alone and adjuvant chemotherapy plus ovarian ablation.

In spite of the cautious views of the overviews of 1992, ovarian ablation deserves re-examination as an adjuvant therapy for premenopausal women, and should be considered as an alternative to adjuvant chemotherapy in premenopausal women (*125*). It is currently a subject of intense clinical investigation (*122*). Kathryn Strasser-Weippl and Paul E Goss extensively discuss this subject in Chapter 9 in this book. The use of ovarian ablation should be weighed against the prolonged side effect of ovarian failure, including disabling osteoporosis. There is, however, no evidence to support the routine use of a combination of ovarian ablation and chemotherapy. The current practice is combination of chemotherapy with Tamoxifen, whereby Tamoxifen is usually started after chemotherapy rather than administered currently with chemotherapy, so as to minimise the risk of thromboembolic complications (*126,127*).

16. TREATMENT SIDE EFFECTS AND QUALITY OF LIFE

Treatment side effects can be short-term or long-term. By convention, short-term side effects include acute side effects such as are encountered during chemotherapy or shortly thereafter, as well as others lasting up to the end of a treatment course of 6-months or slightly longer. They usually dissipate on completion of therapy (*128*). The short-term side effects usually include nausea, vomiting, mucositis/stomatitis, and diarrhoea. Other side effects in this category include hair loss, neutropaenia, myalgia, neuropathy, fatigue and myelosuppression in general. Short-term side effects tend to produce significant impairment on the quality of life of the patient and a source of demoralisation, psychological stress and loss of compliance on the side of the patient. The frequency and usual degree of severity of short-term side effects associated with the various adjuvant breast cancer chemotherapy regimens are outlined in Table12. The degree of psychological impairment tends to be related to treatment intensity. Thus, in the National Cancer Institute of Canada Clinical Trials Group study (MA.5) of CEF versus CMF, the quality of life of patients on the dose-intensive Epirubicin based test regimen was significantly worse in the earlier treatment cycles. However, there was no difference in the quality of life of patients in both arms in the later segment of scheduled six treatment cycles. This has been interpreted as signifying a significant interaction between time and treatment side effects, and concurrent psychological and physical adaptation to the trauma of disease diagnosis, and its subsequent therapies (*75*). Recovery is usually complete by about one year when most patients are able to return to full active life (*128*).

Table 12. Frequency and usual severity of short-term side effects associated with adjuvant breast cancer chemotherapy regimens*

Regimen	Nausea	Vomiting	Diarrhea	Stomatitis	Alopecia	Neutropenia	Febrile neutropenia or infection	Thrombo-cytopenia	Neuropathy	Myalgias
CMF (oral Cyclophosphamide)	Frequent, +/++	Common,	Common, +	Common, +	Frequent, Partial-total	Frequent, ++/+++	Rare	Frequent, +	Almost Never †	Almost Never †
CMF (all Intravenous)	Common, +/++	Frequent, +	Common, +	Uncommon, +	Frequent, Partial-total	Frequent, ++/+++	Rare	Uncommon, +	Almost Never †	Almost Never †
MF	Common, +	Common, +	Common, +/++	Uncommon, +	Uncommon, minimal	Rare, +	Almost never	Almost never †	Almost Never †	Almost never†
AC	Frequent, +/++	Common, +/++	Uncommon, +	Common, +/++	Almost always, total	Frequent, ++/+++	Rare	Uncommon, +	Almost never	Almost Never †
CEF/FAC (oral Cyclophosphamide)	Frequent, +/+++	Frequent, +/++	Common +/++	Frequent, +/+++	Almost always, total	Almost always	Common	Frequent, +/++	Uncommon, +	Uncommon +
CAF/FAC/FEC 100 (all intravenous)	Common, ++/+++	Common, +/++	Common, +/++	Frequent, +/++	Almost always Total	Frequent, +++	Common	Frequent, +/++	Uncommon +	Uncommon +

* Frequency: almost never = less than 1%: rare = 1%-5%; uncommon = 6%-20%; common = 21%-50%; frequent -51%--95%; almost always = more than 95%. Severity (for all toxic effects excluding alopecia): + = mild; ++ = moderate; +++ = severe: =. CMF = cyclophosphamide, methotrexate, and 5-fluorouracil; AC = doxorubicin and cyclophosphamide; CAF = cyclophosphamide. doxorubicin, and 5-fluorouracil; FEC = 5-fluorouracil. epirubicin. and cyclophosphamide; MF = methotrexate and 5-fluorouracil; CEF = cyclophosphamide. epirubicin. and 5-fluorouracil; FAC = 5-fluorouracil, doxorubicin and cyclophosphamide.

†Not recorded in trials *(9-21)*

It is often assumed that older patients (>65 years) tolerate chemotherapy less well than younger ones, and that short-term toxicities of chemotherapy tend to occur more frequently among them, and that they are more prone to the adverse impact of chemotherapy toxicities on quality of life. Crivellari and colleagues (*129*) analysed the International Breast Cancer Study Group Trial VII in order to study the effects of CMF chemotherapy and Tamoxifen in the management of early-stage breast cancer. They reported higher grades of haematological, mucosal and toxicity in older than younger women. However, the impact of the side effects on quality of life measurements (performance status, coping, physical well being, and mood and appetite) was similar in the older and younger study population. This study and others, such as that of Dees et al (*130*) tend to reassure that older women appear to tolerate chemotherapy nearly as well as the younger women (*128*). Effective supportive care agents have become available over the last two decades, which have made most of these acute side effects manageable. Most of the agents used in adjuvant chemotherapy today, including Cyclophosphamide, Methotrexate, 5-fluorouracil, Doxorubicin, Epidoxorubicin (Epirubicin) and Paclitaxel could be classified as intermediate to highly emetogenic effects (*131,132*). This degree of chemotherapy-associated emesis is manageable prophylactically or therapeutically with Dexamethasone alone or in combination with serotonin antagonists such as Ondansetron, Granisetron or Dolasetron (*133-135*). Higher grades of neutropaenia could be life threatening if not promptly recognised and managed. The availability of effective antimicrobial agents has reduced the risk of serious consequences of this common side effect. However, their use should be limited to complicated neutropenic situations (*136*), and according to prescribed guidelines (*137,138*). Alopecia, another frequent side effect of chemotherapy, is a challenge of cosmesis rather than of life itself. Long-term side effects, on the other hand, are those that occur weeks, months and even years after the conclusion of chemotherapy treatment program. They include premature menopause/infertility, weight gain, cardiac dysfunction, leukaemia/myelodysplastic syndromes, and cognitive impairment (*128*). The incidence of anthracycline associated cardiotoxicity increases with increasing cumulative doses of the agents, which are usually Adriamycin and Epirubicin in terms of adjuvant management of breast cancer. The incidence of congestive heart failure in association with anthracycline based breast cancer adjuvant chemotherapy varies between 0.8% (*139*) and 2% (*140*). The relatively high incidence rate in the study of Albain and her colleagues has been attributed to the patient population being older than is commonly found in most adjuvant trials. In general, it has been suggested that physicians can counsel women without pre-existing cardiac disease that the incidence of symptomatic cardiac problems with anthracycline–based regimen is extremely rare (*128*). Premature ovarian failure, leading to premature menopause and infertility is a common long-term side effect of systemic adjuvant therapy. The risk of its development is related to the age of the exposed, as well as the treatment regimen used and the total dose administered. The vast majority of women older than 40 years would experience menopause following treatment with CMF (81%-96%), FEC (83%-98%) and AC (89%). The corresponding risks of premature menopause among women younger than 40 years 31%-38%, 51%-77% and 23% (*128*). Weight gain, complicating adjuvant chemotherapy, appears to be more common in premenopausal than postmenopausal women. The amount of weight gained is usually in the range of 2.5-5.0 kg, although it could be as high as 10-20 kg (*128*). The underlying mechanism for the side effect is unknown. Some of the most dreaded long-term side effect of adjuvant chemotherapy of early-stage breast cancer, acute leukaemia and myelodysplastic syndromes, are fortunately relatively rare. The risk appears to be related to the total dose of Cyclophosphamide use as well as to exposure to anthracycline. The CMF regimen has been linked to a risk level of 5 cases of leukaemia in 10,000 women over a period of 10 years. The risk appears to be greater with anthracycline-containing regimens, with incidence rates of 1 case in 1000 women to 15 cases in 1000 women have been cited (*75,100*). Secondary leukaemias complicating alkylating agent chemotherapy (such as CMF) develop typically about 5-7 years of exposure, while those resulting from exposure to topoisomerase inhibitors (such as the anthracylines) occur within 6 months to 5 years after expose. Cognitive impairment, which manifests with a number of symptoms such as difficulty with concentration, memory impairment and difficulty with word finding, is a frequent complaint in patients who have undergone adjuvant chemotherapy for early-stage breast cancer. This is often referred to in colloquial parlance as “chemo-brain”. However, neither the anecdotes nor the research studies done far to investigate its background, permit any firm conclusions (*128*). The benefit of the use of Tamoxifen for five years by women who have oestrogen receptor positive cancer is no longer in doubt. It is now common practice following the completion of adjuvant chemotherapy. However, the decision by women whether or not to accept treatment with Tamoxifen is often difficult because of the various anecdotes that have been linked to the use of the drug. These include weight gain, hair loss, joint pain, fatigue, depression, vaginal dryness, hot flashes and sweats as well as sexual dysfunction. Ganz (*141*) has argued that many of the symptoms are directly related to chemotherapy-induced menopause or to withdrawal of hormone replacement therapy. Bines et al (*142*) do not think that Tamoxifen therapy statistically significantly increases the risk of menopause onset in premenopausal women. Goodwin et al (*143*) have demonstrated in a multivariate model studying the effect of chemotherapy and hormonal therapy in 183 premenopausal breast cancer patients that age, chemotherapy and Tamoxifen each were statistically significantly associated with the onset of menopause. Thus, the risk of development of menopause after the age of 35 years is statistically significantly increased when chemotherapy is used as opposed to Tamoxifen or no adjuvant systemic therapy. The risk of a premenopausal woman developing menopause from Tamoxifen is not substantial (*141,143*). Adjuvant therapy with Tamoxifen was not observed to be associated with an increased risk of weight gain by Goodwin and her colleagues (*144*). It seems that there is consensus that Tamoxifen is associated with increased risk of genital symptoms such as vaginal discharge or dryness as well as vasomotor symptoms, such as hot flashes as compared with patients receiving no systemic therapy. However, such symptoms do not constitute a statistically significant impairment of the patient's quality of life (*145,146*). Furthermore, Tamoxifen related symptoms tend to increase with age with the oldest age group having the greatest risk of serious adverse medical events. The absolute benefits of Tamoxifen adjuvant therapy are however equivalent and independent of age group. The decision by a woman whether or not to use Tamoxifen should be based on the realities of her cancer, her age and her individual value judgement (*141*).

Simes and Coates (*147*) have also surmised that given the modest survival benefits from adjuvant cytotoxic therapy for most women with early-stage breast cancer, individual preferences are important when weighing trade-offs between survival and adverse treatment effects. For younger women at high risk, the improvements in survival will usually be sufficient to justify therapy. However, for older women and for those at low risk of recurrence, assessment of individual preferences may be critical to optimal decision-making (*147*).

17. COST CONSIDERATIONS AND INFLUENCE ON BREAST CANCER MANAGEMENT

The developments in biomedical research over the last three decades have resulted in unprecedented degree of acquisition of knowledge about the nature of cancer, its early recognition, diagnosis and treatment. The period also coincides with major technological developments in the field of radiology, nuclear medicine and radiotherapeutics. These various developments have enabled the emergence of an era of major progress in cancer management. Unfortunately, availability of numerous expensive diagnostic and treatment options translates into skyrocketing health care cost in the developed countries of North America and Europe. In recent years, the need to evolve a mechanism of cost containment has led, for instance, to serious political discussions about reforming the American health-care system beginning with the Presidential Elections of 1992 (*148*). The political wrangling in the entrepreneurial environment of the United States of America has led to the emergence of managed care institutions. In many ways, and unfortunately, the developments in health care policies in the U.S. has not resulted in a reformed health care system as had been expected. Rather, it has resulted in an alteration in the system of payment (*148*). Furthermore, more Americans have become uninsured than there had been before the onset of the process, thus further complicating the subject of affordability of cancer care by the ordinary American. Developed countries that run publicly funded health care systems have been spared this type of crisis in health care funding. According to Weeks (*149*), not only was it recognised in the 1980's that cost crisis was related mainly to increasing volume and intensity of medical services, but also that there was no evidence in support of associated differences in medical outcomes. The situation has been responsible for the emergence of the "Third Revolution in Medical Care" and the "Era of Assessment and Accountability" (*149,150*). Thus, economic analysis of treatment regimens goes beyond actual cost of the process. It may involve other considerations such as patient outcomes, quality of life, cost-effectiveness and cost-benefit and cost-utility. The goal of any health economic analysis is to determine whether the cost of a particular intervention is justified by the health benefits that it produces. Under such considerations, the appropriate question is not "how much does it cost", but rather "how much more does it cost to provide the treatment than the most reasonable alternative" (*149,151*). A comprehensive cost-effectiveness analysis requires data on how the alternative treatments being compared affect the length of life, the quality of life, and cost (*149*). It measures the benefits of health-care interventions in units of medical effect. For example, the cost-effectiveness of combination chemotherapy, compared to single agent therapy, for a given disease could be assessed by calculating the additional cost (in dollars) per additional patient reaching the 5-year disease-free survival mark. The health benefit may be expressed in units, such as "years of life saved" or "increase in the quality-adjusted life expectancy". Cost-effectiveness ratios are, therefore, usually expressed in terms of dollars per years of life saved (*149*). Table 13 shows the cost-effectiveness of some breast cancer related interventions in comparison to other disease states. Some of these estimates are "quality-adjusted", i.e. adjusted by a numerical description that combines expected survival and expected quality of life (*152*). The cost-effectiveness varies between $15,000 for premenopausal breast cancer patients treated with chemotherapy and Tamoxifen (*153*) and $305,000 (*154*) for patient receiving Pamidronate for metastatic breast cancer. This compares favourably with $51,000 for unrelated bone marrow transplantation for chronic myeloid leukaemia (*155*) and $2,400,000 for diagnosis and treatment of cancer of unknown primary site (*156*). There is no absolute standard for what constitutes a reasonable number of dollars per year of life saved. Generally, interventions costing approximately $50,000 per year of life saved are usually considered cost effective, while that costing $100,000 are considered cost ineffective. In general, however, cancer therapies that result in clinically meaningful improvements in patients' outcome are considered cost effective, regardless of cost (*149*).

Table 13. Cost-Effectiveness Analyses In the Management of Breast and Other Cancers.

Diagnosis	Intervention	Alternative	Cost-Effectiveness Ratio
Premenopausal node-positive breast cancer[#]	Chemotherapy and Tamoxifen	Tamoxifen	$15,000*
Node-negative breast cancer[$]	Chemotherapy	No chemotherapy	$50,000*
Metastatic breast cancer[&]	Pamidronate	Hormonal therapy	$305,000*
Chronic myeloid leukaemia[¢]	Unrelated donor bone marrow transplantation	Alfa-interferon	$51,300*
Acute myelogenous leukaemia[§]	Bone marrow transplantation	Conventional chemotherapy	$15,000
Carcinoma of unknown primary site[¶]	Diagnosis and treatment	Supportive care	$2,400,000

* Quality adjusted
Adapted with modification from Weeks, JC: Outcomes assessment. In: Cancer Medicine (eds. Bast, RC); p:1039-1044.
#Smith, J et al: J Clin Oncol 1993;11:771-776.
$ Hillner, BE et al: N Engl J Med 1991;324:160-168.
& Hillner, BE et al: J Clin Oncol 2000;28:72-79.
¢ Lee, SJ et al: Blood 1988;92:4047-4052.
§ Welch, HG and Larson, EB: N Engl J Med 1989;321:807-812.
¶ Levine, MN et al: Can Med Assoc J 1985;133;977-983.

The policy implications of cost-effectiveness analysis for those who pay for cancer care have been thoroughly examined by Hillner and Smith (*152*). It is understandable that for the individual patient, the position is usually: "treat to prolong life despite the cost." In an era of ever escalating cost of health care and limited health care resources, the approach of "cure at all cost" seems to be getting increasing untenable. The challenge in this area could be particularly sobering when health care resources are severely limited and there are competing health care priorities. How much is society willing to spend for a small potential gain, since the effect of adjuvant chemotherapy on survival is either small or unknown? The answer to that question is bound to be subject to the varying prevalent values in societies not only around the globe, but also from one part of a country to the other, particularly where there are no legislative directives that guide access to health care and its delivery. The examples of cost-effectiveness analysis provided in Table 13 illustrates the wide range of estimates of cost that can be incurred in the management of common but varied cancer problems in oncologic practice. There appears to be a need to have a mechanism to guide the oncologists in the use of the available diagnostic procedures and chemotherapeutic interventions, so as to ensure rationale, fair and judicious usage. Several organisations have assumed leadership roles in developing guidelines for oncologists with the end-effect of creating a background for the use of available resources fairly among their deserving patients. Such bodies include the National Institutes of Health of the U.S., which regularly organises Consensus Conferences at which experts on subjects provide guidance on generally acceptable principles of management. The consensus of opinion is subsequently disseminated to all physicians in the field. The American Society of Clinical Oncology (ASCO) also from time to time issues guidelines for the use of new and expensive drugs or procedures in order to guide its members and their patients. One example of this is the development of guidelines for the use of haemopoietic growth factors (*157*) in the management of neutropaenia associated with cancer therapy. Finally, institutions are also developing their own guidelines for the treatment of common cancers. The British Columbia Cancer Agency in Vancouver, Canada, for instance, has site-specific Tumour Groups, whose function is to develop treatment protocols and management guidelines (Tables 6 and 7) (including follow-up protocols) based on available evidence in the literature for guidance of its oncologists in four cancer centres under its jurisdiction. The overall interest in the Canadian context is to ensure that all citizens requiring care for a specific cancer type receive an internationally acceptable standard of care at a reasonable cost to the health-care provider, the provincial Ministry of Health.

18. FOLLOW-UP AFTER ADJUVANT CHEMOTHERAPY

The various forms of treatment employed in the management of early-stage breast cancer constitute sources of diverse health problems for breast cancer survivors, increasing rates of survival of whom is in support of the success of modern treatment modalities for this common health challenge of women. Thus, breast cancer mortality has declined in the 1990's by the largest amount in over 65 years, with the 5-year survival rate reaching 97%

for women diagnosed at an early stage of the disease (*158*). Of the 180,000 women who receive a diagnosis of invasive breast cancer in the United States every year, more than 80% can expect to survive for at least five years (*159*). There are more than 2 million female breast cancer survivors in USA alone (*160,161*). This large population of women continues to be at risk for recurrence of breast cancer as well as the development of a second primary breast cancer in the contralateral breast. In addition, they need to be followed-up for long-lasting treatment related symptoms including premature menopause and other gynaecological and reproductive complications as well as a myriad of psychological and social problems precipitated by their breast cancer experience. Other problems include the long-term effect on the haemopoietic system and cardiovascular complications.

Many long-term follow-up studies have shown that more than 75% of recurrences are heralded by symptoms or findings on physical examinations (*162,163*). Patients should therefore be carefully evaluated for symptoms and by clinical examination on follow-up visit. The role of intensive surveillance of the well patient is doubtful because only 15%-25% of metastatic disease cases would present without symptoms, and would have been diagnosed on routine physical examination or by radiological/laboratory testing (*164*). There is no evidence therefore that routine close monitoring, or performance of surveillance diagnostic procedures would lead to earlier diagnosis as to make a difference in survival (*165-167*), although it results in a slightly earlier detection of recurrence. The risk of breast cancer mortality in the contralateral breast is estimated at 0.5% to 1.0% per year (*168*). The risk is higher in women diagnosed at a younger age (*169*). A panel of experts of the American Society of Clinical Oncology (ASCO) met and made recommendations for follow-up care of breast cancer patients in 1997 (*170*). Another ASCO panel met the following year and updated their guidelines (*171*). The latter panel concluded that data were sufficient to recommend: a) monthly breast self examination, b) annual mammography of the preserved and contralateral breast, c) careful history and physical examination every 3 to 6 months for 3 years, then every 6 to 12 months for 2 years, then annually. The panel felt that data were not sufficient to recommend routine bone scans, chest radiographs, haematological blood counts, tumour markers, such as carcinoembryonic antigen (CEA), or other cancer antigen (CA 15-3, and CA 27.29), liver ultrasonograms, or computed tomography scans. The recommendations, summarised in Table 14, were evaluated and supported by the ASCO Health Services Research Committee reviewers and the ASCO Board of Directors.

Table 14. ASCO Recommendations for Breast Cancer Follow-up Care.

Test	Frequency	Remarks
Recommended Breast Cancer Surveillance		
History/eliciting of symptoms; physical examination	Every 3 to 6 months for the first 3 years; then 6-12 months for next 2 years; then annually	
Breast Self Examination	Monthly	
Mammography	Annually	Women treated with breast conserving surgery should have their first posttreatment mammogram 6 months after completion of radiotherapy.
Patient Education Regarding Symptoms of Recurrence		Since the majority of recurrences occur between visits, it is prudent to inform women about symptoms of recurrence
Coordination of care		
The majority of breast cancer recurrences will have occurred within the first 5 years after primary therapy. Subsequent care of the patient following primary treatment should be coordinated and not duplicated. In addition, continuity of care should be encouraged and conducted by a physician experienced in the surveillance of cancer patients and in the examination of women with both irradiated and normal contralateral breasts.		
Pelvic examination	Regular intervals.	Longer intervals may be appropriate for women who have had total abdominal hysterectomy and oophorectomy
Breast cancer Surveillance Testing – Not recommended		
Complete Blood Cell count		Data insufficient to support routine use of complete blood count
Automated Chemistry Studies		Data insufficient to support routine use of automated chemistry studics (including protcin, albumin, and calcium levels)
Chest roentgenography		Data insufficient to support routine use of chest radiographs
Bone scans		Data insufficient to support routine use of bone scans
Ultrasound of the liver		Data insufficient to support routine use of liver ultrasounds
Computed Tomography		Data insufficient to support routine use of computed tomography
Breast Cancer Tumour		The routine use of the CA

Marker CA 15-3		15-3 tumour marker for breast cancer surveillance is not recommended.
Breast Cancer Tumour Marker Carcinoembryonic Antigen (CEA)		The routine use of the tumour marker CEA for breast cancer surveillance is not recommended

Adapted from Smith, TJ et al: American Society of Clinical Oncology 1998 Update of Recommended Breast Cancer Surveillance Guidelines. Journal of Clinical Oncology 1999;17:1080-1082.

Health problems that are prevalent among breast cancer survivors include late complications of chemotherapy. Among these are the rare occurrences of leukaemia and myelodysplastic syndromes (MDS), cardiomyopathies (see Long-term side effects of chemotherapy). Routine screening for these disorders is not recommended (*159*). Highly prevalent, however, among patients who have had chemotherapy for breast cancer are gynaecological and reproductive complications. These include amenorrhoea, infertility, and premature menopause in younger woman who continue to menstruate after the completion of chemotherapy. For many of these women, a recurring theme of discussion with health care providers is that of the use of hormone replacement therapy (HRT). A Steering Committee of Health Canada recently reported on the outcome of an extensive research study of this difficult question (*172*). The committee advised on the safety of HRT as follows: "Routine use of HRT (either oestrogen or oestrogen and progesterone) is not recommended for women who have had breast cancer. Randomised controlled trials are required to guide recommendations for this group of women. Women who have had breast cancer are at risk of recurrence and contralateral breast cancer. The potential effect of HRT on these outcomes in women with breast cancer has not been determined in methodologically sound studies. However, in animal and in vitro studies, the development and growth of breast cancer are known to be oestrogen dependent. Given the demonstrated increased risk of breast cancer associated with HRT in women without a diagnosis of breast cancer, it is possible that the risk of recurrence and contralateral breast cancer associated with HRT in women with breast cancer could be of a similar magnitude." Given the magnitude of the problems that breast cancer survivors face in terms of menopausal symptoms, the committee further recommends: "Postmenopausal women with a previous diagnosis of breast cancer who request HRT should be encouraged to consider alternatives to HRT. If menopausal symptoms are particularly troublesome and do not respond to alternative approaches, a well-informed woman may choose to use HRT to control these symptoms after discussing the risks with her physician. In these circumstances, the dose and the duration of treatment should be minimised" (*172*). Table 15 lists the agents that have proven efficacy in the management of hot flashes, a particularly troublesome menopausal symptom complicating chemotherapy. Sexual dysfunction is a common problem among breast cancer survivors. The prevalence of the problem ranges between 15% for reduced physiological arousal to a high of 64% for reduced sexual desire (*158*). Patients are known to be very reluctant to discuss this sensitive issue with health-care providers. There is therefore a need for a proactive probing for the existence of the problem for appropriate intervention. Penson and colleagues (*173*) recommend the "PLISSIT" model. This is a four-step model consisting of "Permission", "Limited Information", "Specific Suggestions" and "Intensive Therapy". This approach leads to gathering of useful information for a purposeful intensive intervention.

Table 15. Agents Used to Ameliorate Hot Flashes.

AGENT	DOSE	COMMENTS WITH LEVEL OF EVIDENCE ()
Vitamin E[#]	800 IU daily	Marginal improvement in clinical symptoms compared to placebo (1)
Megestrol Acetate[$]	20 mg daily or 500 mg i.m. every 2 weeks	50%-90% of patients report 50% decrease in frequency of hot flashes (1). Concern about use of hormonal agent in breast cancer survivor status
Fluoxetine[%]	20 mg daily	Reduction in frequency and intensity of hot flashes compared to placebo (1)
Venlafaxine[&]	75 mg daily (or 37.5 mg in sustained release preparation[@])	Reduction in frequency and intensity of hot flashes as compared to placebo (1)
Paroxetine[¢]	20 mg daily	67% reduction in number of episodes, and 75% reduction in intensity score
Clonidine[£]	Oral or patch 0.1 mg daily	10%-20% reduction in symptoms as compared to placebo (1); substantial side effects; not preferred over placebo by patients at the end of study
Ergotamine- and phenobarbital-based preparations	Various	No benefit after 8 weeks over placebo (1)
Raloxifene[¥]	60 mg daily	No difference in incidence of hot flashes as compared to placebo.
Soy phytoestrogens	Daily tablets, each containing 50 mg of soy isoflavones	No improvement in symptoms as compared to placebo (1)

#: Barton et al: J Clin Oncol 1998;16:495-500.
$: Loprinzi et al: N Engl J Med 1994;331:347-352.
%: Loprinzi et al: Breast Cancer Res Treat 1999;57: 34 (abstract).
&: Loprinzi et al: Proc Am Soc Clin Oncol 2000;19:2a (abstract).
@: Loprinzi et al: Lancet 2000;356: 2059-2063.
¢: Stearns et al: Breast Cancer Res Treat 1998;50: 308 (abstract).
£: Goldberg et al: J Clin Oncol 1994;12: 155-158.
¥: Delmas et al: N Engl J Med 1997;;337:1641-1647.
§: Quella et a: J Clin Oncol 2000;18:1086-1074.
Adapted from Burstein, HJ and Winer, EP:N Engl J Med 2000;343: 1086-1094 with modifications based on Pritchard et al: Can Med Assoc J: 2002;166:1017-1022

Diagnosis and treatment of breast cancer are life-altering events, which, along with the experience of a life-threatening illness such as cancer predispose to psychosocial disorder. Physicians looking after breast cancer survivors should inquire about mood disorders, including fatigue, anxiety, impaired cognitive performance and symptoms of sexual dysfunction. Pre-existent psychosocial problems and family stress could aggravate the situation. Psychosocial problems are worst in the first year of diagnosis, but tend to improve in later years. The long-term quality of life of most breast cancer survivors is excellent. Health care providers should be aware that the use of complementary alternative medicine is common among breast cancer survivors. It could serve as a marker of impaired quality of life.

The well being of breast cancer survivors is challenged on several fronts, thus creating a situation that could lead to "multiple doctoring". Burstein and Winer (*159*) suggest follow-up by a cancer specialist and a primary care physician, and that the latter should look after the gynaecological aspects of the patient's health. The ASCO guidelines for follow-up recommend co-ordination rather than duplication of care. Follow-up by a primary care physician, who is familiar with the patient, is probably to be preferred. A study has shown that primary care follow-up of women with breast cancer who are free of disease, is not associated with increase in time to diagnosis of recurrence, increase in anxiety, or deterioration in health-related quality of life. The study also showed that 69% of recurrences presented between follow-up visits and 44% of the recurrences seen by specialists presented first to the primary care physician (*174*). A cost minimisation analysis of the study revealed that measures of the quality of care such as frequency and length of visits were superior in primary care. Costs to patients and to the health service were lower in primary care. There was no difference in total costs of diagnostic tests, with particular tests being performed more frequently in primary care than in specialist care (*175*). Patients, however, prefer follow-up with the specialist, and perceive the negative aspects of the specialist follow-up – the waiting, the rushed consultations, and the lack of continuity of care – as the trade-off for guarding against a relapse. For patients, an acceptable model of community based care would include clinics staffed by someone with specialised knowledge of cancer and a fast track route back to hospital care when this is required (*176*).

19. RECENT TRENDS TO IMPROVE OUTCOMES IN ADJUVANT CHEMOTHERAPY

The gradual refinement of adjuvant treatments of breast cancer over the last three decades has led to incremental improvements in outcomes, including improvement in mortality of the disease especially in the last decade (*177*). Several directions are currently being pursued to augment the achievement of recent years. These include newer ways of administering chemotherapeutic agents that have been in use for years as well as trial of new agents. Norton (*36*) has proposed the concept that dose dense method of administration of chemotherapy is likely to be successful where other concepts such as dose-augmentation or dose-intensity based approaches have failed. In a recent report of the Intergroup Trial C9741/CALGB 9741, evidence has been adduced to the effect that dose density improves clinical outcomes significantly (*178*). The prognostic and predictive characteristics of the tumour marker HER-2/neu have previously been discussed in this chapter (*27-29,179-180*). HER-2/neu belongs to a family of transmembrane receptor tyrosine kinases, which play significant roles in the signal transduction pathway and contribute in the process of carcinogenesis. Several promising new agents, which are directed against these molecules, have recently been developed. Their potential values in cancer control are being evaluated. The molecule against HER-2/neu is Trastuzumab (Herceptin), which is a humanized IgG_1 monoclonal antibody that binds to the extracellular domain of HER-2/neu. This agent is now being studied in a number of adjuvant treatment regimens including NSABP B31, Intergroup N9831, BCIRG 006 and HERA Trial (NCIC-CTG MA.25) (*181*). The clinical role of prophylactic bisphosphonates in early breast cancer has been evaluated by three studies with different outcomes. Two were positive (*182-183*), while one was negative (*184*). Larger studies of these agents in the adjuvant management of breast cancer are in progress (*177*).

REFERENCES

(1) Margolese R, Fisher B, Hortobagyi G, Bloomer W: Neoplasms of the breast. Hamilton and London: B.C. Decker; 2000.

(2) Cruz E, McDonald G, W. C: Prophylactic treatment of cancer: the use of chemotherapeutic agents to prevent tumor metastasis. Surgery 1956;40:291.

(3) McDonald G, Cruz E, Cole W: The effect of cancer inhibitor drugs on the "take" of Walker carcinosarcoma 256 in rats. Surg Forum 1956.;7:486.

(4) Shapiro D, Fugmann R: A role of chemotherapy as an adjunct to surgery . :. Cancer Res 1957;17:1098.

(5) Parkin DM, Bray F, Ferlay J, Pisani P: Estimating the world cancer burden: Globocan 2000. International Journal of Cancer. 2001;94:153-6.

(6) Greenlee RT, Hill-Harmon MB, Murray T, Thun M: Cancer statistics, 2001.[erratum appears in CA Cancer J Clin 2001 Mar-Apr;51(2):144]. Ca: a Cancer Journal for Clinicians. 2001;51:15-36.

(7) Anonymous: Treatment of Early-Stage Breast Cancer. Consensus Development Conference. Washington, DC, 1990, p. 1-19.

(8) Winchester DP, Osteen RT, Menck HR: The National Cancer Data Base report on breast carcinoma characteristics and outcome in relation to age. Cancer. 1996;78:1838-43.

(9) Carter CL, Allen C, Henson DE: Relation of tumor size, lymph node status, and survival in 24,740 breast cancer cases. Cancer. 1989;63:181-7.

(10) Joslyn SA: Hormone receptors in breast cancer: racial differences in distribution and survival. Breast Cancer Research & Treatment. 2002;73:45-59.

(11) Trock BJ: Breast cancer in African American women: epidemiology and tumor biology. Breast Cancer Research & Treatment. 1996;40:11-24.

(12) Boyer-Chammard A, Taylor TH, Anton-Culver H: Survival differences in breast cancer among racial/ethnic groups: a population-based study. Cancer Detection & Prevention. 1999;23:463-73.

(13) Joslyn SA, West MM: Racial differences in breast carcinoma survival. Cancer. 2000;88:114-23.

(14) Dignam JJ: Efficacy of systemic adjuvant therapy for breast cancer in African-American and Caucasian women. Journal of the National Cancer Institute. Monographs. 2001:36-43.

(15) Muss HB: Factors used to select adjuvant therapy of breast cancer in the United States: an overview of age, race, and socioeconomic status. Journal of the National Cancer Institute. Monographs. 2001:52-5.

(16) English WP, Cleveland KE, Barber WH: There is no difference in survival between African-American and white women with breast cancer. American Surgeon. 2002;68:594-7.

(17) Shavers VL, Harlan LC, Stevens JL: Racial/ethnic variation in clinical presentation, treatment, and survival among breast cancer patients under age 35. Cancer 2003;97:134-47.

(18) Newman LA, Mason J, Cote D, Vin Y, Carolin K, Bouwman D, et al: African-American ethnicity, socioeconomic status, and breast cancer survival: a meta-analysis of 14 studies involving over 10,000 African-American and 40,000 White American patients with carcinoma of the breast. Cancer. 2002;94:2844-54.

(19) El-Tamer MB, Homel P, Wait RB: Is race a poor prognostic factor in breast cancer? Journal of the American College of Surgeons. 1999;189:41-5.

(20) Joslyn SA: Racial differences in treatment and survival from early-stage breast carcinoma. Cancer. 2002;95:1759-66.

(21) Simon MS, Severson RK: Racial differences in breast cancer survival: the interaction of socioeconomic status and tumor biology. American Journal of Obstetrics & Gynecology. 1997;176:S233-9.

(22) Wojcik BE, Spinks MK, Optenberg SA: Breast carcinoma survival analysis for African American and white women in an equal-access health care system. Cancer. 1998;82:1310-8.

(23) Bradley CJ, Given CW, Roberts C: Race, socioeconomic status, and breast cancer treatment and survival.[comment]. Journal of the National Cancer Institute. 2002;94:490-6.

(24) Newman LA, Bunner S, Carolin K, Bouwman D, Kosir MA, White M, et al: Ethnicity related differences in the survival of young breast carcinoma patients. Cancer. 2002;95:21-7.

(25) Dignam JJ, Redmond CK, Fisher B, Costantino JP, Edwards BK: Prognosis among African-American women and white women with lymph node negative breast carcinoma: findings from two randomized clinical trials of the National Surgical Adjuvant Breast and Bowel Project (NSABP). Cancer. 1997;80:80-90.

(26) Mansour EG, Ravdin PM, Dressler L: Prognostic factors in early breast carcinoma. Cancer. 1994;74:381-400.

(27) Thor AD, Liu S, Edgerton S, Moore D, 2nd, Kasowitz KM, Benz CC, et al: Activation (tyrosine phosphorylation) of ErbB-2 (HER-2/neu): a study of incidence and correlation with outcome in breast cancer. Journal of Clinical Oncology. 2000;18:3230-9.

(28) Thor AD, Berry DA, Budman DR, Muss HB, Kute T, Henderson IC, et al: erbB-2, p53, and efficacy of adjuvant therapy in lymph node-positive breast cancer.[comment]. Journal of the National Cancer Institute. 1998;90:1346-60.

(29) Paik S, Bryant J, Tan-Chiu E, Yothers G, Park C, Wickerham DL, et al: HER2 and choice of adjuvant chemotherapy for invasive breast cancer: National Surgical Adjuvant Breast and Bowel Project Protocol B-15. Journal of the National Cancer Institute. 2000;92:1991-8.

(30) Wenger CR, Clark GM: S-phase fraction and breast cancer--a decade of experience. Breast Cancer Research & Treatment. 1998;51:255-65.

(31) Goldhirsch A, Glick JH, Gelber RD, Senn HJ: Meeting highlights: International Consensus Panel on the Treatment of Primary Breast Cancer.[comment]. Journal of the National Cancer Institute. 1998;90:1601-8.

(32) Newman LA, Theriault R, Clendinnin N, Jones D, Pierce I: Treatment choices and response rates in African-American women with breast carcinoma. Cancer 2003;97:246-52.

(33) Franzini L, Williams AF, Franklin J, Singletary SE, Theriault RL: Effects of race and socioeconomic status on survival of 1,332 black, Hispanic, and white women with breast cancer.[comment]. Annals of Surgical Oncology. 1997;4:111-8.

(34) Haybittle J, Houghton J, Baum M: Social class and weight as prognostic factors in early breast cancer. British Journal of Cancer. 1997;75:729-33.

(35) Fisher B, Dignam J, Wolmark N, DeCillis A, Emir B, Wickerham DL, et al: Tamoxifen and chemotherapy for lymph node-negative, estrogen receptor-positive breast cancer.[comment]. Journal of the National Cancer Institute. 1997;89:1673-82.

(36) Norton L: Evolving concepts in the systemic drug therapy of breast cancer. Seminars in Oncology. 1997;24:S10-3-S10-10.

(37) Gilewski T, Dang C, Surbone A, Norton L: Cytokinetics. In Bast R, Kufe D, Pollock R, Weichselbaum R, Holland J, Frei E, editors. Cancer Medicine. Hamilton and London: B.C. Decker, 2000, p. 509-38.

(38) Golding AJ, Venditti JM, Humphreys SR, al. e: Modification of treatment schedules in the management of advanced mouse leukemia with amethopterin. JNCI 1956;17:203-12.

(39) Bruce W, Lau L: Prediction at the cellular level - a study with the C3H mammary tumor. Baltimore: Williams & Wilkins; 1975.

(40) Hryniuk W, Levine MN: Analysis of dose intensity for adjuvant chemotherapy trials in stage II breast cancer. Journal of Clinical Oncology. 1986;4:1162-70.

(41) Abeloff MD, Mellits ED, Baumgardner R, Wilcox P, Watkins S: Prospective trialof standard vs low dose cytoxan, methotrexate, 5-FU (CMF) in adjuvant therapy of breast cancer - assessment of therapeutic efficacyand toxicity. Annual meeting of the American Association of Cancer Research and American Society of Clinical Oncology, 1981, p. 440.

(42) Tannock IF, Boyd NF, DeBoer G, Erlichman C, Fine S, Larocque G, et al: A randomized trial of two dose levels of cyclophosphamide, methotrexate, and fluorouracil chemotherapy for patients with metastatic breast cancer. Journal of Clinical Oncology. 1988;6:1377-87.

(43) Wood WC, Budman DR, Korzun AH, Cooper MR, Younger J, Hart RD, et al: Dose and dose intensity of adjuvant chemotherapy for stage II, node-positive breast carcinoma. New England Journal of Medicine. 1994;330:1253-9.

(44) Bonadonna G, Valagussa P, Moliterni A, Zambetti M, Brambilla C: Adjuvant cyclophosphamide, methotrexate, and fluorouracil in node- positive breast cancer: the results of 20 years of follow-up. N Engl J Med 1995;332:901-6.

(45) Budman DR, Berry DA, Cirrincione CT, Henderson IC, Wood WC, Weiss RB, et al: Dose and dose intensity as determinants of outcome in the adjuvant treatment of breast cancer. The Cancer and Leukemia Group B. Journal of the National Cancer Institute. 1998;90:1205-11.

(46) Fisher B, Anderson S, DeCillis A, Dimitrov N, Atkins JN, Fehrenbacher L, et al: Further evaluation of intensified and increased total dose of cyclophosphamide for the treatment of primary breast cancer: findings from National Surgical Adjuvant Breast and Bowel Project B-25. Journal of Clinical Oncology. 1999;17:3374-88.

(47) Hudis C, Seidman A, Raptis G, Fennelly D, Gilewski T, Baselga J, et al: Sequential adjuvant therapy: the Memorial Sloan-Kettering Cancer Center experience. Seminars in Oncology. 1996;23:58-64.

(48) Anonymous: Adjuvant chemotherapy of breast cancer. National Institutes of Health Concensus Development Conference. Washington, DC: National Institutes of Health, 1985, p. 1-19.

(49) Anonymous: Highlights of the 2000 NIH Consensus Conference on Adjuvant Therapy of Breast Cancer. In Sledge GW, editor. NIH Consensus Conference on Adjuvant Therapy for Breast Cancer, 2000. Washington, DC, 2000, p. 1-8.

(50) Anonymous: Adjuvant therapy of breast cancer. NIH Consensus Development Conference. Washington, DC, 2000, p. 1-35.

(51) Fisher B, Ravdin RG, Ausman RK, Slack NH, Moore GE, Noer RJ: Surgical adjuvant chemotherapy in cancer of the breast: results of a decade of cooperative investigation. Ann Surg 1968;168:337-56.

(52) Fisher B, Slack N, Katrych D, Wolmark N: Ten year follow-up results of patients with carcinoma of the breast in a co-operative clinical trial evaluating surgical adjuvant chemotherapy. Surg Gynecol Obstet 1975;140:528-34.

(53) Kjellgren K, Nissen-Meyer R, Norin T: Perioperative adjuvant chemotherapy in breast cancer. The Scandinavian Adjuvant Chemotherapy Study 1. Acta Oncologica. 1989;28:899-901.

(54) Mendelsohn DL: The growth fraction: a new concept applied to tumors. Science. 1960;132:1496.

(55) Skipper HE: Kinetics of mammary tumor cell growth and implications for therapy. Cancer. 1971;28:1479-99.

(56) Skipper HE, Schabel FM: Quantitative and cytokinetic studies in experimental tumor models. In Holland J, Frei E, editors. Cancer Medicine. Philadelphia: Lea and Febiger, 1973, p. 629.

(57) Holland J: Clinical studies of unmaintained remissions in acute lymphocytic leukemia. 21st Annual Symposium on Fundamental Cancer Research 1967. Baltimore, MD: Williams and Wilkins, 1968, p. 453-62.

(58) Schabel FM, Jr.: Concepts for systemic treatment of micrometastases. Cancer. 1975;35:15-24.

(59) Bonadonna G, Brusamolino E, Valagussa P, Rossi A, Brugnatelli L, Brambilla C, et al: Combination chemotherapy as an adjuvant treatment in operable breast cancer. N Engl J Med 1976;294:405-10.

(60) Hortobagyi GN: Progress in systemic chemotherapy of primary breast cancer: an overview. Journal of the National Cancer Institute. Monographs. 2001:72-9.

(61) A'Hern RP, Smith IE, Ebbs SR: Chemotherapy and survival in advanced breast cancer: the inclusion of doxorubicin in Cooper type regimens. British Journal of Cancer. 1993;67:801-5.

(62) Fossati R, Confalonieri C, Torri V, Ghislandi E, Penna A, Pistotti V, et al: Cytotoxic and hormonal treatment for metastatic breast cancer: a systematic review of published randomized trials involving 31,510 women.[comment]. Journal of Clinical Oncology. 1998;16:3439-60.

(63) Mouridsen HT, Rose C, Brincker H, Thorpe SM, Rank F, Fisherman K, et al: Adjuvant systemic therapy in high-risk breast cancer: the Danish Breast Cancer Cooperative Group's trials of cyclophosphamide, or CMF in premenopausal and tamoxifen in postmenopausal patients. Heidelberg: Springer-Verlag; 1984.

(64) Fisher B, Redmond C, Wickerham DL, Bowman D, Schipper H, Wolmark N, et al: Doxorubicin-containing regimens for the treatment of stage II breast cancer: The National Surgical Adjuvant Breast and Bowel Project experience. Journal of Clinical Oncology. 1989;7:572-82.

(65) Fisher B, Brown AM, Dimitrov NV, Poisson R, Redmond C, Margolese RG, et al: Two months of doxorubicin-cyclophosphamide with and without interval reinduction therapy compared with 6 months of cyclophosphamide, methotrexate, and fluorouracil in positive-node breast cancer patients with tamoxifen-nonresponsive tumors: results from the National Surgical Adjuvant Breast and Bowel Project B-15. Journal of Clinical Oncology. 1990;8:1483-96.

(66) Misset JL, Gil-Delgado, Chollet P, Belpomme D, Fargeot Pm, Fumoleau P, et al: Ten years results of the French trial comparing adriamycin, vincristin, 5-fluorouracile and cyclophosphamide to standard CMF as adjuvant therapy for node positive breast cancer. Twenty-eighth Annual Meeting of the American Society of Clinical Oncology. San Diego, CA, 1992, p. 54 (abstract 41).

(67) Carpenter JT, Velez-Garcia E, Aron BS, Salter M, Stahl DL, Estes N, et al: Five-year results of a randomized comparison of cyclophosphamide, doxorubicin (Adriamycin) and fluorouracil (CAF) vs cyclophosphamide, methotrexate and fluorouracil (CMF) for node positive breast cancer. Annual Meeting of the American Society of Clinical Oncology, 1994, p. 66.

(68) Marty M, Bliss JM, Coombes RC, Wils JM, Amadori D, Gambrosier P, et al: Cyclophosphamide, methotrexate, fluorouracil (CMF) vs F-epirubicin-C (FEC) chemotherapy in premenopausal women with node positive breast cancer: results of a randomized trial (abstract). Annual Meeting of the American Society of Clinical Oncology, 1994, p. 62.

(69) Coombes RC, Bliss JM, Wils J, Morvan F, Espie M, Amadori D, et al: Adjuvant cyclophosphamide, methotrexate, and fluorouracil versus fluorouracil, epirubicin, and cyclophosphamide chemotherapy in premenopausal women with axillary node-positive operable breast cancer: results of a randomized trial. The International Collaborative Cancer Group. Journal of Clinical Oncology. 1996;14:35-45.

(70) Galligioni E, Cetto G, Nascimben O, Buonadonna A, Crivellari D, Molino A, et al: Adjuvant chemotherapy with high-dose epirubicin (EC) and cyclophosphamide versus cyclophosphamide, methotrexate, and fluorouracil (CMF) in high risk premenopausal breast cancer patients. A prospective randomized trial. Annual Meeting of the American Society of Clinical Oncology, 1997, p. 145a.

(71) Hutchins L, Green S, Ravdin P, Lew P, Martino S, Abeloff M, et al: CMF versus CAF with and without tamoxifen in high-risk node-negative breast cancer patients and a

natural history follow-up study in low-risk node-negative patients: first results of Intergroup trial INT 0102. Thirty-fourth Annual Meeting of the American Society of Clinical Oncology. Los Angeles, CA, 1998, p. 1a.

(72) Di Leo A, Larcimont D, Beauduin M, Vindevoghel A, Michel J, Focan C, et al: CMF or anthracycline-based adjuvant chemotherapy for node-positive breast cancer patients: 4 year results of a Belgian randomised clinical trial with predictive marker analysis. Thirty-fifth annual meeting of the American Society of Clinical Oncology. Atlanta, GA, 1999, p. 69a, abstract 258.

(73) Galligioni E, Cetto G, Crivellari D, Nascimben O, Buonadonna A, Molino A, et al: High dose epirubicin and cyclophosphamide (EC) vs cyclophosphamide, methotrexate, fluorouracil (CMF) as adjuvant chemotherapy in high risk premenopausal breast cancer patients: 5-year results of a prospective randomized trial (abstract). Breast Cancer Research & Treatment. 2000;64:63.

(74) Fisher B, Anderson S, Tan-Chiu E, Wolmark N, Wickerham DL, Fisher ER, et al: Tamoxifen and chemotherapy for axillary node-negative, estrogen receptor-negative breast cancer: findings from National Surgical Adjuvant Breast and Bowel Project B-23. Journal of Clinical Oncology. 2001;19:931-42.

(75) Levine MN, Bramwell VH, Pritchard KI, Norris BD, Shepherd LE, Abu-Zahra H, et al: Randomized trial of intensive cyclophosphamide, epirubicin, and fluorouracil chemotherapy compared with cyclophosphamide, methotrexate, and fluorouracil in premenopausal women with node-positive breast cancer. National Cancer Institute of Canada Clinical Trials Group. Journal of Clinical Oncology. 1998;16:2651-8.

(76) Anonymous: Systemic treatment of early breast cancer by hormonal, cytotoxic, or immune therapy. 133 randomised trials involving 31,000 recurrences and 24,000 deaths among 75,000 women. Early Breast Cancer Trialists' Collaborative Group. Lancet. 1992;339:1-15.

(77) Anonymous: Systemic treatment of early breast cancer by hormonal, cytotoxic, or immune therapy. 133 randomised trials involving 31,000 recurrences and 24,000 deaths among 75,000 women. Early Breast Cancer Trialists' Collaborative Group. Lancet. 1992;339:71-85.

(78) Anonymous: Polychemotherapy for early breast cancer: an overview of the randomised trials. Early Breast Cancer Trialists' Collaborative Group. Lancet. 1998;352:930-42.

(79) Piccart MJ, Di Leo A, Beauduin M, Vindevoghel A, Michel J, Focan C, et al: Phase III trial comparing two dose levels of epirubicin combined with cyclophosphamide with cyclophosphamide, methotrexate, and fluorouracil in node-positive breast cancer. Journal of Clinical Oncology. 2001;19:3103-10.

(80) Cortes JE, Pazdur R: Docetaxel. Journal of Clinical Oncology. 1995;13:2643-55.

(81) Hortobagyi GN, Holmes FA: Single-agent paclitaxel for the treatment of breast cancer: an overview. Seminars in Oncology. 1996;23:4-9.

(82) Henderson IC, Berry DA, Demetri GD, Cirrincione CT, Goldstein LJ, Martino S, et al: Improved outcomes from adding sequential Paclitaxel but not from escalating Doxorubicin dose in an adjuvant chemotherapy regimen for patients with node-positive primary breast cancer. J Clin Oncol 2003;21:976-83.

(83) Nabholz J-M, Pienkowski T, Mackey M, Pawlicki M, Guastalla C, Vogel C, et al: Phase III trial comparing TAC (docetaxel, doxorubicin, cyclophosphamide) with FAC (5-fluorouracil, doxorubicin, cyclophosphamide) in the adjuvant treatment of node positive breast cancer patients: interim analysis of the BCIRG 001study. Thirty-eighth Annual Meeting of the American Society of Clinical Oncology. Orlando, FL, 2002, p. 36a (abstract 141).

(84) Ravdin PM, Siminoff LA, Davis GJ, Mercer MB, Hewlett J, Gerson N, et al: Computer program to assist in making decisions about adjuvant therapy for women with early breast cancer. Journal of Clinical Oncology. 2001;19:980-91.

(85) Fisher B, Carbone P, Economou SG, Frelick R, Glass A, Lerner H, et al: 1-Phenylalanine mustard (L-PAM) in the management of primary breast cancer. A report of early findings. New England Journal of Medicine. 1975;292:117-22.

(86) Buzdar AU, Gutterman JU, Blumenschein GR, Hortobagyi GN, Tashima CK, Smith TL, et al: Intensive postoperative chemoimmunotherapy for patients with stage II and stage III breast cancer. Cancer. 1978;41:1064-75.

(87) Henderson IC, Gelman RS, Harris JR, Cannelos GP: Duration of therapy in Adjuvant chemotherapy trials. NCI Monographs 1986;1:95-98.

(88) Bonadonna G, Valagussa P, Rossi A, Zucali R, Tancini G, Bajetta E, et al: Are surgical adjuvant trials altering the course of breast cancer? Seminars in Oncology. 1978;5:450-64.

(89) Ahmann DL, O'Fallon JR, Scanlon PW, Payne WS, Bisel HF, Edmonson JH, et al: A preliminary assessment of factors associated with recurrent disease in a surgical adjuvant clinical trial for patients with breast cancer with special emphasis on the aggressiveness of therapy. American Journal of Clinical Oncology. 1982;5:371-81.

(90) Howell A, Rubens RD, Bush H, George WD, Howat JMT, Crowther D, et al: A controlled trial of adjuvant chemotherapy with melphalan versus cyclophosphamide, methotrexate, and fluorouracil for breast cancer. Berlin, Heidelberg: Springer-Verlag; 1984.

(91) Anonymous: Effects of adjuvant tamoxifen and of cytotoxic therapy on mortality in early breast cancer. An overview of 61 randomized trials among 28,896 women. Early Breast Cancer Trialists' Collaborative Group.[comment]. New England Journal of Medicine. 1988;319:1681-92.

(92) Glucksberg H, Rivkin SE, Rasmussen S, Tranum B, Gad-el-Mawla N, Costanzi J, et al: Combination chemotherapy (CMFVP) versus L-phenylalanine mustard (L-PAM) for operable breast cancer with positive axillary nodes: a Southwest Oncology Group Study. Cancer. 1982;50:423-34.

(93) Redmond C, Fisher B, Wieand HS: The methodologic dilemma in retrospectively correlating the amount of chemotherapy received in adjuvant therapy protocols with disease-free survival. Cancer Treatment Reports. 1983;67:519-26.

(94) Hryniuk W, Bush H: The importance of dose intensity in chemotherapy of metastatic breast cancer. Journal of Clinical Oncology. 1984;2:1281-8.

(95) Cooper RG. Combination chemotherapy in hormone resistant breast cancer. Annual meeting of the American Association for Cancer Research, 1969, p. 15 (abstract).

(96) Gelman RS, Henderson IC: A reanalysis of dose intensity for adjuvant chemotherapy trial in stage II breast cancer. SAKK Bull 1987;1:10-12.

(97) Henderson IC, Berry D, Demetri G, Cirrincione C, Goldstein L, Martino S, et al: Improved disease-free and overall survival from the addition of sequential paclitaxel but not from escalation of doxorubicin dose level in the adjuvant chemotherapy of patients with node-positive primary breast cancer. Thirty-fourth annual meeting of the American Society of Clinical Oncology. Los Angeles, CA, 1998, p. 101a, abstract no. 390A.

(98) Korzun A, Norton L, Perloff M, Wood W, Carey R, Rice M, et al: Clinical equivalence of despite dosage differences of two schedule of cyclophosphamide, methotrexate, 5-fluorouracil, vincristine, and prednisone for adjuvant therapy of node-positive stage II breast cancer. Twenty-fourth annual meeting of the American Society of Clinical Oncology, 1988, p. 12.

(99) Anonymous: A randomized trial of adjuvant combination chemotherapy with or without prednisone in premenopausal breast cancer patients with metastases in one to three axillary lymph nodes. Ludwig Breast Cancer Study Group. Cancer Research. 1985;45:4454-9.

(100) Fisher B, Anderson S, Wickerham DL, DeCillis A, Dimitrov N, Mamounas E, et al: Increased intensification and total dose of cyclophosphamide in a doxorubicin-cyclophosphamide regimen for the treatment of primary breast cancer: findings from National Surgical Adjuvant Breast and Bowel Project B-22. Journal of Clinical Oncology. 1997;15:1858-69.

(101) Bremond A, Kerbrat P, Fumoleau P, De Gislain C, Namer M, Schraub S, et al: Five year follow-up results of a randomized trial testing the role of the dose intensity and duration of chemotherapy in node positive premenopausal breast cancer patients. Thirty-second Annual Meeting of the American Society of Clinical Oncology. Philadelphia, PA, 1996, p. 113, Abstract no. 19.

(102) Fumoleau P, Bremond A, Kerbrat P, Fargeot M, Namer M, Montcuquet J, et al: Better outcome of premenopausal node-positive breast cancer patients treated with 6 cycles vs 3 cycles of adjuvant chemotherapy: eight year follow-up results of FASG 01. Thirty-fifth annual meeting of the American Society of Clinical Oncology. Atlanta, GA, 1999, p. 67a, abstract no. 252.

(103) Bonneterre J, Roche H, Bremond A, Kerbrat P, Namer M, Fumoleau P, et al: Results of a randomized trial of adjuvant chemotherapy with FEC 50 vs FEC 100 in high risk node-positive breast cancer patients. Thirty-fourth annual meeting of the American Society of Clinical Oncology. Los Angeles, CA, 1998, p. 124a, abstract no. 473.

(104) Buzdar AU, Smith TL, Marcus CE, Hortobagyi GN, Blumenschein GR: Significance of drug dose, timing and radiotherapy in adjuvant therapy of breast cancer. Recent Results in Cancer Research. 1984;96:141-7.

(105) Pronzato P, Campora E, Amoroso D, Bertelli G, Botto F, Conte PF, et al: Impact of administration-related factors on outcome of adjuvant chemotherapy for primary breast cancer. American Journal of Clinical Oncology. 1989;12:481-5.

(106) Cleator S, Parton M, Dowsett M: The biology of neoadjuvant chemotherapy for breast cancer. Endocrine-Related Cancer 2002;9:183-95.

(107) Jacquillat CI, Baillet F, Blondon J, al. e: Preliminary results of neoadjuvant chemotherapy in initial management of breast cancer. Annual Meeting of the American Society of Clinical Oncology, 1983, p. 112 (abstract).

(108) Jacquillat CI, Weil M, Auclerc G, al. e, editors: Neoadjuvant chemotherapy in the conservative management of breast cancers: Study of 143 patients. New York, NY: Springer-Verlag; 1986.

(109) Wolff AC, Davidson NE: Preoperative therapy in breast cancer: lessons from the treatment of locally advanced disease. Oncologist. 2002;7:239-45.

(110) Fisher B, Bryant J, Wolmark N, Mamounas E, Brown A, Fisher ER, et al: Effect of preoperative chemotherapy on the outcome of women with operable breast cancer. Journal of Clinical Oncology. 1998;16:2672-85.

(111) Mauriac L, Durand M, Avril A, Dilhuydy JM: Effects of primary chemotherapy in conservative treatment of breast cancer patients with operable tumors larger than 3 cm. Results of a randomized trial in a single centre. Annals of Oncology. 1991;2:347-54.

(112) Scholl SM, Fourquet A, Asselain B, Pierga JY, Vilcoq JR, Durand JC, et al: Neoadjuvant versus adjuvant chemotherapy in premenopausal patients with tumours considered too large for breast conserving surgery: preliminary results of a randomised trial: S6. European Journal of Cancer. 1994;30A:645-52.

(113) Semiglazov VF, Topuzov EE, Bavli JL, Moiseyenko VM, Ivanova OA, Selevnev IK, et al: Primary (neoadjuvant) chemotherapy and radiotherapy compared with primary radiotherapy compared with primary radiotherapy alone in stage IIb-IIIa breast cancer. Annals of Oncology. 1994;5:591-95.

(114) Powles TJ, Hickish TF, Makris A, Ashley SE, O'Brien ME, Tidy VA, et al: Randomized trial of chemoendocrine therapy started before or after surgery for treatment of primary breast cancer. Journal of Clinical Oncology. 1995;13:547-52.

(115) Makris A, Powles TJ, Ashley SE, Chang J, Hickish T, Tidy VA, et al: A reduction in the requirements for mastectomy in a randomized trial of neoadjuvant chemoendocrine therapy in primary breast cancer.[comment]. Annals of Oncology. 1998;9:1179-84.

(116) Broet P, Scholl SM, de la Rochefordiere A, Fourquet A, Moreau T, De Rycke Y, et al: Short and long-term effects on survival in breast cancer patients treated by primary chemotherapy: an updated analysis of a randomized trial. Breast Cancer Research & Treatment. 1999;58:151-6.

(117) van der Hage JA, van de Velde CJ, Julien JP, Tubiana-Hulin M, Vandervelden C, Duchateau L: Preoperative chemotherapy in primary operable breast cancer: results from the European Organization for Research and Treatment of Cancer trial 10902.[comment]. Journal of Clinical Oncology. 2001;19:4224-37.

(118) Kuerer HM, Newman LA, Smith TL, Ames FC, Hunt KK, Dhingra K, et al: Clinical course of breast cancer patients with complete pathologic primary tumor and axillary lymph node response to doxorubicin-based neoadjuvant chemotherapy.[comment]. Journal of Clinical Oncology. 1999;17:460-9.

(119) Mamounas EP: NSABP Protocol B-27. Preoperative doxorubicin plus cyclophosphamide followed by preoperative or postoperative docetaxel. Oncology (Huntington). 1997;11:37-40.

(120) Richards MA, O'Reilly SM, Howell A, George WD, Fentiman IS, Chaudary MA, et al: Adjuvant cyclophosphamide, methotrexate, and fluorouracil in patients with axillary node-positive breast cancer: an update of the Guy's/Manchester trial. Journal of Clinical Oncology. 1990;8:2032-9.

(121) Bianco AR, Del Mastro L, Gallo C, Perrone F, Matano E, Pagliarulo C, et al: Prognostic role of amenorrhea induced by adjuvant chemotherapy in premenopausal patients with early breast cancer. British Journal of Cancer. 1991;63:799-803.

(122) Anonymous: Ovarian ablation in early breast cancer: overview of the randomised trials. Early Breast Cancer Trialists' Collaborative Group.[comment]. Lancet. 1996;348:1189-96.

(123) Anonymous: Tamoxifen for early breast cancer: an overview of the randomised trials. Early Breast Cancer Trialists' Collaborative Group.[comment]. Lancet. 1998;351:1451-67.

(124) Emens LA, Davidson NE: Adjuvant hormonal therapy for premenopausal women with breast cancer. Clinical Cancer Research 2003;9:486s-94s.

(125) Anonymous: Ovarian ablation in early breast cancer: phoenix arisen? Lancet. 1992;339:95-6.

(126) Pritchard KI, Paterson AH, Paul NA, Zee B, Fine S, Pater J: Increased thromboembolic complications with concurrent tamoxifen and chemotherapy in a randomized trial of adjuvant therapy for women with breast cancer. National Cancer Institute of Canada Clinical Trials Group Breast Cancer Site Group. Journal of Clinical Oncology. 1996;14:2731-7.

(127) Pritchard KI, Paterson AH, Fine S, Paul NA, Zee B, Shepherd LE, et al: Randomized trial of cyclophosphamide, methotrexate, and fluorouracil chemotherapy added to

tamoxifen as adjuvant therapy in postmenopausal women with node-positive estrogen and/or progesterone receptor-positive breast cancer: a report of the National Cancer Institute of Canada Clinical Trials Group. Breast Cancer Site Group. Journal of Clinical Oncology. 1997;15:2302-11.

(128) Partridge AH, Burstein HJ, Winer EP: Side effects of chemotherapy and combined chemohormonal therapy in women with early-stage breast cancer.[erratum appears in J Natl Cancer Inst 2002 Jun 5;94(11):866]. Journal of the National Cancer Institute. Monographs. 2001:135-42.

(129) Crivellari D, Bonetti M, Castiglione-Gertsch M, Gelber RD, Rudenstam CM, Thurlimann B, et al: Burdens and benefits of adjuvant cyclophosphamide, methotrexate, and fluorouracil and tamoxifen for elderly patients with breast cancer: the International Breast Cancer Study Group Trial VII. Journal of Clinical Oncology. 2000;18:1412-22.

(130) Dees EC, O'Reilly S, Goodman SN, Sartorius S, Levine MA, Jones RJ, et al: A prospective pharmacologic evaluation of age-related toxicity of adjuvant chemotherapy in women with breast cancer. Cancer Investigation. 2000;18:521-9.

(131) Anonymous: Prevention of chemotherapy- and radiotherapy-induced emesis: results of Perugia Consensus Conference. Antiemetic Subcommittee of the Multinational Association of Supportive Care in Cancer (MASCC). Annals of Oncology. 1998;9:811-9.

(132) Gralla RJ, Osoba D, Kris MG, Kirkbride P, Hesketh PJ, Chinnery LW, et al: Recommendations for the use of antiemetics: evidence-based, clinical practice guidelines. American Society of Clinical Oncology.[erratum appears in J Clin Oncol 1999 Dec;17(12):3860]. Journal of Clinical Oncology. 1999;17:2971-94.

(133) Gralla RJ, Navari RM, Hesketh PJ, Popovic W, Strupp J, Noy J, et al: Single-dose oral granisetron has equivalent antiemetic efficacy to intravenous ondansetron for highly emetogenic cisplatin-based chemotherapy. Journal of Clinical Oncology. 1998;16:1568-73.

(134) Perez EA, Hesketh P, Sandbach J, Reeves J, Chawla S, Markman M, et al: Comparison of single-dose oral granisetron versus intravenous ondansetron in the prevention of nausea and vomiting induced by moderately emetogenic chemotherapy: a multicenter, double-blind, randomized parallel study. Journal of Clinical Oncology. 1998;16:754-60.

(135) Gralla RJ: Antiemetic therapy. In Bast RC, Kufe DW, Pollock RE, Weichselbaum RR, Holland JF, Frei III E, editors. Cancer Medicine. Hamilton, London: B.C. Decker Inc., 2000, p. 2243-50.

(136) Hartmann LC, Tschetter LK, Habermann TM, Ebbert LP, Johnson PS, Mailliard JA, et al: Granulocyte colony-stimulating factor in severe chemotherapy-induced afebrile neutropenia.[comment]. New England Journal of Medicine. 1997;336:1776-80.

(137) Goodnough LT, Anderson KC, Kurtz S, Lane TA, Pisciotto PT, Sayers MH, et al: Indications and guidelines for the use of hematopoietic growth factors. Transfusion. 1993;33:944-59.

(138) Bennett CL, Smith TJ, Weeks JC, Bredt AB, Feinglass J, Fetting JH, et al: Use of hematopoietic colony-stimulating factors: the American Society of Clinical Oncology survey. The Health Services Research Committee of the American Society of Clinical Oncology. Journal of Clinical Oncology. 1996;14:2511-20.

(139) Valagussa P, Zambetti M, Biasi S, Moliterni A, Zucali R, Bonadonna G: Cardiac effects following adjuvant chemotherapy and breast irradiation in operable breast cancer.[comment]. Annals of Oncology. 1994;5:209-16.

(140) Albain K, Green S, Ravdin P, Cobau C, Levine E, Ingle J, et al: Overall survival after cyclophosphamide, adriamycin, 5-FU, and Tamoxifen (CAFT) is superior to T alone in postmenopausal, receptor(+), node(+) breast cancer: new findings from phase III Southwest Oncology Group Intergroup Trial S8814 (INT-0100). Annual Meeting of the American Society of Clinical Oncology, 2001, p. 24a.

(141) Ganz PA: Impact of tamoxifen adjuvant therapy on symptoms, functioning, and quality of life. Journal of the National Cancer Institute. Monographs. 2001:130-4.

(142) Bines J, Oleske DM, Cobleigh MA: Ovarian function in premenopausal women treated with adjuvant chemotherapy for breast cancer.[comment]. Journal of Clinical Oncology. 1996;14:1718-29.

(143) Goodwin PJ, Ennis M, Pritchard KI, Trudeau M, Hood N: Risk of menopause during the first year after breast cancer diagnosis. Journal of Clinical Oncology. 1999;17:2365-70.

(144) Goodwin PJ, Ennis M, Pritchard KI, McCready D, Koo J, Sidlofsky S, et al: Adjuvant treatment and onset of menopause predict weight gain after breast cancer diagnosis. Journal of Clinical Oncology. 1999;17:120-9.

(145) Ganz PA, Rowland JH, Meyerowitz BE, Desmond KA: Impact of different adjuvant therapy strategies on quality of life in breast cancer survivors. Recent Results in Cancer Research. 1998;152:396-411.

(146) Day R, Ganz PA, Costantino JP, Cronin WM, Wickerham DL, Fisher B: Health-related quality of life and tamoxifen in breast cancer prevention: a report from the National Surgical Adjuvant Breast and Bowel Project P-1 Study. Journal of Clinical Oncology. 1999;17:2659-69.

(147) Simes RJ, Coates AS: Patient preferences for adjuvant chemotherapy of early breast cancer: how much benefit is needed? Journal of the National Cancer Institute. Monographs. 2001:146-52.

(148) Raber MN, Bailes JS: Clinical oncology in a changing health care environment. In Bast RC, Kufe DW, Pollock RE, Weichselbaum RR, Holland JF, Frei E, editors. Cancer Medicine. Hamilton and London: B.C. Decker, Inc., 2000, p. 1035-38.

(149) Weeks JC: Outcomes Assessment. In Bast RC, Kufe DW, Pollock RE, Weichselbaum RR, Holland JF, Frei E, editors. Cancer Medicine. Hamilton and London: B.C. Decker, Inc., 2000, p. 1039-44.

(150) Relman AS: Assessment and accountability: the third revolution in medical care. New England Journal of Medicine. 1988;319:1220-2.

(151) Detsky AS, Naglie IG: A clinician's guide to cost-effectiveness analysis.[comment]. Annals of Internal Medicine. 1990;113:147-54.

(152) Hillner BE, Smith TJ: Efficacy and cost effectiveness of adjuvant chemotherapy in women with node-negative breast cancer. A decision-analysis model. New England Journal of Medicine. 1991;324:160-8.

(153) Smith TJ, Hillner BE: The efficacy and cost-effectiveness of adjuvant therapy of early breast cancer in premenopausal women. Journal of Clinical Oncology. 1993;11:771-6.

(154) Hillner BE, Weeks JC, Desch CE, Smith TJ: Pamidronate in prevention of bone complications in metastatic breast cancer: a cost-effectiveness analysis.[comment]. Journal of Clinical Oncology. 2000;18:72-9.

(155) Lee SJ, Anasetti C, Kuntz KM, Patten J, Antin JH, Weeks JC: The costs and cost-effectiveness of unrelated donor bone marrow transplantation for chronic phase chronic myelogenous leukemia. Blood. 1998;92:4047-52.

(156) Levine MN, Drummond MF, Labelle RJ: Cost-effectiveness in the diagnosis and treatment of carcinoma of unknown primary origin. Cmaj. 1985;133:977-87.

(157) Ozer H, Armitage JO, Bennett CL, Crawford J, Demetri GD, Pizzo PA, et al: 2000 update of recommendations for the use of hematopoietic colony-stimulating factors: evidence-based, clinical practice guidelines. American Society of Clinical Oncology Growth Factors Expert Panel.[comment]. Journal of Clinical Oncology. 2000;18:3558-85.

(158) Thors CL, Broeckel JA, Jacobsen PB: Sexual functioning in breast cancer survivors. Cancer Control. 2001;8:442-8.

(159) Burstein HJ, Winer EP: Primary care for survivors of breast cancer.[comment]. New England Journal of Medicine. 2000;343:1086-94.

(160) Ries LAG, Kosary CL, Hankey BF, Miller BA, Clegg LX, Edwards BK, editors: Cancer prevalence and survivorship issues: analyses of the 1992 National Health Interview Survey. (NIH Publication no. 99-2789). Bethesda, MD,; 1996.

(161) Hewitt M, Breen N, Devesa S: Cancer prevalence and survivorship issues: analyses of the 1992 National Health Interview Survey. Journal of the National Cancer Institute. 1999;91:1480-6.

(162) Winchester DP, Sener SF, Khandekar JD, Oviedo MA, Cunningham MP, Caprini JA, et al: Symptomatology as an indicator of recurrent or metastatic breast cancer. Cancer. 1979;43:956-60.

(163) Pandya KJ, McFadden ET, Kalish LA, Tormey DC, Taylor SGt, Falkson G: A retrospective study of earliest indicators of recurrence in patients on Eastern Cooperative Oncology Group adjuvant chemotherapy trials for breast cancer. A preliminary report. Cancer. 1985;55:202-5.

(164) Schapira DV, Urban N: A minimalist policy for breast cancer surveillance.[comment]. Jama. 1991;265:380-2.

(165) Anonymous: Impact of follow-up testing on survival and health-related quality of life in breast cancer patients. A multicenter randomized controlled trial. The GIVIO Investigators.[comment]. Jama. 1994;271:1587-92.

(166) Rosselli Del Turco M, Palli D, Cariddi A, Ciatto S, Pacini P, Distante V: Intensive diagnostic follow-up after treatment of primary breast cancer. A randomized trial. National Research Council Project on Breast Cancer follow-up.[comment]. Jama. 1994;271:1593-7.

(167) Anonymous: Clinical practice guidelines for the use of tumor markers in breast and colorectal cancer. Journal of Clinical Oncology 1996;14:2843-77.

(168) Dawson LA, Chow E, Goss PE: Evolving perspectives in contralateral breast cancer. European Journal of Cancer. 1998;34:2000-9.

(169) Bernstein JL, Thompson WD, Risch N, Holford TR: The genetic epidemiology of second primary breast cancer. American Journal of Epidemiology. 1992;136:937-48.

(170) Anonymous: Recommended breast cancer surveillance guidelines. American Society of Clinical Oncology. Journal of Clinical Oncology. 1997;15:2149-56.

(171) Smith TJ, Davidson NE, Schapira DV, Grunfeld E, Muss HB, Vogel VG, 3rd, et al: American Society of Clinical Oncology 1998 update of recommended breast cancer surveillance guidelines. Journal of Clinical Oncology. 1999;17:1080-2.

(172) Pritchard KI, Khan H, Levine M, Steering Committee on Clinical Practice Guidelines for the C, Treatment of Breast C: Clinical practice guidelines for the care and treatment of breast cancer: 14. The role of hormone replacement therapy in women with a previous diagnosis of breast cancer. Cmaj. 2002;166:1017-22.

(173) Penson RT, Gallagher J, Gioiella ME, Wallace M, Borden K, Duska LA, et al: Sexuality and cancer: conversation comfort zone. Oncologist. 2000;5:336-44.

(174) Grunfeld E, Mant D, Yudkin P, Adewuyi-Dalton R, Cole D, Stewart J, et al: Routine follow up of breast cancer in primary care: randomised trial.[comment]. Bmj. 1996;313:665-9.

(175) Grunfeld E, Gray A, Mant D, Yudkin P, Adewuyi-Dalton R, Coyle D, et al: Follow-up of breast cancer in primary care vs specialist care: results of an economic evaluation. British Journal of Cancer. 1999;79:1227-33.

(176) Maher J, Bradburn J, Adewuyi-Dalton R: Follow up in breast cancer. Patients prefer specialist follow up.[comment]. Bmj. 1995;311:54.

(177) Coleman RE: Current and future status of adjuvant therapy for breast cancer. Cancer 2003;97:880-86.

(178) Citron ML, Berry DA, Cirrincione C, Hudis C, Winer EP, Gradishar WJ, et al: Randomized Trial of Dose-Dense Versus Conventionally Scheduled and Sequential Versus Concurrent Combination Chemotherapy as Postoperative Adjuvant Treatment of Node-Positive Primary Breast Cancer: First Report of Intergroup Trial C9741/Cancer and Leukemia Group B Trial 9741. J Clin Oncol 2003;25:25.

(179) Slamon DJ, Clark GM, Wong SG, Levin WJ, Ullrich A, McGuire WL: Human breast cancer: correlation of relapse and survival with amplification of the HER-2/neu oncogene. Science. 1987;235:177-82.

(180) Thor A: Are patterns of HER-2/neu amplification and expression among primary tumors and regional metastases indicative of those in distant metastases and predictive of Herceptin response? [letter; comment.]. Journal of the National Cancer Institute. 2001;93:1120-1.

(181) Tan-Chiu E, Piccart M: Moving forward: Herceptin in the adjuvant setting. Oncology. 2002;63:57-63.

(182) Diel IJ, Solomayer EF, Costa SD, Gollan C, Goerner R, Wallwiener D, et al: Reduction in new metastases in breast cancer with adjuvant clodronate treatment.[comment]. New England Journal of Medicine. 1998;339:357-63.

(183) Powles TJ, Paterson AH, Nevantaus A, Legault S, Pajunen VA, Tidy VA, et al: Adjuvant clodronate reduces the incidence of bone metastases in patients with primary operable breast cancer. Annual Meeting of the American Society of Clinical Oncology. Los Angeles, 1998, p. 123a (abstract).

(184) Saarto T, Blomqvist C, Virkkunen P, Elomaa I: Adjuvant clodronate treatment does not reduce the frequency of skeletal metastases in node-positive breast cancer patients: 5-year results of a randomized controlled trial. Journal of Clinical Oncology. 2001;19:10-7.

Chapter 9

ENDOCRINE THERAPY OF BREAST CANCER

[1]Kathrin Strasser-Weippl, MD and [2]Paul E. Goss, MD, FRCPC, FRCP (UK), PhD

[1]First Medical Department with Medical Oncology, Wilhelminen Hospital, Montleartstr. 37, A-1160 Vienna, Austria, and [2]Department of Medical Oncology and Hematology, Princess Margaret Hospital, University Health Network, Toronto, ON, Canada and Harvard Medical School, Boston, MA, USA

1. INTRODUCTION

Breast cancer is one of the most common cancers in the Western world. Although the incidence of breast cancer in white women is higher than in black women, mortality from breast cancer is relatively higher in African-Americans. This has been shown in several large cancer surveillance studies (*1-2*). The survival disadvantage of African-American women is due to numerous factors, the most important of which is stage of disease at the time of diagnosis. It has been shown in multiple studies, that black women are more likely to present with more advanced disease (*3-4*), including more patients with inoperable breast cancer and with nodal involvement. The differences in breast cancer stage and biology at presentation between Caucasian and African-American patients are reviewed extensively in Chapters 2 and 3.

It is clear from most studies that some disparity in prognosis exists between racial/ethnic groups after accounting for stage differences. This might in part be due to differences in treatment. For example, black patients are less likely to receive surgical treatment, even after adjusting for age, stage and histology of the tumour (*5*). Several studies have shown that when Caucasian and African American breast cancer patients, matched for age, stage of disease and tumour characteristics, are treated similarly in the context of clinical studies, their outcomes are similar (*6-7*).

C. Williams et al. (eds.), Breast Cancer in Women of African Descent, 221-247.

Endocrine therapy has provided important gains in the outcome of women with breast cancer. In the metastatic setting it has provided clinically meaningful responses and palliation of symptoms. In the adjuvant setting, Tamoxifen has improved survival of all age groups considerably. Furthermore, Tamoxifen has been shown to reduce the incidence of invasive and pre-invasive breast lesions in the short term and offered hope that endocrine strategies will be increasingly important in the prevention of this disease in the future. In about 70% of breast cancers, oestrogen or progesterone receptors can be found on more than 10% of the tumour cells (ER/PR positive), providing the rationale for anti-estrogenic treatment of these tumours. Importantly, there are racial differences in tumour ER and PR content, with breast cancer patients of African descent having a smaller likelihood to present with hormone sensitive disease (*8-9*). The differences between racial groups in biological tumour characteristics including ER/PR status are also discussed in Chapters 2 and 3.

The topic of this chapter is to examine emerging trends in endocrine therapy in breast cancer patients presenting with ER/PR positive disease, and to review these data with respect to women of African descent. In line with traditional divisions of clinical management of breast cancer, the treatment of breast cancer with cytotoxic agents with or without hormonal manipulation is discussed in Chapters 8. In this chapter we have elected to begin by reviewing Tamoxifen because it is the standard upon which all-novel endocrine research and clinical care is based.

2. RATIONALE FOR ENDOCRINE TREATMENT OF BREAST CANCER

The association between oestrogen and breast cancer growth has been appreciated for over one hundred years, since George Beatson and A. Schinzinger produced remissions of advanced breast cancer by performing bilateral oophorectomy in premenopausal patients (*10-11*). Since that time, a body of evidence has accumulated indicating that oestrogen and its metabolites are related to both the initiation and promotion of breast cancer. Initial endocrine therapy included oophorectomy and more morbid surgical procedures such as hypophysectomy and adrenalectomy. Surgical, medical and radiation-induced ovarian ablation antagonizes oestrogen's action in premenopausal women by blocking ovarian oestrogen synthesis and thus reducing its levels both in the circulation and in normal and malignant breast tissue. Hypophysectomy and adrenalectomy were abandoned when the first endocrine agent, Tamoxifen citrate, was introduced. Tamoxifen was first approved by the FDA for use in advanced breast cancer in 1978. Since then, more than 40,000 breast cancer patients have been treated with Tamoxifen in the context of clinical research studies, and many more in clinical practice. Apart from numerous single trials, the benefit of Tamoxifen and oophorectomy as adjuvant therapies of hormone dependent breast cancer has been shown in two large meta-analyses of all the trial data (*12-14*).

In parallel with the clinical studies, the mechanism of action of endocrine agents has been extensively investigated in preclinical laboratory models. Additional SERMS (selective oestrogen receptor modulators) and several new classes of endocrine agents, which antagonize the effects of oestrogen, have been identified. SERMS and pure anti-oestrogens antagonize ER function by binding competitively to the receptor. In contrast to SERMS steroidal pure anti-oestrogens have no agonistic effect on ER and they additionally reduce ER concentration by inducing oestrogen receptor degradation. Aromatase inhibitors work by blocking the enzyme complex responsible for the final step in oestrogen synthesis. In premenopausal women aromatase inhibitors given as monotherapy incompletely block ovarian oestrogen production and cause a reflex rise in gonadotropin levels leading to an ovarian hyperstimulation syndrome. Their use in premenopausal women therefore is being investigated in combination with ovarian function suppression. In postmenopausal women the aromatase inhibitors very effectively suppress extra-gonadal peripheral aromatase. In this chapter we review the current recommendations, as well as possible future developments, in the use of endocrine agents in the treatment of pre- and postmenopausal breast cancer patients, in both the adjuvant and the metastatic settings. Typically, novel therapies have been examined first in metastatic disease and then moved into the adjuvant setting. This has been true for Tamoxifen, its analogue Toremifene and the aromatase inhibitors, which have recently been introduced in the adjuvant setting. Newer agents have been looked at only in the metastatic setting to date. To follow the clinical development of these agents we have therefore elected to present the data in the order of metastatic, adjuvant, neoadjuvant and prevention therapy.

The limitations of these data in women of African-American descent are discussed separately in section 7 below.

3. TAMOXIFEN

Since its introduction, Tamoxifen has been the mainstay of endocrine therapy of breast cancer. Since Tamoxifen was approved for the treatment of advanced disease, numerous studies have been conducted investigating its effects in the adjuvant setting. These trials have been conducted both in pre-

and postmenopausal women and for postoperative durations of one to ten years. In premenopausal women Tamoxifen has been given both as monotherapy and in combination with ovarian ablation. Recently, the combination of Tamoxifen with aromatase inhibitors in the postmenopausal setting and the use of aromatase inhibitors as single agents have also been studied. In both pre-and postmenopausal women endocrine therapies have been investigated in combination with or as a substitute for chemotherapies.

The available data on the use of Tamoxifen in the adjuvant setting of ER positive breast cancer have been extensively reviewed by the Early Breast Cancer Trialists' Collaborative Group (EBCTCG) (*12-14*). In their most recent meta-analysis including 37,000 women, adjuvant treatment with 5 years of Tamoxifen resulted in a reduction of 47% in disease recurrence, and a mortality reduction of 26% (*12*). The benefit of Tamoxifen was restricted to the group of women with ER-positive tumours. In this meta-analysis, 5 years of Tamoxifen therapy were better than 1 or 2 years of treatment, and the benefit from 5 years of Tamoxifen therapy persisted through 10 years of follow-up. In the EBCTCG overview, the relative benefit from Tamoxifen was independent of axillary lymph node involvement, age, Tamoxifen dose, menopausal status, or use of chemotherapy. Based on these data, the 2000 US National Institutes of Health Consensus Development Conference recommended the use of adjuvant Tamoxifen therapy for women "regardless of age, menopausal status, involvement of axillary nodes, or tumour size"(*15*). The 2001 St. Gallen Consensus Panel came to a similar conclusion, suggesting the use of Tamoxifen in most women with ER positive breast cancer unless the patient has an absolute contraindication to Tamoxifen (*16*). In both consensus statements, it was pointed out that for women with very low risk hormone responsive disease, no adjuvant medical treatment is an alternative choice to Tamoxifen. The interaction in the adjuvant setting of the effects of chemotherapy and Tamoxifen in pre- and postmenopausal women has also been extensively reported on by EBCTCG. Their findings are discussed in Chapter 8.

In order to evaluate a duration of adjuvant Tamoxifen of more than five years the treatment arm of the NSABP B-14 study, evaluating five years of adjuvant Tamoxifen versus placebo, was extended. When 1,172 women, who had finished five years of adjuvant Tamoxifen within the B-14 protocol, were randomised to another five years of Tamoxifen or placebo, no additional advantage was obtained from continuing Tamoxifen therapy for more than 5 years (*17*). Several explanations of the ineffectiveness of Tamoxifen have been offered by studies in pre-clinical models. Firstly, Tamoxifen resistance develops as indicated by the experiments of MCF-7 cells in long-term culture with Tamoxifen (*18-19*). In addition Masamura et al. showed that breast cancer cells become hypersensitive to oestrogen, when deprived of oestrogen for a long time (*20*). In the clinical setting, the postulated increasing hypersensitivity of normal and residual malignant breast cells to Tamoxifen estrogenic effects over time and the results of the B-14 trial have lead to Tamoxifen use being restricted to five years. The ongoing ATLAS ('Adjuvant Tamoxifen – longer against shorter') and ATOM ('Adjuvant Tamoxifen – Offer More?) studies will clarify this issue, as patients are randomised to 5 versus more than 5 years of adjuvant Tamoxifen in these trials. The benefit of 5 years of adjuvant Tamoxifen was shown to be equal in premenopausal women to that in postmenopausal women in the 1998 EBCTCG overview (*12*).

The side effect profile of Tamoxifen has been examined in the adjuvant studies. In addition, important toxicity data were gained from four large Tamoxifen prevention trials, in which Tamoxifen was compared to placebo and therefore probably most accurately reflect its true toxicities (*21*). In the four prevention trials, Tamoxifen effects on the uterus, bone, lipid metabolism, thromboembolic risk, cataract incidence and quality of life were carefully evaluated.

Participants in the prevention trials receiving Tamoxifen had a 2.4 times greater risk of developing invasive endometrial cancer than those getting placebo (p=.00005) (*21*), and this effect was more pronounced in women over 50 years of age compared to younger women (RR 4.01 vs. 1.21) (*22*). When the data from the adjuvant studies included in the 1998 EBCTCG analysis are taken together, this excess of endometrial cancer risk is confirmed (OR 3.4, p=.00002) (*12*). Tamoxifen also leads to an excess risk of thromboembolic events when compared with placebo. In the NSABP P-1 prevention study, including more than 13,000 women, pulmonary embolism (PE) was three times as common (RR=3.01) and strokes were nearly twice as frequent among women on Tamoxifen >50 years of age compared to younger women taking the drug (RR 1.75) (*22*). Venous thromboembolic events were also increased in all prevention trials, with a relative risk of 1.9 in the Tamoxifen compared to the placebo arms (p<.0001) (*21*). Overall, the increase in vascular events on Tamoxifen was comparable to that seen with hormone replacement therapy.

While in retrospective analyses of three randomised Tamoxifen trials a reduction in coronary heart disease (CHD) was observed, no such benefit was demonstrated in NSABP P-1 (*21-25*). In the EBCTCG overview of 1998, mortality rates for causes "not attributed to breast or endometrial cancer" were nearly identical in patients receiving Tamoxifen or placebo in the adjuvant setting (*12*). In view of the recently published data from the "Women's Health Initiative" (WHI) study (*26*), it is uncertain whether Tamoxifen favourable influence on lipid metabolism translates into a true reduction in CHD.

Tamoxifen has been shown to preserve bone mineral density (BMD) in postmenopausal breast cancer patients (*26-28*). P-1 is the only prospective trial which has evaluated the effect of Tamoxifen on bone fractures versus placebo and it showed a reduction in the risk of long bone and symptomatic vertebral fractures of borderline statistical significance (RR=0.81, 95% CI 0.63-1.05) (*22*). To date, Tamoxifen has not been evaluated in a prospective trial in women with osteoporosis.

Apart from the effects of hormonal agents on hormone-sensitive tissues, there are a number of non-endocrinologic side effects of these agents. These include toxicities such as skin rashes, gastrointestinal toxicity and musculoskeletal disorders.

A small excess risk of cataracts (RR 1.14 95% CI 1.01-1.29) for women on Tamoxifen was identified in the P-1 trial (*22*). Tamoxifen also significantly increased bothersome hot flashes and vaginal discharge, but this did not affect overall physical and emotional well-being (*29*). The reported effects of Tamoxifen on cognition are variable and dependent on the parameters measured. Several ongoing studies are addressing this point.

The toxicity profile of Tamoxifen in premenopausal/younger women is comparable to the use of this agent in the postmenopausal setting. However, in the prevention trials, women less than 50 years of age were shown to have a smaller or no increase in the risk of uterine cancer compared to older women on Tamoxifen (*21*). The relative risk of venous thromboembolic events was equal for pre- and postmenopausal women (HR 1.9, p=.0001), but the absolute risk was smaller in the premenopausal cohort.

At this time regular ophthalmic examination and transvaginal ultrasound assessment of the endometrium are not routinely recommended for women on Tamoxifen therapy.

4. NOVEL ENDOCRINE STRATEGIES IN THE METASTATIC SETTING

In recent years, Tamoxifen has been challenged by a number of new hormonal agents, which have been developed with the intention to improve the outcome of women above that achieved by Tamoxifen. The therapeutic index of treatment can be enhanced either by greater efficacy against the disease and/or by reducing toxicities. In addition, to improve the outcome of patients, new endocrine agents have also been combined with each other and with chemotherapy. Some of these strategies are presented in this chapter.

4.1 Aromatase Inhibitors

In postmenopausal women, aromatase inhibitors are now the first-line treatment of choice in metastatic breast cancer. Anastrozole, Letrozole and Exemestane have all been compared to the former standard of care, Tamoxifen, in this setting. For Anastrozole, data from two independent phase III studies are available, showing equality to Tamoxifen (*30-31*). In a retrospective subgroup analysis, Anastrozole increased time to progression (10.7 vs. 6.4 months; p=.022) and clinical benefit rate (57.1% vs 52.0%) in the ER positive subset of patients (*32*). In both studies, anastrozole was better tolerated than Tamoxifen, causing fewer venous thromboembolic events and episodes of vaginal bleeding. Letrozole was compared to Tamoxifen in a phase III study in the first-line treatment of metastatic breast cancer, in which it was significantly superior to Tamoxifen in terms of time to progression (41 vs. 26 weeks; HR 0.7; p=0.0001), objective response rate (30% vs. 20%; p=.0006) and clinical benefit rate (49% vs. 38%; p=.001) (*33*). For Exemestane, only phase II data are as yet available, showing a trend to greater response rate (PR+CR: 42% vs. 16%) and rate of clinical benefit (PR+CR: 58% vs. 131%) for Exemestane and an ongoing phase three study is being conducted (*34*). Available data from the first-line setting indicate at least equal short-term safety of the aromatase inhibitors compared to Tamoxifen including symptoms of menopause and quality of life.

Taking these data together, the aromatase inhibitors are now considered as the standard of care as first-line treatment of ER/PR positive metastatic breast cancer in postmenopausal women. Anastrozole and Letrozole are now widely approved for this indication. Several studies are ongoing comparing the inhibitors to each other. In the first of these studies, anastrozole and Letrozole were compared as second-line, post Tamoxifen, therapy for locally advanced or metastatic breast cancer in 713 postmenopausal women. The two aromatase inhibitors were equal in time to progression, duration of response and duration of clinical benefit, but Letrozole was significantly superior to anastrozole in terms of overall response (19.1% vs. 12.3%; p=.014) and non-significantly superior in terms of clinical benefit (27% vs. 23%; p=.218) (*35*). Results of other studies comparing the inhibitors are awaited with great interest.

As the most important toxicity data pertaining to the aromatase inhibitors have been and are being collected in adjuvant studies, the toxicity profiles of the inhibitors will be discussed in more detail in section 5.2.

4.2 Novel Approaches to Ovarian Function

As stated earlier for premenopausal women with metastatic breast cancer, ovarian ablation has long been known to be an effective treatment. When compared to Tamoxifen in numerous studies, equality between the two treatment regimens was found in this setting. This was also confirmed in a retrospective analysis of four randomised trials comparing oophorectomy by irradiation or surgery with Tamoxifen (*36*). When surgical oophorectomy was compared to ovarian ablation by Goserelin in a large, prospective, randomised study in patients with advanced breast cancer, Goserelin provided similar benefit to oophorectomy in terms of objective response, failure free survival, and overall survival (*37*). Based on these data, Goserelin and oophorectomy by other means are regarded as equal treatments today. The question whether adding Tamoxifen to ovarian ablation is of benefit, was also addressed by several studies. In a recently published, patient-based meta-analysis conducted by the European Organisation for the Research and Treatment of Cancer (EORTC), Tamoxifen plus an LHRH analogue was better than the LHRH analogue alone in terms of objective response, progression-free survival and overall survival (HR .67;.7;.78 respectively) (*38*). However, in another study comparing ovarian ablation with Goserelin, with or without Tamoxifen, Goserelin was equally effective as oophorectomy by irradiation, but the addition of Tamoxifen seemed to be beneficial only in the Goserelin arm (*39*). In view of the fact that the combination of both agents induced more side effects in this study, it seems unclear whether Tamoxifen should be added to ovarian ablation in the palliative setting, in which quality of life is a major concern.

4.3 Megestrol Acetate

For many years the progestin Megestrol Acetate and the first-generation aromatase inhibitor Aminoglutethimide were the standard of care as second-line hormonal treatment of postmenopausal metastatic breast cancer after Tamoxifen. As other endocrine agents, including the third generation aromatase inhibitors and Faslodex, are now moving to earlier treatment settings, the significance of Megestrol Acetate and Aminoglutethimide is declining. Apart from their inferior efficacy compared to the newer agents, their troublesome side effect profiles frequently lead to toxicity related withdrawal of treatment. In the second-line metastatic breast cancer trials, all third generation aromatase inhibitors showed a superior toxicity profile compared to both Megestrol Acetate and Aminoglutethimide. Today, these agents are only used in patients with metastatic breast cancer, who have benefited from prior endocrine therapies for a long time and are therefore expected to have endocrine sensitive disease even after failure of first- and second-line hormonal treatments.

4.4 Estrogens

Estrogens administered in high doses were widely used for therapy of advanced breast cancer before the introduction of contemporary endocrine therapy. While the mechanism of their antitumour effect is unknown, in vitro investigations have shown estrogens in high concentrations to be toxic to cell growth. In a recently published study, diethylstilbestrol (DES) was given to postmenopausal women with metastatic breast cancer progressing on prior anti-estrogenic treatments (range 2-10). Among 32 patients, the clinical benefit rate was 37.5% (*40*). In another study including 151 postmenopausal women with advanced breast cancer and no prior endocrine treatment, DES was compared to Tamoxifen (*41*). In terms of overall response rate and median duration of response, DES was non-significantly superior to Tamoxifen, and there was equality between the two agents in progression-free survival. However, median survival significantly favoured DES, being 3.0 years for DES vs. 2.4 years for Tamoxifen (p=.039). Treatment with DES was more commonly associated with toxicity such as nausea, oedema, vaginal bleeding, and cardiac problems, whereas hot flashes were more often seen in the Tamoxifen arm. This trial, which was conducted before the era of the third generation aromatase inhibitors, underscores the fact that estrogens have activity and remain in the armamentarium for treatment of selected patients with metastatic breast cancer.

4.5 New SERMS and SERDS

Raloxifene is a SERM differing slightly from Tamoxifen in its estrogenic and anti-estrogenic properties. Raloxifene was originally developed for the prevention of osteoporosis because of its pro-estrogenic effect on bone.

In the Multiple Outcomes of Raloxifene Evaluation (MORE) study, 7705 postmenopausal women with osteoporosis and no history of breast or endometrial cancer were randomised to receive either Raloxifene (60 mg or 120 mg) or placebo daily for 3 years. In the MORE trial, which was primarily designed to test the bone-preserving effects of Raloxifene, the occurrence of breast cancer was a secondary endpoint. After a median follow-up of 40 months, there were substantially fewer invasive cancers in women receiving Raloxifene (RR=0.24; 95% CI 0.13-0.44; P<0.001) than in those taking placebo. This difference was entirely attributable to a 90% reduction in ER positive invasive breast cancers (RR=0.1; 95% CI 0.04-

0.24) by Raloxifene with no difference in the occurrence of ER negative tumours (*42*). A partially overlapping analysis reviewing data from nine Raloxifene trials (including MORE) has confirmed a significant reduction in newly diagnosed breast cancers in women on Raloxifene, making the MORE data likely to be correct (*43*). Based on these data, which suggest a similar efficacy of Raloxifene and Tamoxifen in reducing the incidence of receptor positive breast tumours, a confirmatory study comparing Raloxifene to Tamoxifen is underway. Raloxifene has not been evaluated in metastatic breast cancer in the context of large clinical studies. When it was given to 14 patients with Tamoxifen resistant or refractory metastatic breast cancer in one small trial, only one patient showed a minor response (*44*). Raloxifene is not approved for the treatment of breast cancer.

A relatively new drug in the armamentarium of endocrine therapies in metastatic breast cancer is the selective oestrogen receptor downregulator (SERD) Faslodex, differing from the SERMS by its ability to degrade and destroy the oestrogen receptor. Having no pro-estrogenic effects, Faslodex is therefore called a 'pure oestrogen antagonist.' In 2002, two studies were published in which Faslodex was compared to anastrozole in postmenopausal women with advanced breast cancer progressing on prior endocrine therapy. In these trials, including 851 patients altogether, Faslodex was as effective as anastrozole, with a non-significant difference favouring Faslodex in time to progression and clinical benefit rate in one study (*45-46*). Based on these data, Faslodex has been approved for the treatment of postmenopausal women with metastatic breast cancer after failure of Tamoxifen in the USA. Several studies evaluating Faslodex in the first-line metastatic setting and in the adjuvant setting are ongoing.

5. NOVEL ENDOCRINE STRATEGIES IN THE ADJUVANT SETTING

5.1 Novel Approaches to Ovarian Function Suppression

In contrast to the use of Tamoxifen in the adjuvant setting, the role of ovarian ablation in premenopausal women is less clear. Ovarian ablation can be achieved by surgical removal or irradiation of the ovaries, or by treatment with GNRH-analogues, which are all equal in their effectiveness of suppressing ovarian oestrogen production. In the EBCTCG overview of 1996, ovarian ablation as a monotherapy in premenopausal breast cancer patients (any ER status of the primary tumour) was found to reduce the risk of disease recurrence by 25% and the risk of death by 24% (*47*). It is unclear, however, whether ovarian ablation, with or without Tamoxifen, can replace the use of chemotherapy in premenopausal women. Up to now, eight trials have been reported addressing this question, most of them in preliminary form (*48*-(*49*). Taken together, these trials show that ovarian ablation, with or without Tamoxifen, is equal to "CMF-type" chemotherapy. However, there are several weaknesses of this dataset. First, in none of the studies was Tamoxifen offered in the chemotherapy arm, despite the well-known benefit of Tamoxifen added to adjuvant chemotherapy. Second, the "CMF-type" chemotherapy used in these trials was less effective than the types of chemotherapies considered optimal in this setting today. Therefore, the endocrine arm in these studies was shown to be equal to a less-than-optimal chemotherapy approach. Nevertheless, both the 2000 NIH and the 2001 St. Gallen consensus panels included endocrine therapy without chemotherapy in the treatment options of premenopausal breast cancer patients (*15-16*). A new trial set to begin soon will address the question of the value of chemotherapy in addition to ovarian function suppression combined with an endocrine agent. The trial, known as the PERCHE ('Premenopausal Endocrine-Responsive Chemotherapy trial') study, will clarify the question whether the addition of chemotherapy to ovarian function suppression in combination with an endocrine agent in premenopausal patients with ER-positive disease will be of added value. In the PERCHE trial, 1,750 premenopausal breast cancer patients with ER-positive disease will be treated with ovarian ablation plus Tamoxifen or Exemestane for five years. In addition, they will be randomised to receive chemotherapy or not.

Two adjuvant breast cancer trials in premenopausal women are currently being launched which include ovarian blockade plus an aromatase inhibitor as part of their design. In the SOFT ('Suppression of Ovarian Function Trial') study, women who remain premenopausal, as determined by plasma estradiol levels, after surgery and/or after neoadjuvant or adjuvant chemotherapy are randomised to Tamoxifen for 5 years, versus ovarian ablation plus Tamoxifen for 5 years, versus ovarian ablation plus Exemestane for 5 years. One very important feature of this trial is the fact that patients are evaluated for their hormonal status after completion of chemotherapy, so that the benefit of ovarian function suppression will be evaluated in a cohort of truly premenopausal patients. In the TEXT ('Tamoxifen and Exemestane Trial') study, women will receive an LHRH agonist (Triptorelin) immediately after surgery. They will be randomised to combine Triptorelin with Tamoxifen versus Exemestane. In both the SOFT and the TEXT studies, the new combination of ovarian ablation plus an aromatase inhibitor will be analysed.

From the available data, the optimal duration of ovarian function suppression as an alternative to complete ovarian ablation is still unclear. In the trials using GNRH-analogues, durations between one and five years have been used. The 2003 St. Gallen consensus panel recommended the use of one, two or five years of ovarian ablation. In the new TEXT, SOFT and PERCHE trials beginning internationally, GNRH use for 5 years has been chosen by the international investigators who have designed those trials and therefore provide a new "defensible" duration of therapy for patients outside the context of clinical trials.

As ovarian ablation renders premenopausal patients postmenopausal, the side effects of this intervention include risks and symptoms of menopause. One of the adjuvant trials that carefully evaluated the side effects of ovarian ablation and compared them to chemotherapy was the ZEBRA ('Zoladex Early Breast Cancer Research Association') study (*50*). In this adjuvant breast cancer study, 1,640 pre or perimenopausal patients were randomised to receive CMF chemotherapy for 6 cycles or the LHRH agonist Goserelin for two years. As expected, the side effects of the LHRH agonist included menopausal symptoms such as hot flushes (60.4%), vaginal dryness (25.9%) and amenorrhoea (>95%). After cessation of treatment, more than 90% of patients below 40 years of age had a return of menstruation. One additional important adverse event occurring while on an LHRH agonist is loss of bone mineral density (BMD). In the ZEBRA study, the mean percentage of loss in BMD for Goserelin after 2 years was 10.5% for lumbar spine and 6.4% for neck and femur. The loss in BMD on the LHRH agonist was significantly greater than in patients receiving chemotherapy. After cessation of Goserelin, a partial recovery of BMD was seen at the 3-year-assessment.

5.2 Aromatase Inhibitors

Since the publication of the recent ATAC ('Anastrozole, Tamoxifen, Alone and Combined') trial, the primacy of Tamoxifen in the adjuvant, postmenopausal setting is being questioned. In this trial, 9,366 postmenopausal women with ER/PR positive or unknown breast cancer were randomised to receive Tamoxifen (20 mg/d), anastrozole (1 mg/d), or the combination, for 5 years. The median age of the study population was 63 years, and 74% of patients had ER or PR positive tumours. After a median follow-up of 47 months, anastrozole led to a 17% greater reduction in the risk of recurrence compared to Tamoxifen (HR 0.83, p=.015), and a 22% greater reduction in the ER positive subgroup (HR 0.78, p=.007) (*51-52*). The first overall survival data from this trial are due to be presented in early 2004. As the combination of Tamoxifen and anastrozole was not better than Tamoxifen monotherapy in this study, the combination arm was stopped after the first interim analysis. The differences between Tamoxifen and anastrozole with respect to their side effect profiles will be discussed below. The ATAC trial is the largest adjuvant breast cancer study ever performed, and its results are in line with data from the preclinical and metastatic clinical settings, all showing that the third generation aromatase inhibitors are superior to Tamoxifen in ER/PR positive breast cancer. Based on the published results from ATAC, the 2003 St. Gallen consensus panel included the option of giving anastrozole to postmenopausal women in the adjuvant breast cancer setting, "if Tamoxifen is contraindicated" in their recommendations [unpublished]. In this context, a contraindication to Tamoxifen could be a history of deep vein thrombosis, pulmonary embolism, cerebrovascular event or vaginal bleeding. A similar recommendation was published by the American Society of Clinical Oncology Technology assessment (*53*). In their position statement, the only adjuvant setting in which the use of anastrozole was considered appropriate was in patients with a "relative or absolute contraindication to adjuvant Tamoxifen," however, the possible contraindications were not further specified (*53*). Despite the position taken by the ASCO technology assessment group, the Food and Drug Administration approved anastrozole for use in adjuvant therapy in 2002 and likewise it has been approved by regulatory authorities widely elsewhere in the world. Although it is conceivable that the curves of disease-free survival in the ATAC trial may yet converge, it seems unlikely based on past experience of endocrine therapies in general. Another adjuvant aromatase inhibitor trial recently published is the MA.17 study comparing five years of Letrozole versus placebo following five years of adjuvant Tamoxifen *(54)*. This study was discontinued at the first interim analysis because of a highly significant reduction in breast cancer recurrences in the Letrozole arm. After a median follow-up of 2.4 years, the hazard ratio for local and distant recurrence or new contralateral breast cancer in the Letrozole group as compared with the placebo group was 0.57 (95% CI 0.43-0.75, p=.00008).

ATAC is the first of a number of large adjuvant breast cancer studies comparing the aromatase inhibitors anastrozole, Letrozole and Exemestane to Tamoxifen. The designs of these trials include monotherapy arms, the combination of Tamoxifen plus an aromatase inhibitor, and the sequential use of the two groups of agents. The question of whether one of the third generation inhibitors is superior to the others and whether their efficacy can be improved by combining them with other agents, will also be addressed by future trials. The new NCIC CTG (National Cancer Institute of Canada Clinical Trials Group) study MA.27 is a large North American adjuvant endocrine trial being launched in the second quarter of 2003. It is the first head-to-head comparison of two aromatase inhibitors, anastrozole versus Exemestane, in the adjuvant setting. Its goal is to show superior efficacy of the steroidal inhibitor as well as an improved profile on end-organ function such as bone and lipid metabolism. In addition, the value of adding a COX-2 inhibitor, with pleiotropic putative anti-cancer actions, to anastrozole or Exemestane will be evaluated in this trial.

The short-term, common toxicity profiles of the third-generation inhibitors are similar, with the most common adverse events being nausea, vomiting, hot flashes, musculoskeletal discomfort, fatigue and headaches. The available data comparing aromatase inhibitors in the first-line metastatic setting to Tamoxifen show a trend in favour of the inhibitors for important toxicities such as thromboembolism (*55*). Although the aromatase inhibitors are often discussed together, there might be important differences in their side effect profiles, particularly between the steroidal and non-steroidal inhibitors.

As discussed above, anastrozole is the only agent for which results in the adjuvant setting have been presented. In the ATAC trial (*51-52*), women taking anastrozole were at a significantly lower risk of developing endometrial cancer than those taking Tamoxifen (p=.02). The non-stimulatory effect of anastrozole on the endometrium was also reflected by a significantly decreased rate of vaginal bleeding during the trial (4.5% vs 8.2%, p<.0001). In addition, both ischaemic cerebrovascular and venous thromboembolic events (including DVT) were rarer in the anastrozole group (1.0% vs 2.1% and 2.1% vs 3.5% respectively, both p=.00006). By contrast, women taking anastrozole were more likely to suffer from musculoskeletal disorders (27.8% vs 21.3%, p<.00001), in particular fractures (5.9% vs 3.7%, p<.0001). In a subanalysis of the ATAC study evaluating treatment effects on quality of life during the first two years, no differences were seen between the three treatment arms (*29*). In the ATAC trial, influences of either anastrozole or Tamoxifen on the serum lipid profile have not been reported. In another small study, no influence of anastrozole on the lipid profile was seen (*56*). In the MA.17 study comparing Letrozole to placebo, low-grade hot flashes, arthritis, arthralgia and myalgia were more frequent in the Letrozole group, but vaginal bleeding was less frequent. Similar to the ATAC trial, there were more new diagnoses of osteoporosis in women in the placebo group, but this was not statistically significant (54).The effects of Exemestane on organs other than the breast have not been as comprehensively studied to date. However, similar to anastrozole, Letrozole has in several studies been shown to significantly increase parameters of bone resorption (*56*). When Letrozole was given to healthy women for three months, no influence was seen on their lipid profiles (*57*). In another study including 20 women with breast cancer, Letrozole significantly increased total and LDL cholesterol levels, as well as the atherogenic risk ratios total/HDL and LDL/HDL cholesterol (*58*). More detailed data on the effects of Letrozole on bone metabolism and parameters of cardiovascular risk were extensively studied in the MA.17-trial (Figure 1) and will be published soon.

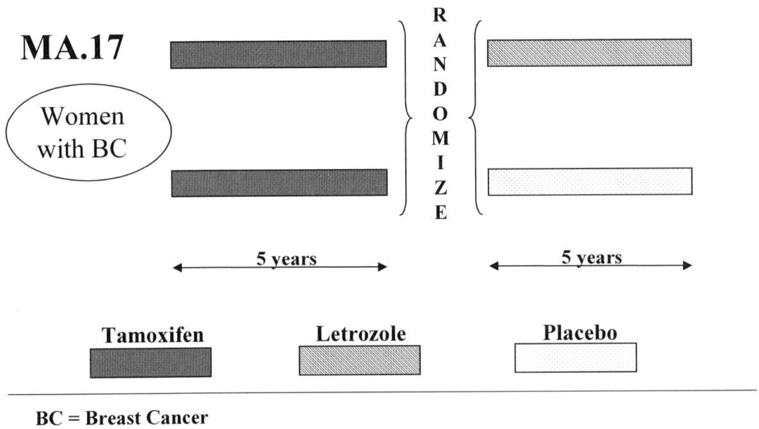

Figure 1. Five Years of adjuvant Tamoxifen, then Letrozole vs. Placebo for 5 years.

The effect of Exemestane on a marker of bone metabolism (excretion of pyridinoline) was studied in two preclinical studies of ovariectomised rats. In the first (*59*), control rats were oophorectomised (OVX), and OVX-treated animals received Exemestane. The aromatase inhibitor was able to prevent the increase in pyridinoline excretion compared to non-ovariectomized rats by 96% ($p<.0001$ vs OVX control) and the profound reduction in BMD caused by oophorectomy was obviated in the Exemestane treated animals. A follow-up experiment of five groups of animals included two lower doses of Exemestane, as well as a group treated with 17-hydroexemestane, the principal metabolite of Exemestane, and another treated with the anti-tumour dose of Letrozole in this rat model (*60*). The bone sparing effects of all three doses of Exemestane were demonstrated in a dose-dependent manner and an equally positive effect of 17-hydroexemestane was demonstrated. Letrozole had no effect on bone metabolism in the castrated animals, which lack peripheral aromatase.

In a further follow-up study in postmenopausal women, bone biomarkers of formation and resorption were studied in groups of women given placebo, Letrozole 2.5 mg/day and Exemestane 25 mg/day. In this experiment, Exemestane-treated women appeared similar to placebo in terms of early markers of bone turnover whereas Letrozole increased bone resorption and reduced bone formation (*61*).

These studies therefore support the notion that Exemestane has a superior safety profile compared to the other third generation inhibitors with regard to bone metabolism. Likewise, Exemestane seems to have effects converse to the other aromatase inhibitors in terms of lipid metabolism. In an EORTC companion study, Exemestane was compared to Tamoxifen in breast cancer patients. After a treatment period of 24 weeks, Exemestane had beneficial effects on triglycerides and a stabilizing effect on HDL and total cholesterol levels (*62*). Similarly in the two rat experiments cited above, Exemestane improved the serum lipid profile in treated animals compared with either oophorectomy or Letrozole treatment. Overall, it seems that being an androgenic steroid may make Exemestane superior to the other inhibitors in terms of side effects caused by oestrogen depletion of target tissues other than the breast.

Apart from the toxicities concerning endocrine-responsive tissues such as the endometrium, bone and lipid profile, the therapeutic index of aromatase inhibitors is also influenced by their general tolerability. The side effect profiles reported from the metastatic studies may not be applicable in the preventive setting and data from the adjuvant setting are only available for anastrozole (*51-52*). In the ATAC trial, anastrozole caused significantly fewer vasomotor symptoms than Tamoxifen as reported by the investigators although in a parallel quality of life study this was not confirmed (*29*). The incidence of cataracts, nausea/vomiting, fatigue and mood disturbances were the same in both groups. In the case of Exemestane and Letrozole, data from studies in advanced breast cancer are all that are available at present to ascertain their tolerability. In a study in the first-line setting of metastatic breast cancer, Exemestane induced fewer hot flashes and peripheral oedema than Tamoxifen (*63*). In terms of nausea and sweating, Exemestane was also better tolerated than Tamoxifen. When Letrozole was compared to Tamoxifen in advanced breast cancer, the tolerability of the two agents was similar with respect to nausea and hot flashes (*33*).

6. COMBINATIONS OF ENDOCRINE AGENTS

6.1 Combination with Chemotherapy

6.1.1 Tamoxifen Plus Chemotherapy

A frequently asked question in the adjuvant treatment of postmenopausal breast cancer is, whether chemotherapy should be added to Tamoxifen in women with a particularly high risk for disease recurrence. This question has not been answered by the two consensus panels mentioned above. The 1998

EBCTCG overview provides some important data on this issue by showing that the benefit of adjuvant chemotherapy seems to decline with increasing age (*64*). In the overview, the relative reduction of disease recurrence achieved by chemotherapy was 22% in women aged 50-59 years, but only 18% in those 60-69 years. There are few data on the effect of chemotherapy in women older than 69 years, but it seems likely that their benefit will be equal to or smaller than that seen in women 60-69 years. The corresponding reductions in mortality were considerably lower (14% and 8% respectively), presumably because of the high prevalence of other co-morbidities in these age groups. When chemotherapy was added to Tamoxifen, the EBCTCG overview showed a 19% proportional reduction in disease recurrence and an 11% reduction in mortality in women 50-69 years. Taken together these data make it likely that the benefit from adjuvant chemotherapy, added to Tamoxifen in women with hormone responsive disease, is greatest in younger postmenopausal women with high-risk disease. A more detailed review on the use of chemotherapy in the adjuvant setting can be found in Chapter 8.

When Tamoxifen is used in women receiving chemotherapy, it was unclear for a long time whether endocrine treatment should start at the time of diagnosis, or after chemotherapy has been completed. In 2002, this issue was probably resolved, at least for Tamoxifen, by the publication of the preliminary results of the Intergroup 0100 (SWOG 8814) trial. In this trial, 1,477 postmenopausal breast cancer patients were randomised to Tamoxifen, chemotherapy plus Tamoxifen (CAFT-T), or chemotherapy followed by Tamoxifen (CAF-T) (*65*). When the two chemotherapy arms were compared after a median follow-up of 8 years, disease free survival (DFS) was significantly better in the CAF-T arm, in which Tamoxifen was started late (HR 1.18, 95% CI 0.98-1.43). The estimated 10-year DFS (OS) rates for T, CAFT-T, and CAF-T were 48% (60%), 53% (62%) and 60% (68%), respectively. This means that the early start of Tamoxifen concurrently with chemotherapy reduces the benefit from chemotherapy by half. These data are in line with several preclinical studies, showing that anti-oestrogens are able to abrogate the effectiveness of chemotherapy (*66*). The optimal sequencing of newer endocrine agents will need to be evaluated in future studies, but in the absence of data it seems sensible for now to start endocrine treatment only after the completion of chemotherapy.

6.1.2 Ovarian Function Suppression plus Chemotherapy

An important question in the premenopausal setting is whether treatment with ovarian ablation should follow chemotherapy in patients with ER-positive disease. In the 1996 EBCTCG meta-analysis, the benefit of ovarian ablation was smaller in women who were also treated with chemotherapy (*47*). This may be due to the fact that many women were postmenopausal after chemotherapy, therefore diluting the effect of ovarian ablation. Three recently published trials have further explored this issue. In the Intergroup 0101 study, CAF chemotherapy given for six cycles was compared to CAF followed by 5 years of Goserelin with or without Tamoxifen. In more than 1,500 randomised patients, CAF plus ovarian ablation was not superior to CAF alone. In the IBCSG VIII study, 1063 breast cancer patients were randomised to six cycles of CMF versus ovarian ablation or the combination of both (*67*). Patients with ER-positive disease achieved similar results with any of the three arms. However, in the subgroup of patients less than 40 years old, the combination arm was significantly better than the two other arms, possibly because fewer women became postmenopausal during chemotherapy in the younger age group. In the third, recently published study investigating the role of ovarian ablation, the ZIPP study, Goserelin was given alone or to women also receiving chemotherapy or Tamoxifen or both. The published results showed Goserelin to be beneficial in the overall study population, but a smaller effect was seen in women also receiving chemotherapy or Tamoxifen. At the recent 2003 American Society of Clinical Oncology (ASCO) meeting, Davidson presented the long-term analysis of the Eastern Cooperative Oncology Group Phase III Intergroup Trial (E5188 INT 0101). This trial randomised premenopausal women to CAF, CAF followed by Z (Goserelin) or CAFT followed by Z. Overall there was no advantage to adding Z to chemotherapy although in a retrospective analysis women < 40 years of age, those who did not become amenorrhoea from chemotherapy and those with premenopausal estradiol levels after chemotherapy benefited from the addition of Z. Taking all these data together, it is still not clear whether treatment with ovarian ablation after chemotherapy is beneficial in certain cohorts of women. The ongoing SOFT study is currently addressing this important issue (see also 5.1). In this trial, 3,000 patients with ER-positive tumours will be randomised to Tamoxifen, Tamoxifen plus ovarian ablation, or ovarian ablation plus Exemestane after chemotherapy or immediately after surgery, if no chemotherapy is given. Importantly, patients entering this trial are required to be premenopausal after the completion of chemotherapy as measured by serum estradiol determinations.

6.2 Strategies to Overcome Endocrine Resistance

It has been known since the early development of endocrine therapies that some patients with ER/PR positive breast cancer do not benefit from endocrine therapy. In addition, even patients whose tumours initially respond to endocrine treatment will eventually experience disease progression. In recent years, substantial progress has been made in understanding specific mechanisms of endocrine resistance. One major mechanism for tumour insensitivity to hormonal agents seems to be cross-talk between the oestrogen receptor and other growth factor signalling pathways (*68*). The studies investigating this interaction have revealed at least three different levels of cross-talk between signal transduction pathways and steroid hormone receptors (*69-72*). In breast cancer patients, overexpression or aberrant activation of epidermal growth factor receptor 2 (EGFR2/erbB-2/Her-2) has been widely demonstrated in breast tumours and linked to an adverse prognosis and endocrine resistance (*73*). Consequently, combining endocrine therapies with signal transduction inhibitors might be a strategy to overcome the development of endocrine resistance. In preclinical studies, signal transduction inhibitors such as the tyrosine kinase inhibitor (TKI) of EGFR1, ZD-1839 ('Iressa'), seem to be only marginally successful if used as monotherapy. However, in a recently published study in a breast cancer xenograft mouse model, significant synergism was shown between ZD-1839 and Tamoxifen (*74*). In this study, mice bearing ER-positive MCF-7 breast tumours were treated with Tamoxifen, ZD-1839 or the combination of both. ZD-1839 improved the antitumour effect of Tamoxifen, and markedly delayed the emergence of acquired resistance from 2-3 months to over 6 months. Importantly, in mice treated with oestrogen deprivation, there was no demonstrable benefit from adding ZD-1839, indicating that activation of the EGFR pathway enhances Tamoxifen agonist effects. Apart from ZD-1839, at least three other growth factor tyrosine kinase inhibitors are in different stages of preclinical and clinical development, including OSI-744 ('Tarceva'), PKI-166, GW-572016 and CI-1033, which is an irreversible pan-erbB tyrosine kinase inhibitor. These compounds are being studied as monotherapy in numerous tumour types and clinical studies evaluating them in combination with endocrine agents are awaited with great interest.

Downstream of the growth factor receptor tyrosine kinases, the Ras proteins, for which aberrant function has been demonstrated in breast cancer, play a major role in intracellular signal transduction. The enzyme farnesyl transferase, which is needed for posttranslational processing of Ras, can be inhibited with the newly developed farnesyl transferase inhibitors (FTIs) (*75*). One of these compounds, R115777 ('Zarnestra'), is active in breast cancer models (*76*), and has already been studied in women with advanced, hormone resistant breast cancer (*77*). Among 76 patients, 24% of patients derived clinical benefit from R115777. The FTIs have as yet not been studied in combination with endocrine therapy.

Other signal transduction inhibitors in clinical development include the Raf kinase and MEK inhibitors, cell cycle inhibitors, and the mTOR inhibitors, of which CCI-779 ('Rapamycin') has entered clinical development in advanced breast cancer (*78*). For CCI-779, improved efficacy has been demonstrated in combination with endocrine therapy and a phase II trial combining the compound with aromatase inhibitors is being planned (*69*).

In addition to the signal transduction inhibitors, synergism with endocrine agents has been shown for inhibitors of the enzyme cyclooxygenase (COX-inhibitors). Celecoxib is an inhibitor of the COX-2 pathway, which is an approved cancer prevention drug (for familial polyposis of the colon) and apparently synergises significantly with Exemestane in preclinical animal models (*79*). Due to its pleiotropic mechanisms of action and its inhibitory effect on receptor negative breast cancer cells in-vitro, celecoxib may also achieve a reduction in oestrogen receptor negative breast cancers. In the adjuvant MA.27 study (see 5.2) the value of adding celecoxib to anastrozole or Exemestane will be investigated. The design of MA.27 adjuvant trial is shown in Figure 2.

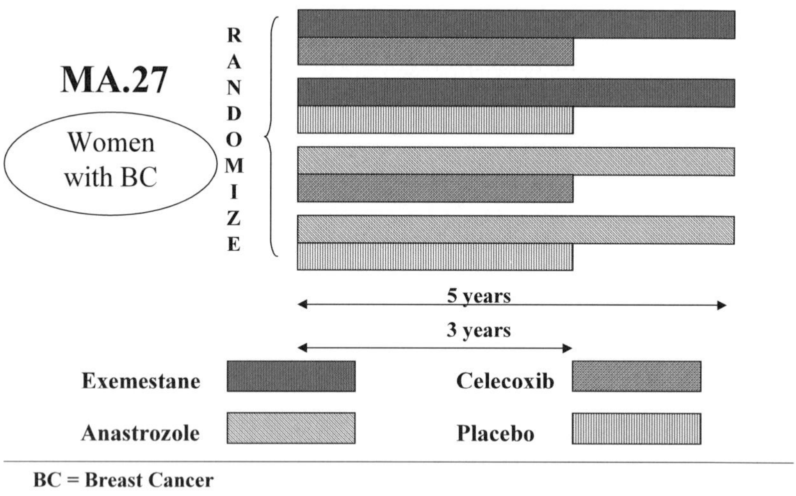

Figure 2. Exemestane vs. Anastrozole and Celecoxib vs. Placebo, 2x2 Factorial Design.

7. LIMITATIONS OF DATA IN WOMEN OF AFRICAN-AMERICAN DESCENT

The question as to whether blacks and whites similarly benefit from Tamoxifen has been investigated by a study analysing the NSABP B-13 and B-14 protocols (*80*). In the B-14 study, surgery plus placebo was compared with surgery plus 5 years of Tamoxifen among patients with ER-positive tumours. When the participants in B-14 were analysed by racial status, black women were younger and were more likely to have larger tumours. However, among all ER-positive patients, event-free survival at 5 years, overall survival and risk of recurrence were equal between blacks and whites. These data show an equal degree of benefit from adjuvant Tamoxifen for Caucasian patients and those of African descent.

Accrual of women of African descent into breast cancer trials has been low in the past (*81*), so that knowledge about breast cancer in this cohort gained within studies might not necessarily reflect the situation in all patients of African descent. Although a study evaluating the inclusion of racial minorities in NCI-sponsored treatment trials show that there is proportional racial/ethnic representation in these studies (*82*), a more recent analysis contradicts this view (*81*). In Chapter 11, the issue of clinical trials research in African-American women will be reviewed in detail. Of importance is that current treatment and prevention paradigms are automatically applied to women of African descent and in the absence of additional data this seems the most reasonable approach. However, just as adopting treatment guidelines for men to women may be sub-optimal the same may be said for doing so among women of different ethic origin including women of African descent. Future studies of the risk factors, pathogenesis, tumour biology and genetic influences on breast cancer in African women are needed in order to establish optimal treatment strategies for these women.

8. CONCLUSION

In this chapter we have summarised the emerging endocrine strategies and treatment options for women with hormone receptor positive breast cancer. With the introduction of the aromatase inhibitors in particular the outcome of women is being improved. It can be seen from the data discussed however, that, although numerous studies have been conducted in women with hormone responsive breast cancer, none have specifically addressed the issue of women of African descent. The fact that accrual of African American women into studies has been low in the past may have multiple reasons. This short fall together with possible strategies to overcome the lack of published data is addressed in more detail in Chapter 11.

In addition to enhancing accrual of women of African descent to ongoing breast cancer studies, future studies should specifically address women from other ethnic minorities. The results of such studies might result in endocrine treatment of breast cancer being tailored according to the features of disease in cohorts of different ethnic origin. Studies of genetic polymorphisms in women at risk for developing breast cancer may highlight inherent differences in the pathogenesis of breast cancer between individuals and possibly between women of different ethnicity. This in turn may lead ultimately to the question as to whether endocrine breast cancer prevention strategies, which are effective in Caucasian women, will be equally successful in women of African descent.

REFERENCES

(1) Axtell LM, Myers MH: Contrasts in survival of black and white cancer patients, 1960-73. J Natl Cancer Inst 1978;60:1209-15.

(2) Young JL, Jr., Ries LG, Pollack ES: Cancer patient survival among ethnic groups in the United States. J Natl Cancer Inst 1984;73:341-52.

(3) Natarajan N, Nemoto T, Mettlin C, Murphy GP: Race-related differences in breast cancer patients. Results of the 1982 national survey of breast cancer by the American College of Surgeons. Cancer 1985;56:1704-9.

(4) Eley JW, Hill HA, Chen VW, Austin DF, Wesley MN, Muss HB, et al: Racial differences in survival from breast cancer. Results of the National Cancer Institute Black/White Cancer Survival Study. Jama 1994;272:947-54.

(5) McWhorter WP, Mayer WJ: Black/white differences in type of initial breast cancer treatment and implications for survival. Am J Public Health 1987;77:1515-7.

(6) Yood MU, Johnson CC, Blount A, Abrams J, Wolman E, McCarthy BD, et al: Race and differences in breast cancer survival in a managed care population. J Natl Cancer Inst 1999;91:1487-91.

(7) Heimann R, Ferguson D, Powers C, Suri D, Weichselbaum RR, Hellman S: Race and clinical outcome in breast cancer in a series with long-term follow-up evaluation. J Clin Oncol 1997;15:2329-37.

(8) Mohla S, Sampson CC, Khan T, Enterline JP, Leffall L, Jr., White JE: Estrogen and progesterone receptors in breast cancer in Black Americans: Correlation of receptor data with tumor differentiation. Cancer 1982;50:552-9.

(9) Beverly LN, Flanders WD, Go RC, Soong SJ: A comparison of estrogen and progesterone receptors in black and white breast cancer patients. Am J Public Health 1987;77:351-3.

(10) Beatson G: On the treatment of inoperable cases of the mamma: suggestions for a new method of treatment with illustrative cases. Lancet 1896;2:104-07.

(11) Schinzinger A: Centralblatt fur Chirurgie. Beilage 1889;18:55-56.

(12) Tamoxifen for early breast cancer: an overview of the randomised trials. Early Breast Cancer Trialists' Collaborative Group. Lancet 1998;351:1451-67.

(13) Systemic treatment of early breast cancer by hormonal, cytotoxic, or immune therapy. 133 randomised trials involving 31,000 recurrences and 24,000 deaths among 75,000 women. Early Breast Cancer Trialists' Collaborative Group. Lancet 1992;339:1-15.
(14) Systemic treatment of early breast cancer by hormonal, cytotoxic, or immune therapy. 133 randomised trials involving 31,000 recurrences and 24,000 deaths among 75,000 women. Early Breast Cancer Trialists' Collaborative Group. Lancet 1992;339:71-85.
(15) Eifel P, Axelson JA, Costa J, Crowley J, Curran WJ, Jr., Deshler A, et al: National Institutes of Health Consensus Development Conference Statement: adjuvant therapy for breast cancer, November 1-3, 2000. J Natl Cancer Inst 2001;93:979-89.
(16) Thurlimann B: Breast Cancer. International Consensus Meeting of the Treatment of Primary Breast Cancer. Switzerland: St. Gallen, 2001, p. 294-97.
(17) Fisher B, Dignam J, Bryant J, DeCillis A, Wickerham DL, Wolmark N, et al: Five versus more than five years of tamoxifen therapy for breast cancer patients with negative lymph nodes and estrogen receptor-positive tumors. J Natl Cancer Inst 1996;88:1529-42.
(18) Morrow M, Jordan VC: Molecular mechanisms of resistance to tamoxifen therapy in breast cancer. Arch Surg 1993;128:1187-91.
(19) Gottardis MM, Jordan VC: Development of tamoxifen-stimulated growth of MCF-7 tumors in athymic mice after long-term antiestrogen administration. Cancer Res 1988;48:5183-7.
(20) Masamura S, Santner SJ, Heitjan DF, Santen RJ: Estrogen deprivation causes estradiol hypersensitivity in human breast cancer cells. J Clin Endocrinol Metab 1995;80:2918-25.
(21) Cuzick J, Powles T, Veronesi U, Forbes J, Edwards R, Ashley S, et al: Overview of the main outcomes in breast-cancer prevention trials. Lancet 2003;361:296-300.
(22) Fisher B, Costantino JP, Wickerham DL, Redmond CK, Kavanah M, Cronin WM, et al: Tamoxifen for prevention of breast cancer: report of the National Surgical Adjuvant Breast and Bowel Project P-1 Study. J Natl Cancer Inst 1998;90:1371-88.
(23) McDonald CC, Stewart HJ: Fatal myocardial infarction in the Scottish adjuvant tamoxifen trial. The Scottish Breast Cancer Committee. Bmj 1991;303:435-7.
(24) Costantino JP, Kuller LH, Ives DG, Fisher B, Dignam J: Coronary heart disease mortality and adjuvant tamoxifen therapy. J Natl Cancer Inst 1997;89:776-82.
(25) Rutqvist LE, Mattsson A: Cardiac and thromboembolic morbidity among postmenopausal women with early-stage breast cancer in a randomized trial of adjuvant tamoxifen. The Stockholm Breast Cancer Study Group. J Natl Cancer Inst 1993;85:1398-406.
(26) Rossouw JE, Anderson GL, Prentice RL, LaCroix AZ, Kooperberg C, Stefanick ML, et al: Risks and benefits of estrogen plus progestin in healthy postmenopausal women: principal results From the Women's Health Initiative randomized controlled trial. Jama 2002;288:321-33.
(27) Love RR, Mazess RB, Tormey DC, Barden HS, Newcomb PA, Jordan VC: Bone mineral density in women with breast cancer treated with adjuvant tamoxifen for at least two years. Breast Cancer Res Treat 1988;12:297-302.
(28) Love RR, Mazess RB, Barden HS, Epstein S, Newcomb PA, Jordan VC, et al: Effects of tamoxifen on bone mineral density in postmenopausal women with breast cancer. N Engl J Med 1992;326:852-6.
(29) Fallowfield L CD: Assessing the quality of life (QOL) of postmenopausal (PM) women radomized into the ATAC ('Arimidex' tamoxifen, alone or in combination) adjuvant breast cancer (BC) trial. Proc. Am. Soc. Clin. Oncol 2002;21.

(30) Bonneterre J, Thurlimann B, Robertson JF, Krzakowski M, Mauriac L, Koralewski P, et al: Anastrozole versus tamoxifen as first-line therapy for advanced breast cancer in 668 postmenopausal women: results of the Tamoxifen or Arimidex Randomized Group Efficacy and Tolerability study. J Clin Oncol 2000;18:3748-57.

(31) Nabholtz JM, Buzdar A, Pollak M, Harwin W, Burton G, Mangalik A, et al: Anastrozole is superior to tamoxifen as first-line therapy for advanced breast cancer in postmenopausal women: results of a North American multicenter randomized trial. Arimidex Study Group. J Clin Oncol 2000;18:3758-67.

(32) Bonneterre J, Buzdar A, Nabholtz JM, Robertson JF, Thurlimann B, von Euler M, et al: Anastrozole is superior to tamoxifen as first-line therapy in hormone receptor positive advanced breast carcinoma. Cancer 2001;92:2247-58.

(33) Mouridsen H, Gershanovich M, Sun Y, Perez-Carrion R, Boni C, Monnier A, et al: Superior efficacy of letrozole versus tamoxifen as first-line therapy for postmenopausal women with advanced breast cancer: results of a phase III study of the International Letrozole Breast Cancer Group. J Clin Oncol 2001;19:2596-606.

(34) Paridaens R, Dirix L, Beex L, Nooij MN, Cufer T, Lohrisch C, et al: Promising Results with Exemestane in the First-Line Treatment of Metastatic Breast Cancer: A Randomized Phase II EORTC Trial with a Tamoxifen Control. Clin Breast Cancer 2000;1 Suppl 1:S19-21.

(35) Rose C, Vtoraya O, Pluzanska A, Neave F, Clemens M, Chaudri-Ross HA, et al: Letrozole (Femara) vs. anastrozole (Arimidex). Proc. Am. Soc. Clin. Oncol 2002;21.

(36) Crump M, Sawka CA, DeBoer G, Buchanan RB, Ingle JN, Forbes J, et al: An individual patient-based meta-analysis of tamoxifen versus ovarian ablation as first line endocrine therapy for premenopausal women with metastatic breast cancer. Breast Cancer Res Treat 1997;44:201-10.

(37) Taylor CW, Green S, Dalton WS, Martino S, Rector D, Ingle JN, et al: Multicenter randomized clinical trial of goserelin versus surgical ovariectomy in premenopausal patients with receptor-positive metastatic breast cancer: an intergroup study. J Clin Oncol 1998;16:994-9.

(38) Klijn JG, Blamey RW, Boccardo F, Tominaga T, Duchateau L, Sylvester R: Combined tamoxifen and luteinizing hormone-releasing hormone (LHRH) agonist versus LHRH agonist alone in premenopausal advanced breast cancer: a meta-analysis of four randomized trials. J Clin Oncol 2001;19:343-53.

(39) Boccardo F, Rubagotti A, Perrotta A, Amoroso D, Balestrero M, De Matteis A, et al: Ovarian ablation versus goserelin with or without tamoxifen in pre- perimenopausal patients with advanced breast cancer: results of a multicentric Italian study. Ann Oncol 1994;5:337-42.

(40) Lonning PE, Taylor PD, Anker G, Iddon J, Wie L, Jorgensen LM, et al: High-dose estrogen treatment in postmenopausal breast cancer patients heavily exposed to endocrine therapy. Breast Cancer Res Treat 2001;67:111-6.

(41) Peethambaram PP, Ingle JN, Suman VJ, Hartmann LC, Loprinzi CL: Randomized trial of diethylstilbestrol vs. tamoxifen in postmenopausal women with metastatic breast cancer. An updated analysis. Breast Cancer Res Treat 1999;54:117-22.

(42) Cauley JA, Norton L, Lippman ME, Eckert S, Krueger KA, Purdie DW, et al: Continued breast cancer risk reduction in postmenopausal women treated with raloxifene: 4-year results from the MORE trial. Multiple outcomes of raloxifene evaluation. Breast Cancer Res Treat 2001;65:125-34.

(43) Jordan VC, Glusman JE, Eckert S: Raloxifene reduces incident primary breast cancer. Breast Cancer Res Treat 1998;50:227.

(44) Buzdar AU, Marcus C, Holmes F, Hug V, Hortobagyi G: Phase II evaluation of Ly156758 in metastatic breast cancer. Oncology 1988;45:344-5.
(45) Osborne CK, Pippen J, Jones SE, Parker LM, Ellis M, Come S, et al: Double-blind, randomized trial comparing the efficacy and tolerability of fulvestrant versus anastrozole in postmenopausal women with advanced breast cancer progressing on prior endocrine therapy: results of a North American trial. J Clin Oncol 2002;20:3386-95.
(46) Howell A, Robertson JF, Quaresma Albano J, Aschermannova A, Mauriac L, Kleeberg UR, et al: Fulvestrant, formerly ICI 182,780, is as effective as anastrozole in postmenopausal women with advanced breast cancer progressing after prior endocrine treatment. J Clin Oncol 2002;20:3396-403.
(47) Ovarian ablation in early breast cancer: overview of the randomised trials. Early Breast Cancer Trialists' Collaborative Group. Lancet 1996;348:1189-96.
(48) Davidson NE: Ovarian ablation as adjuvant therapy for breast cancer. J Natl Cancer Inst Monogr 2001;30:67-71.
(49) Dees EC, Davidson NE: Ovarian ablation as adjuvant therapy for breast cancer. Semin Oncol 2001;28:322-31.
(50) Jonat W, Kaufmann M, Sauerbrei W, Blamey R, Cuzick J, Namer M, et al: Goserelin versus cyclophosphamide, methotrexate, and fluorouracil as adjuvant therapy in premenopausal patients with node-positive breast cancer: The Zoladex Early Breast Cancer Research Association Study. J Clin Oncol 2002;20:4628-35.
(51) Cuzick J: The ATAC ('Arimidex', Tamoxifen, Alone or in Combination) trial in postmenopuasal women with early breast cancer - updated efficacy results based on a median follow-up of 47 months. The Breast 2003;12:47.
(52) Baum M: The ATAC (Arimidex, Tamoxifen, Alone or in Combination) adjuvant breast cancer trial in postmenopausal patients: factors influencing the success of patient recruitment. Eur J Cancer 2002;38:1984-6.
(53) Winer EP, Hudis C, Burstein HJ, Chlebowski RT, Ingle JN, Edge SB, et al: American Society of Clinical Oncology technology assessment on the use of aromatase inhibitors as adjuvant therapy for women with hormone receptor-positive breast cancer: status report 2002. J Clin Oncol 2002;20:3317-27.
(54) Goss PE, Ingle JN, Martino S et al. A randomized trial of letrozole in postmenopausal women after five years of tamoxifen therapy for early-stage breast cancer. N Engl J Med 2003;349:1793-1802.
(55) Goss PE, Strasser K: Aromatase inhibitors in the treatment and prevention of breast cancer. J Clin Oncol 2001;19:881-94.
(56) Dewar J, Nabholtz JM, Bonneterre J, Buzdar A, Robertson JF, Thurlimann B, et al: The effect of anastrozole (Arimidex) on serum lipids - data from a randomized comparison of anastrozole (AN) vs. tamoxifen (TAM) in postmenopausal (PM) women with advanced breast cancer (ABC). Breast Cancer Res Treat 2000;64.
(57) Harper-Wynne C, Ross G, Sacks N, Salter J, Nasiri N, Iqbal J, et al: Effects of the aromatase inhibitor letrozole on normal breast epithelial cell proliferation and metabolic indices in postmenopausal women: a pilot study for breast cancer prevention. Cancer Epidemiol Biomarkers Prev 2002;11:614-21.
(58) Elisaf MS, Bairaktari ET, Nicolaides C, Kakaidi B, Tzallas CS, Katsaraki A, et al: Effect of letrozole on the lipid profile in postmenopausal women with breast cancer. Eur J Cancer 2001;37:1510-3.
(59) Goss PE, Grynpas M, Qui S, Hu H: The effects of exemestane on bone and lipids in the ovariectomized rat. Breast Cancer Res Treat 2001;69.

(60) Goss PE, Cheung AM, Lowery C: Comparison of the efffects of exemestane, 17-hydroexemestane and letrozole on bone and lipid metabolism inthe ovariectomized rat. Breast Cancer Res Treat 2002;76.

(61) Goss PE, Thomsen T, Banke-Bochita J: A randomized, placebo-controlled, explorative study to investigate the effect of low estrogen plasma levels on markers of bone turnover in healthy postmenopausal women during the 12 week treatment with exemestane or letrozole. Breast Cancer Res Treat 2002;76.

(62) Lohrisch C, Paridaens R, Dirix L, Beex L, Nooij MN, Cameron D, et al: No adverse impact on serum lipids of the irreversible aromatase inactivator Aromasin (exemestane [E]) in lst line treatment of metastatic breast cancer (MBC): companion study to a European Organization of Research and Treatment of Breast Cancer (Breast Group) trial with Pharmacias' Upjohn. Proc. Am. Soc. Clin. Oncol 2001;20.

(63) Dirix L, Piccart M, Lohrisch C, Beex L, Nooij MN, Biganzoli L, et al: Efficacy of and tolerance to exemestane (E) versus tamoxifen (T) in flst line hormone therapy (HT) of postmenopausal metastatic breast cancer (MBC) patients (pts): A European Organisation for the Research and Treatment of Cancer (EORTC Breast Group) phase II trial with Pharmacia and Upjohn,. Proc. Am. Soc. Clin. Oncol 2001;20.

(64) Polychemotherapy for early breast cancer: an overview of the randomised trials. Early Breast Cancer Trialists' Collaborative Group. Lancet 1998;352:930-42.

(65) Albain K, Green S, Ravdin P, Cobau C, Levine EG, Ingle JN: Adjuvant chemohormonal therapy for primary breast cancer should be sequential instead of concurrent: initial results from intergroup trial 0100(SWOG-8814). Proc. Am. Soc. Clin. Oncol 2002;21.

(66) Osborne CK, Kitten L, Arteaga CL: Antagonism of chemotherapy-induced cytotoxicity for human breast cancer cells by antiestrogens. J Clin Oncol 1989;7:710-7.

(67) Castiglione-Gertsch M, Oneill A, Gelber RD, Nasi ML: Is the addition of adjuvant chemotherapy always necessary in node negative (N-) pre/perimenopausal breast cancer patients (pts) who receive goserelin? First results of IBCSG trial VIII. Proc. Am. Soc. Clin. Oncol 2002;21.

(68) Nicholson RI, McClelland RA, Robertson JF, Gee JM: Involvement of steroid hormone and growth factor cross-talk in endocrine response in breast cancer. Endocr Relat Cancer 1999;6:373-87.

(69) Johnston SR, Head J, Pancholi S, Detre S, Martin LA, Smith IE, et al: Integration of signal transduction inhibitors with endocrine therapy: an approach to overcoming hormone resistance in breast cancer. Clin Cancer Res 2003;9:524S-32S.

(70) Bunone G, Briand PA, Miksicek RJ, Picard D: Activation of the unliganded estrogen receptor by EGF involves the MAP kinase pathway and direct phosphorylation. Embo J 1996;15:2174-83.

(71) Simoncini T, Hafezi-Moghadam A, Brazil DP, Ley K, Chin WW, Liao JK: Interaction of oestrogen receptor with the regulatory subunit of phosphatidylinositol-3-OH kinase. Nature 2000;407:538-41.

(72) Webb P, Lopez GN, Uht RM, Kushner PJ: Tamoxifen activation of the estrogen receptor/AP-1 pathway: potential origin for the cell-specific estrogen-like effects of antiestrogens. Mol Endocrinol 1995;9:443-56.

(73) Wright C, Nicholson S, Angus B, Sainsbury JR, Farndon J, Cairns J, et al: Relationship between c-erbB-2 protein product expression and response to endocrine therapy in advanced breast cancer. Br J Cancer 1992;65:118-21.

(74) Massarweh S, Shou J, DiPietro M, Mohsin SK, Hilsenbeck SG: Targeting the epidermal growth factor receptor pathway improves the ant-tumor effect of tamoxifen

and delays acquired resistance in xenograft model of breast Cancer. Breast Cancer Res Treat 2002;33.
(75) Johnston SR: Farnesyl transferase inhibitors: a novel targeted tnerapy for cancer. Lancet Oncol 2001;2:18-26.
(76) Kelland LR, Smith V, Valenti M, Patterson L, Clarke PA, Detre S, et al: Preclinical antitumor activity and pharmacodynamic studies with the farnesyl protein transferase inhibitor R115777 in human breast cancer. Clin Cancer Res 2001;7:3544-50.
(77) Johnston SR, Hickish T, Houston S, Ellis PA: Efficacy and tolerability of two dosing regimens of R115777 (Zarnestra), a farnesyl protein transferase inhibitor, in patients with advanced breast cancer. Proc. Am. Soc. Clin. Oncol 2002;21.
(78) Chan A, Johnston SR, Scheulen ME, Mross K: First report: a phase 2 study of the safety and activity of CCI-779 for patients with locally advanced or metastatic breast cancer failing prior chemotherapy. Proc. Am. Soc. Clin. Oncol 2002;21.
(79) Pesenti E, Masferrer JL, Di Salle E: Effect if exenestane and celecoxib alone or in combingation on DMBA-induced mammary carcinoma in rats. Breast Cancer Res Treat 2003;69.
(80) Dignam JJ: Differences in breast cancer prognosis among African-American and Caucasian women. CA Cancer J Clin 2000;50:50-64.
(81) Christian MC, Trimble EL: Increasing participation of physicians and patients from underrepresented racial and ethnic groups in National Cancer Institute- sponsored Clinical Trials. Cancer Epidemiol Biomarkers Prev 2003;12:277S-83S.
(82) Tejeda HA, Green SB, Trimble EL, Ford L, High JL, Ungerleider RS, et al: Representation of African-Americans, Hispanics, and whites in National Cancer Institute cancer treatment trials. J Natl Cancer Inst 1996;88:812-6.

Chapter 10

PALLIATIVE CARE FOR BREAST CANCER IN THE AFRICAN-AMERICAN POPULATION

Joseph O'Donnell, MD
Dartmouth Medical School and the Norris Cotton Cancer Center, Hanover, NH, USA.

1. INTRODUCTION

Breast cancer is the most common cancer among African-American women, but its incidence is 13% lower than that in Caucasian women. The incidence slowly rose in the decade of the 1970's but was fairly stable during the 1990's. The death rate increased between 1973 and 1993 but it has declined since. In spite of this decline, the death rate is still 28% higher in African-American than in Caucasian women. There are likely to be many factors, which contribute to this higher death rate, including access to quality screening and treatment, late diagnosis and perhaps even more aggressive tumour biology (*1*).

The gap in care also occurs in end-of-life care and this chapter will review the data and suggest possible solutions.

2. PALLIATIVE CARE

The Hastings Center (*2*) has developed a useful outline for the goals of medicine. They have defined these goals as:

a) The prevention of disease and injury and the promotion and maintenance of health.
b) The relief of pain and suffering caused by illness.
c) The care and cure of those with a malady and the care of those who cannot be cured.

C. Williams et al. (eds.), Breast Cancer in Women of African Descent, 249-266.

d) The avoidance of premature death and the pursuit of a peaceful death.

There still remains much to be done to achieve all these goals but there has been an especially great level of dissatisfaction with care provided to patients at the end of life and the relief of pain and suffering, the care of those who cannot be cured and the achievement of a peaceful death. One of the early prime movers in the efforts to improve end-of-life care was the Robert Wood Johnson Foundation, which funded the SUPPORT study (Study to Understand Prognosis and Preferences for Outcomes and Risks of Treatment) (*3*). The SUPPORT study documented many of the issues including inadequate pain management, communication, systems for delivery of care and inattention to patient preferences, and despite attempted intervention to change these inadequate outcomes, showed how difficult it was to change the system and provide excellence in care. The results helped spawn numerous corrective programs, including the Soros Foundation's "Project Death in America" (*4*). Reports from the Institute of Medicine, (*5,6*) the President's Cancer Council (*7*). The Medicare Payment Advisory Committee, (*8*) the American Society for Clinical Oncology (*9*) and many other organizations suggested solutions. The public was engaged, especially through a six-hour public television special on palliative and end-of-life care: ("On Our Own Terms: Moyers on Dying" in September 2000) (*10*). All these efforts are slowly producing positive change, but there is still a long way to go. Part of the difficulty with this issue involves our great ambivalence about death and dying well documented in "Death is that Man Taking Names" by Burt (*11*).

3. ORGANIZED EFFORTS

The modern hospice movement began in England and was led by Dame Cicely Saunders. A former nurse and social worker, Dr Saunders realized that to be change agent for this aspect of healthcare, she had to become a doctor. Her example has revolutionized medical care and it all began when she encountered a very special patient, a young Pole dying from war injuries, who said to Dr Saunders the words that drove her to start St Christopher's Hospice: "I only want what is in your mind and in your heart." Hospice started as a movement that was peripheral to mainstream medicine, but as medicine evolves and embraces those aspects in the Hastings Center message above, end of life care has become more central and there is a great deal of effort and success in making it better. The discipline of Palliative Medicine has evolved to be "... an interdisciplinary approach to the study and care of patients with active, progressive, far advanced disease for whom the prognosis is limited and the focus of care is on quality of life." Palliative medicine involves the comprehensive management of the physical, psychological, social, spiritual and existential needs of patients (*12*).

This type of medicine involves what the patient asked of Dr Saunders: the skills of the mind, embodied in competent medical care and the friendship of the heart, with its caring, acceptance, vulnerability and reciprocity. It is not just handholding, but includes an ever evolving and sophisticated science of symptom management. The most common physical symptoms at end of life are weakness, loss of appetite, pain, nausea, constipation, shortness of breath and confusion, but we have come to recognize that something called "suffering" may be the most common symptom of all. Suffering occurs when there is a threat, real or perceived, to the intactness of whole person. (What will happen to me? To my family? Who will remember me? Has my life been worthwhile? Etc). Palliative medicine must deal with both physical symptoms like pain and existential/spiritual ones like suffering.

The author, Wendall Berry, says in a piece called "Health is Membership" (*13*) that "patients come from a world of love. In the hospital what I would call the world of love meets the world of efficiency--the world that is, of specialization, machinery, and abstract procedure. Or rather I should say that these two worlds come together in the hospital but do not meet". Palliative medicine attempts to help these worlds to meet.

Six major skill sets comprise palliative medicine: communication, decision making, management of complications of treatment and disease, symptom control, psychosocial care of the patient and family and care of the dying (*14*).

Hospice care implies a philosophy of care at the end of life for patients who are dying and their families. The focus of care is on the relief of distressing symptoms, including emotional, psychosocial and spiritual ones. Care includes not only the patient, but also the loved ones and is usually given by an interdisciplinary team. Hospice care can occur in a dedicated facility, in any healthcare institution, but especially in the home. Medicare has established a hospice benefit. Eligibility requires a documentED prognosis of less than 6 months, goals of palliation and an identified physician of record. Some agencies require an established "do not resuscitate" order, no planned use of hydration, total parental nutrition and a primary caregiver in place in the home. Hospice agencies usually provide physical symptom control, home health aide services for help with bathing, dressing, feeding, psychological counselling services, help with preparations for death, spiritual support and bereavement programs (*15*).

4. BARRIERS TO END-OF LIFE CARE IN AFRICAN AMERICANS

4.1 Barriers in General

Many of the reports mentioned above detail the barriers to excellent end-of-life care in general. These include such things as: the separation of end-of-life from curative care and the late referrals (if referred at all to palliative services); the inadequate training and number of trained health care professionals and the silos in which we work that often undermine effective teamwork and communication with important partners like chaplains, social workers and home care agencies: the inadequate funding for these services; the lack of effective or accessible patient information; the inadequate or poorly known standards of care; the under-use or lack of useful tools for measurement of things like quality of life; the stigma of this type of care as "giving-up"; the high variability of this type of care by geography; and the inadequate attention to this by the funders of research. Steps are being taken to reduce all these barriers (*16*).

4.2 Barriers in African Americans

African Americans comprise 14% of the population but only 5-7% of those who utilize hospice services. 93% of those using the Medicare hospice benefit are Caucasian and only 7% African-American (*17*). Less than 10% of those on the National Hospice and Palliative Care Organization's database are African Americans and less than 10% using hospice services in managed care organizations come from minority groups. African American patients are more likely to die in hospitals, less likely to have advance directives or do not resuscitate orders, and more likely to have fatalistic views about cancer (*18,19*). The cost for care in the last year of life is likely to be higher in African Americans than Caucasians ($32,000 vs $25,000) (*20*).

All of this should be taken with the perspective that it is difficult to generalize about any group of people and such generalizations may risk stereotyping. The "African" American group (which refers to blacks in the United States) includes people from many different and varied cultures (eg those with family origins in Africa, in the United States and the Caribbean) and it is dangerous to generalize. Whenever one is dealing with an African American patient or family, it is important to ask how they define their cultural tradition and how closely they adhere to traditional practices. For instance, the affluent in general are more likely to move away from the traditions of origin than the poor, who are more likely to follow traditional practices (*21*).

That being said, some generalizations include the following: spirituality is more likely to be highly important, belief in an afterlife and the continuation of a "presence" of an individual even after death are more common, belief in the sanctity of life and "not giving up" are important, decisions are much more likely to be deferred to a group or family for consensus and all members should have input, family members should be physically present with their loved one whenever possible, and suffering is ennobling. Language like "passing on" or "transition" to a new period of existence, not still with us but among us is used rather than dying. A wake, or "sit-up" may also be called a “home going service" in which a loved one goes home to the spirit world. There is a belief in a circle of existence, that death is part of the natural rhythm of life and that whenever someone dies, a baby is born someplace else (*21*).

As noted at the beginning of this chapter, patients from minority groups have higher incidence and death rates at younger ages for certain cancers, more exposure to risk factors predisposing to cancer, tend to be diagnosed later with more advanced stage tumours, may experience less aggressive care (*1*) and seem to have fewer opportunities for excellent end-of-life care. The tragedy is that the gap seems to be widening rather than closing in the U.S.

There have been numerous reports and calls to action because of the great disparities in healthcare outcomes that occur throughout the United States (*22*). The Clinton Administration in their Initiative on Race in America (http://www.wsws.org/public_html/iwb6-30/edit.htm) instituted a number of ongoing initiatives including the upgrading of the office of Minority Health to a Center for Minority Health and Health Disparities at the National Institutes of Health (NIH). This signalled an important message: a center is more influential in setting policy, in influencing budgetary expenditures and can award grants. For example, the National Cancer Institute's Office of Education is updating its materials on pain and symptom management to make them more accessible, culturally and educationally appropriate and useful to under-served groups (*22*). The American Cancer Society has as one of its primary goals to eliminate disparities in cancer outcomes. The Intercultural Cancer Council is bringing together organizations and individuals to collaborate around issues of minority health and disparities in outcomes (http://icc.bcm.tmc.edu/), and the Institute of Medicine (IOM) recently issued a landmark report on eliminating disparities, which has documented inequities and sought solutions (*23*). The IOM also has a task force reviewing the structure of the NIH. The NIH has less than 1% of its grants going to investigators from under-served minority groups and only 23 million of a 3 billion NCI budget goes to issues of resolving disparities (*24*).

4.3 Barriers in the Management of Pain

There has been more written about barriers to effective pain management than any other symptom. A particularly important symptom in cancer and one that is especially common in end-of-life situations, i.e. pain is much more likely to be inadequately assessed, treated and controlled in African American patients (*25*).

Multiple studies have shown inadequate treatment of cancer related pain in minority populations. Cleeland and colleagues (*26*) found that outpatients with cancer pain who were treated in clinics that served ethnic and racial minority patients were three times more likely to be under-medicated with analgesics when compared to patients in other settings. In a later study, this group (*27*) reported that 65% of minority patients left an office visit with an inadequate medication prescription for control of their pain versus 50% of non-minority patients (the latter statistic a large enough problem in its own right). The dose of narcotics prescribed in patient controlled anaesthesia varied in different ethnic groups (*28*). Minority patients with bone fractures were twice as likely to receive inadequate doses of analgesics in the emergency room compared to non-minority patients (*29*). These inequities also have been documented in nursing homes and cancer centres (*30*). Numerous studies document inadequate end-of-life care in minority populations (*22*). Lessons from the approach to this problem can be applied to overcoming barriers to the control of other end-of-life issues. Many of the reasons for inadequate pain control and the barriers to adequate pain relief have been identified and reviewed in the pain literature (*5*). Barriers include those attributed to the healthcare professional, to the patient, and to the system. These same barriers are present in minority populations to an even greater degree.

Although as has been said, it is inappropriate to promote stereotypes for a group, there are certain characteristics that do seem to be more common in African American individuals. They tend to come from poorer socio-economic groups, to be more often un- or underinsured, to have less formal educational achievement, and to come from poorer neighbourhoods. However, numerous studies have shown that such patients may not be offered the same level of intervention no matter what their health or economic status, so these issues go beyond those associated with just poverty (*31*).

Patient and family barriers include a reluctance to report pain in the fear of jeopardizing aggressive disease oriented treatment, fear that pain means the disease is progressing, concern about not being perceived as a "good" patient, reluctance to take medications for fear of addiction or stigmatisation, worries about side effects and tolerance. Minority patients tend to experience even greater barriers to adequate pain control. Language, underlying cultural belief systems and the ability to understand "the other" remain challenging. Patients from some minority populations may show less outward behaviour related to pain causing observers to underestimate the pain (*21*).

African Americans tend to believe in the "dignity" that is associated with suffering and may report later in the course of their illnesses to providers or tend to minimize discomfort (*32*). These patients tend to lack trust in a system that has not treated them fairly (*33*). This is felt to be due to an historical accumulation of social injustices making the system and its motives seem suspect. Patients know the statistics of morbidity and mortality in minorities and may fear being "under-treated" rather than receiving aggressive disease specific care. For many, "quantity of life equals quality of life." Minority patients are more likely to live in poverty and to have poorer health in general. They are more likely to have to attend to the stresses and survival needs of daily existence than to have time and energy for self-care. Their insurance coverage and coverage is likely to be less. Maslow found that 31% of minority patients had no health insurance versus 14% of Caucasians (*23*). Even when employed, minority workers were less likely to be covered (56% versus 66%) (*23*). Fifteen percent of minority patients openly said that they would get better care if they were members of a different race (*23*).

African American patients tend to be less knowledgeable about their symptoms and about federal, state and local programs that may benefit them.

Another set of barriers are those of healthcare professionals and they include a lack of up-to-date knowledge, inadequate assessments in time limited encounters, fear of prescribing powerful drugs with regulatory concerns about their licenses, fear of promoting addiction in patients, or producing difficult to manage side effects (*34*). Minorities again face more significant barriers. The maldistribution of services makes areas populated by large numbers of minority patients have fewer practitioners and fewer choices. Overall access to care is harder and even when in the "system", minority patients make fewer visits. African American patients have fewer available practitioners and fewer choices. Clinics in African American neighbourhoods are more likely to be understaffed and busier. African American patients have less access to specialists (8% of minorities vs. 18% of non-minorities get referrals) (*35*) and less participatory visits with providers (*36*).

Most importantly, racial stereotypes still exist. The IOM report described these in great detail. Non-minorities are more likely to believe that minorities; "live off welfare, are more prone to violence, are less intelligent and lazy"(*37*). This may cause projection of a stereotype that undervalues a patient's participation in medical decision-making. Often this is thought to be unintentional and unconscious but the patients readily perceive it. Studies have shown that physicians make different and less aggressive recommendations to minority standardized patients who are playing roles with the same levels of illness as non-minorities (*38*). There are also not enough minority physicians in the workforce despite valiant efforts to raise the numbers (*39*). Presently, there are ongoing efforts in medical schools and residencies to address issues of cultural competency training, but there is still a long way to go in this area (*40*).

African American physicians account for only 2.9% of the physician workforce, but care for 30% of the African American population. These physicians tend to be strong advocates for aggressive therapies, less likely to support withdrawal of treatment (e.g. tube feedings), or be supportive of physician-assisted suicide. African American physicians themselves say they would be more likely to push for aggressive therapy for themselves if they were in a persistent vegetative state.

System issues that have been identified include a low priority given to pain management, inadequate reimbursement, restrictive regulation policies and problems with availability of the medications, and limited access to pain specialists. Despite large gains in equity, institutional racism still exists. Racial stereotypes are listed above and views like these may pervade institutions. Experts now say that the old version of "Jim Crow" overt racism has turned to a subtle "laissez faire" version and that people aren't even conscious of it (*41*). It can be identified in hiring policies, leadership policies (who are the advocates for these patients?), the fact that managed care shies away from minority patients with greater risk, and is less likely to employ and more likely to terminate minority providers (*42*). Our system tolerates poorer health outcomes in the poor, and we see hospitals and clinics either not locating in or moving from minority neighbourhoods. Even with cost and lack of access as huge barriers and even if a minority patient gets "in" the system and gets a prescription for pain medicines, the pharmacies in their neighbourhoods are less likely to stock the drugs because of fear of robbery and violence (*43*). In addition, we don't have enough data now, nor do we collect it, to know the extent of this problem. Experts suggest that regulatory organizations require reporting of such health care disparities as a first step in beginning to correct them (*44*).

There is also a lack of educational materials aimed at African American populations to address the issues of end-of-life care (*6*).

5. BREAST CANCER AND COMMUNICATION ABOUT PAIN, SUFFERING AND OTHER PALLIATIVE CARE ISSUES

Stage 1V breast cancer is incurable and will eventuate in death. At present, there is no effective curative therapy for this stage but patents should be afforded the most effective anti-neoplastic therapy to prolong life (as long as the quality of life is acceptable to the patient) and control symptoms. Meticulous attention to the control of symptoms should occur along the whole trajectory of the disease course. Practice guidelines for the control of pain, distress, delirium, dyspnoea, anxiety, psychosocial issues, spiritual issues, fatigue, constipation, incontinence and nausea are available.

Barriers to the control of physical pain have been described above. The expression of pain is a form of communication, and cultural differences in emotional expression and stereotypes about pain also influence the clinician's interpretation and expectations regarding pain severity. This is possibly due to the conscious and unconscious bias on the part of the clinician and patient, or the subtle differences in cultural expressive cues (i.e. body language) that the individual can more readily recognize emotion expressed in the style in which they are accustomed (*45-47*). Given the problems with communication and treating cancer-related pain, guidelines for effective pain communication are receiving more attention than ever in medical education. It has been designated one of the core competencies that the ACGME (Accreditation Council for Graduate Medical Education) has said that all residency graduates must possess (www.acgme.org/outcome/comp/compFull.asp).

Communication may be verbal, in writing, or even by "sign" and it may utilize techniques from simple words or gestures to sophisticated media. Body language (or para communication) can portray even more information than is carried in words. No matter what the medium, each message involves a sender, a receiver, and a code and is done in a particular context. Such things as age, sex, culture, language, education, physical impairments, and emotional states can affect the message of the sender and what the receiver hears.

Communication about pain involves words, expectations and behaviours. Words may have different meanings (e.g., words like "pain, hurt or ache" may be used and interpreted differently by both the sender and receiver). Body language and behaviours indicating pain may be much less pronounced in certain minority groups causing providers to underestimate their pain (*25*).

A typical history of pain should include the quality, quantity, duration or frequency, onset, location, radiation to other sites, precipitating factors, aggravating and alleviating factors and associated symptoms. It is important to ascertain the time line of the pain and when the patient last was pain free, how it interferes with daily life, how it has been treated (including any over the counter or "folk" remedies) and what are the complications of the pain itself and what it is being treated with. It is also important to ascertain information about co-morbid conditions and any social, emotional or other things contributing (*48-50*).

Useful questions might include: What keeps you from doing what you want to do? Where is the pain? Does it interrupt your sleep? Is it predictable? How long does it last? If it goes away after you take medicines, how long does it take? (This helps determine if shorts acting analgesic doses are adequate.) How long does the relief last? (This helps determine if the dosing interval of the short acting drugs is appropriate or if the dose of long acting drugs is appropriate).

Pain is more than the physical symptom. Dame Cicely Saunders, coined the term "total pain" to define the pain experience (*51*). Pain is more than a simple biological construct since it is more than the biological damage that may occur in a nerve or a tissue or an organ. Pain is the stimulus and the person's response to it since it includes psychological, social, emotional, financial, spiritual and existential dimensions. It varies from person to person from the stoic, "keep it all inside" person to the emotional and expressive person (*50*).

Suffering is a broader term. It is probably one of the most common symptoms that accompany illness. It refers to loss of the personhood, a threat to the identity and future of the person, and it constitutes a treat to the self-image, a perceived lack of options for coping, a sense of personal loss, and a lack of a basis for hope (*52*). Suffering is most eloquently discussed in the work of Eric Cassell. Cassell states that: "the test of medicine should be its adequacy in the face of suffering...modern medicine fails that test. In fact, the central assumptions in which twentieth century medicine is founded provide no basis for an understanding of suffering. For pain, difficulty in breathing, or other difficulties of the body, superbly yes: for suffering, no"(*53*).

Michael Kearney, an Irish palliative care physician, has written extensively about this topic (*54,55*). He states that the "medical model" is excellent in attending to the physical aspects of the pain experience. Physical pain, which is damage or perceived damage to a cell or a tissue or an organ, can often be "controlled" through attention to careful assessment, to treating the underlying process, and by careful applications of interventions from "without", such as medications or procedures with careful assessment and adjustments to make sure they work. The other physical symptoms like anxiety, depression, social concerns, and other co-existent symptoms may also be approached by interventions from without (*54,55*).

Kearney further proposes that the medical model remains unable to deal with the nature of suffering including the threatened loss of integrity of the person. As a remedy, he suggests that we invoke a different model, given that most of our ideas in Western, "evidence based" medicine are inherited from the school of Hippocrates. Kearney informs us that at the time of Hippocrates, there were actually many schools of medicine. One was called the Aescalapian school or "temple medicine" (*55*). Aescalapius is the Greek God of healing and the staff of Aescalapius is the symbol of medicine. The Aesculapian School attended to people with illnesses that often couldn't be dealt with by other schools, including incurable diseases. Patients in this tradition would go to the temples, located in beautiful places and be attended by people called Therapeutes (from which the term therapeutic is derived). They would bathe, rest, meditate, eat well, and at some point in their "stay", go to sleep in the temple. At this time, there they would have some type of personal experience, e.g., a dream, and perhaps, they would leave the temple on the next day "healed". The Aescalapian tradition believed healing wasn't ours to give, it came from within, or from the earth (like the snake crawling up the staff) (*55*).

Kearney suggests we take a similar approach to suffering, since "our job is to create a safe space where healing can happen." We do this best by listening, "witnessing", and validating our patient's plight. He uses terms from psychotherapy. We must "contain and hold" the space, "stay with and support" the patient so their own healing powers from within can have a chance to work (*46*). There is a tremendous benefit to patients to feel "listened to". Moreover, the patients may be able to teach the physicians how to approach the care of suffering. For example, the African-American spiritual tradition of "witnessing" should be very helpful to the relief of suffering. Kearney also uses tools from psychotherapy like dream analysis to draw deep into the psyche to deal with suffering which he also calls "soul pain" (*46*). Much of what is seen in the Aescalapian tradition is also reflected in some aspects of alternative or complementary medicine, which seems to focus more on the "care of the self" with things like nutrition, meditation, aromatherapy, massage, and laying down of the hands (*55*).

It has been said that the most powerful tool in the clinician's armamentarium is the physician him or herself. Kearney teaches the therapeutic use of self and his writings are important tools in the approach to and relief of suffering (*55*). And perhaps it was Dame Cicely Saunders who said it best when she stated that: "The way we care can reach the most hidden places" and "the real presence of another person is a sense of security" (*56*).

One of the key aspects of communication with African American patients is to develop a sense of trust in the relationship. This may be more difficult to do across the often-deep chasms of different backgrounds and with the long history of overt and covert prejudice. Providers of different race should not be surprised if attempts to reach out are met with coolness or even with providers from the same race, there may be suspicion of bending to an institutional position. The best way to combat this is to develop a relationship over time in which words and actions are congruent. Patients come to believe the provider means what he says, understands and values their perspectives and advocates for them. According to Barrett and Heller, in times of crisis or need, people care most about being comforted by people whom they believe genuinely care (*53*).

Developing rapport happens best when there is mutual respect and trust. A model for building this respect comes from the work of Dr Ned Cassem (*57*). Cassem suggests that all illness is a threat to the personhood. He tells his residents who are seeing patients with heart attacks that: "On the same day your patient suffers a myocardial infarction, he or she also suffers an ego infarction and it takes the ego much longer to recover." As a remedy, Cassem suggests that we need to try to "rescue our patients from the anonymity that accompanies illness." We need to find out: "Who is this person? Who were they at the top of their game?" What defines him/her?" Thus rather than subordinating the patient's narrative of illness to the dominant biomedical narrative, we need to connect our stories with theirs.

Suffering in the context of cancer and pain can serve to create a disparity between which the person was who they are and if they have a future. According to Cassem's model, care and repair of the self can begin by asking three categories of questions: a) What are their accomplishments? (E.g. prowess in acting, music, awards won, sports, rank in military, status in the neighbourhood etc. He also suggests we ask about things that they are ambivalent about or even "naughty" stories. b) Who is really important to them? (e.g. who loves them? whom do they love? who are they closest to? what are their big stories? even who are their enemies). c) What are their favorite things? (Cassem's favourite topic)-(e.g. music, books or poems, newspapers, movies, stars, hobbies, sports teams, restaurants, food, and cars. Also, do they have any aversions? Any addictions? Somewhere in these questions may come a closeness, a respect, a "connection" and once the patient knows that we care, he or she is more likely to let us into "hidden" places and may indeed even let us know what might help them.

Stewart (*57*) has argued; "Without some agreement about the nature of what is wrong, it is difficult for a doctor and a patient to agree on a plan of management that is acceptable to both of them. It is not essential for the physician actually to believe that the nature of the problem is as the patient see it, but the doctor's explanation and recommended treatment must at least be consistent with the patient's point of view." These points of view differ not only across ethnic differences but also across socio-economic, gender, sex, age, religious and occupational differences. Perhaps the greatest challenge is to understand patients across cultural barriers. To do so, one must acquire a set of skills, knowledge and attitudes that enhance the understanding of and respect for a patient's values, beliefs and expectations, awareness of one's own assumptions and values (as well as those of the medical system) and the ability to adapt care to be congruent with the patient's expectations (*58*). Some dimensions of culture include; health and illness beliefs (What paradigm is used to explain illness/healing?): decision making style (Does decision making rest with the individual patient, the group/family or community peers?); healing traditions (What are the alternative/complementary approaches used?); locus of control (does the individual believe that they are responsible for their own destiny or is it predetermined (fate)?); status/hierarchy (Is the status of head of household conferred by age, gender or kinship?); privacy (Is privacy at the level of the individual or family?); communication (Is there a preferred mode of communication, e.g. written, verbal, sign? Is there a preferred language? (Is an interpreter needed?); socio-economic status (Is social status in the community conferred based on family, vocation, wealth or education?); and immigrant status (are there acculturation and generational issues at play?).

Kleinman (*59*) suggests asking the following questions to ascertain the patient's explanatory model for their illness:

1) What do you call the problem?
2) What do you think caused the problem?
3) Why do you think it started when it did?
4) What do you think the sickness does? How does it work?
5) How severe is the sickness? Will it have a long or a short course?
6) What kind of treatment do you think you should receive?
7) What are the most important results you hope to receive from this treatment?
8) What are the chief problems the sickness has caused?
9) What do you fear most about the sickness?

Since spirituality is such an important aspect of the African American experience, it is important for the provider to inquire about this and work to provide what is needed. The framework developed by Puchlaski (*58*) may be useful.

She suggests inquiry around 4 topics and uses the acronym FICA to organize this:

F. What is your faith or belief? Do you consider yourself spiritual or religious? What things do you believe in that give meaning to your life?

I: Is it important in your life? What influence does it have on how you take care of yourself? How have your beliefs influenced your behaviour during this illness? What role do your beliefs play in regaining your health?

C: Are you part of a spiritual or religious community? Is this of support to you and how? Is there a person or a group of people you really love or who are really important to you?

A: How would you like me, your healthcare provider, to address these issues in your healthcare?

In the situation at end-of-life, Byock (*60*) has suggested another set of 5 issues that may be helpful to address between loved ones to help provide closure. These can be framed as five statements: "I forgive you. Will you forgive me? Thank you. Good-bye and I love you." It is useful for the provider to try to get the patient and their loved ones to address these issues.

6. MODELS AND SUGGESTIONS

All is not bleak though. There exist jewels of programs that may serve as models to better address these issues. One outstanding example is the Harlem Palliative Care Network, a collaborative project between Memorial Sloan-Kettering Cancer Center, North General Hospital and the Visiting Nurse Service of New York is such a jewel worth emulating (*61*). The program involves patient and community focus groups so that the voice of the patient may be heard, practitioners who work in and know the community and its resources and have a better idea of the values of the community members. There is an ongoing evaluation and effective feedback so that the community sees that their voice is heard. There was institutional by in from the partners with the providers in the community having an equal voice, community support, staffing by community members. Leadership by prominent and respected African Americans and assurance of trust and continuity of care including bereavement services. Another program that provides expert care to a predominantly African American population is the Balm in Gilead Hospice in Birmingham, Alabama. This was featured extensively in the Moyers Series on PBS (*10*). Both of these programs emphasize that the solution to providing good care lies in the community in which the patients live.

At Dartmouth Medical School and the Norris Cotton Cancer Center, we have developed a Robert Wood Johnson supported project called Project Enable (Educate, Nurture, Advise Before Life Ends) (*62*). This project identifies patients and their families with stage IV breast, lung and colon cancers and provides for them educational sessions about end of life issues, and a contact person to help navigate the system. The project helps address end of life issues but earlier in the course of disease than usually occurs. The modules could be adapted to minority populations.

There is another successful innovation called the Advanced Illness Coordinated Care Program (the AICC) developed by Dr Daniel Tobin at the Lifecare Institute in Albany New York (*63*). In this program, patients with life threatening illnesses meet with providers trained to address end of life issues in six structured outpatient visits. This also might be adapted to provide specific interventions for African Americans.

One solution will be to establish more hospice/palliative care units in hospitals with large African American populations. Providers must be educated about palliative medicine, appropriate literature must be developed and community advisory boards established. The NCI must focus more research on this topic and charge cancer centers to find solutions to providing excellent care to vulnerable populations.

Excellent end of life care in African American women with breast cancer will require hearing their voices, their concerns, and their wishes. It will require developing trust and rapport, addressing spirituality issues, involving families and addressing the structural barriers that may inhibit things like access to pain medications. Perhaps the first and most important principle is that in order to provide excellent palliative care, there should be attention to excellence in care all along the disease course with equal access to high quality preventive, detection and treatment services. It is only by doing this that excellence in end of life care can be achieved.

REFERENCES

(1) Brawley OW. Some perspective on black-white cancer statistics. CA Cancer J Clin 2002;52:322-5.

(2) Anonymous. The Hastings Center Project: The Goals of Medicine: Setting New Priorities. Hastings Center Report, Special Supplement,; 1996 p 54

(3) Anonymous. A controlled trial to improve care for seriously ill hospitalized patients. The study to understand prognoses and preferences for outcomes and risks of treatments (SUPPORT). The SUPPORT Principal Investigators. Jama 1995;274:1591-8.

(4) Cassel CK, Foley KM. Principles and care of patients at the end of life: an emerging concensus among specialties of medicine. New York: Milbank Memorial Fund; 1999.

(5) Fields MJ, Cassel CK, editors. Institute of Medicine. Approaching Death: Improving Care at the End-of-Life. Washington DC: National Academy Press; 1999.

(6) Foley KM, Gelband H. Improving Palliative Care for Cancer. Washington, DC: Institute of Medicine (IOM). National Academy Press.; 2001.

(7) Foley KM, Gelband H, editors. President's Cancer Panel. Cancer Care Issues in the United States: Quality of Care, Quality of Life, 1997-98. Washington, DC: National Academy Press; 2001.

(8) Anonymous. MedPac 2000. Medicare Beneficiaries' Cost and Use of Care in the Last Year of Life. Washington, DC: MedPac; 2000.

(9) Anonymous. American Society of Clinical Oncology (ASCO). Cancer care during the last year of life. Journal of Clinical Oncology 1998;16:1986-96.

(10) Moyers W. "On Our Own Terms", on PBS, September 10-13, 2000, www.pbs.org/on our own terms.: PBS; 2000.

(11) Burt RA. Death is That Man Taking Names: Intersections of American Medicine, Law and Culture. Berkeley and Los Angeles: University of California Press; 2002.

(12) Anonymous. Task Force on Palliative Care. Precepts of Palliative Care. Jl of Palliative Med. 1998;12:109-12.

(13) Berry W. Health is membership.: Counterpoint Press (a member of Perseus Books). 1995.

(14) Foley KM, Gelband H, editors. Chapter 1: Background and Recommendations. Washington, DC: National Academy Press; 2001.

(15) Beresford L. The Hospice Handbook - A Complete Guide. Boston: Little Brown and Co.; 1993.

(16) Hewitt M, Simone JV, editors. Ensuring Quality Cancer Care. Washington, DC: National Academy Press; 1999.

(17) Christakis NA, Escarce JJ. Survival of Medicare patients after enrollment in hospice programs. N Engl J Med 1996;335:172-8.

(18) Caralis PV, Davis B, Wright K, Marcial E. The influence of ethnicity and race on attitudes toward advance directives, life-prolonging treatments, and euthanasia. J Clin Ethics 1993;4:155-65.

(19) McKinley ED, Garrett JM, Evans AT, Danis M. Differences in end-of-life decision making among black and white ambulatory cancer patients. J Gen Intern Med 1996;11:651-6.

(20) Payne R. Palliative Care for African Americans and other Vulnerable Populations: Access and Quality Issues. In Foley KM, Gelband H, editors. Institute of Medicine (IOM): Improving Palliative Care. Washington, DC: National Academy Press; 2001. p. 153-60.

(21) Crawley LM. Palliative care in African American communities. J Palliat Med 2002;5:775-9.

(22) Hogue CJR, Hargraves MA, K.S. C. Minority Health in America: Findings and Policy Implications from the Commonwealth Fund Minority Health Survey. Baltimore, MD: Johns Hopkins University Press; 2000.

(23) Anonymous. Institute of Medicine. Unequal Treatment: Confronting Racial and Ethnic Disparities in Health Care. Washington, DC: National Academy Press.; 2002.

(24) Anonymous. Institute of Medicine. The Unequal Burden of Cancer: An Assessment of NIH Research and Programs for Ethnic Minorities. Washington, DC; 1998.

(25) Todd KH. Pain assessment and ethnicity. Ann Emerg Med 1996;27:421-3.

(26) Cleeland CS, Gonin R, Hatfield AK, Edmonson JH, Blum RH, Stewart JA, et al. Pain and its treatment in outpatients with metastatic cancer. N Engl J Med 1994;330:592-6.

(27) Cleeland CS, Gonin R, Baez L, Loehrer P, Pandya KJ, Hatfield AK, et al. Pain and treatment of pain in minority patients with cancer. The Eastern Cooperative Oncology Group Minority Outpatient Pain Study Pain and its treatment in outpatients with metastatic cancer. Ann Intern Med 1997;127:813-6.

(28) McDonald DD. Gender and ethnic stereotyping and narcotic analgesic administration. Res Nurs Health 1994;17:45-9.

(29) Todd KH, Lee T, Hoffman JR. The effect of ethnicity on physician estimates of pain severity in patients with isolated extremity trauma. Jama 1994;271:925-8.

(30) Benabei R, Gambosi G, Lapane K. Management of pain in elderly patients with cancer. Journal of the American Medical Association. 1998;279:1877-82.

(31) Williams DR, Rucker TD. Understanding and addressing racial disparities in health care. Health Care Financ Rev 2000;21:75-90.
(32) Bolling J. Guinea across the water: the African American approach to death and dying. In Perry JK, Ryan AJ, editors. A Cross-Cultural Look at Death, Dying and Religion. Chicago, IL: Nelson-Hall Publishers; 1995. p. 145-9.
(33) DeSpelder L, Strickland A. The Last Dance: Encountering and Dying. Boston, MA: McGraw Hill Higher Education; 2002.
(34) Anderson KO, Richman SP, Hurley J, Palos G, Valero V, Mendoza TR, et al. Cancer pain management among underserved minority outpatients: perceived needs and barriers to optimal control. Cancer 2002;94:2295-304.
(35) Reed WL, Roberson NL, editors. Health and Medical Care of African Americans. Westport, CT: Auburn House; 1993.
(36) Cooper-Patrick L, Gallo JJ, Gonzales JJ, Vu HT, Powe NR, Nelson C, et al. Race, gender, and partnership in the patient-physician relationship. Jama 1999;282:583-9.
(37) Davis JA, Smith TW. General Social Survey 1972-1990. Chicago.: National Opinion Research Center; 1990.
(38) Bonhom V. Race, ethnicity and pain treatment: striving to understand the causes and solutions to the disparities in pain treatment. J Law Med Ethics 2001;2952-68.
(39) Anonymous. Minorities Continue to be Underrepresented in Medicine. Am Med News, Sept. 1998; 1998. p. 11.
(40) Like RC, Steiner RP, Rubel AJ. STFM Core Curriculum Guidelines. Recommended core curriculum guidelines on culturally sensitive and competent health care. Fam Med 1996;28:291-7.
(41) Bobo L, Kluegel JR, smith RA. Laissez-faire racism: the crystallization of a "kinder, gentler" anti-black ideology. In Tuck SA, Martin JK, editors. Racial Attitudes in the 1990s Continuity and Change. Westport, CT: Prager; 1997.
(42) Lavizzo-Mourey R, Clayton LA, Byrd WM, Johnson G, 3rd, Richardson D. The perceptions of African-American physicians concerning their treatment by managed care organizations. J Natl Med Assoc 1996;88:210-4.
(43) Morrison RS, Wallenstein S, Natale DK, Senzel RS, Huang LL. "We don't carry that"--failure of pharmacies in predominantly nonwhite neighborhoods to stock opioid analgesics. N Engl J Med 2000;342:1023-6.
(44) Bierman AS, Lurie N, Collins KS, Eisenberg JM. Addressing racial and ethnic barriers to effective health care: the need for better data. Health Aff (Millwood) 2002;21:91-102.
(45) Galanti G A. Caring for patients from different cultures: case studies from American hospitals. Second Edition ed. Philadelphia: University of Philadelphia Press; 1997.
(46) Dedier J, Penson R, Williams W, Lynch T, Jr. Race, ethnicity, and the patient-caregiver relationship. Oncologist 1999;4:325-31.
(47) Loeser J. Medical evaluation of the patient with pain. In J.D. L, Butler SH, Chapman CR, Turk DC, editors. Bonica's management of pain. Philadelphia: Lippincott Williams & Wilkins; 2001. p. 267-78.
(48) Luckman J. Transcultural communication in health care. Albany, N.Y.: Delmar Thomson Learning; 2000.
(49) Cohen-Cole SA. The Medical Interview:The Three Function Approach. St. Louis: Mosby Year Book; 1991.
(50) Abrams BM. History taking in the patient in pain. In Raj PP, editor. Practical Management of Pain. St. Louis, MA: Mosby; 2000. p. 333-8.
(51) Sauders C. The Management of Terminal Disease. London, England: Edward Arnold.; 1978.

(52) Cassel EJ. The nature of suffering and the goals of medicine. N Engl J Med 1982;306:639-45.
(53) Chapman CR, Gavrin J. Suffering: the contributions of persistent pain. Lancet 1999;353:2233-7.
(54) Kearny M. Mortally Wounded. Stories of Soul Pain, Death and Healing. Dublin, Ireland: Marino Books.; 1996.
(55) Kearny M. A Place of Healing. New York, N.Y.: Oxford Press; 2000.
(56) Holland JC, Cassem EH, Mount B, Simmons P. Dealing with loss, death, and grief in clinical practice: spiritual, meaning in oncology. Alexandria, VA: American Society of Clinical Oncology Educational Book.; 2002. p. 198-205.
(57) Stewart M, Brown JB, Weston WW, McWilliam CL, Freeman TR. Patient Centered Medicine. Transforming the Clinical Method. Thousand Oaks, CA: Sage Publications.; 1995.
(58) Pulchalski CM, Romer AL. Taking a spiritual history allows Clinicians to Understand Patients More Fully. J Pall Med 2000;3:129-37.
(59) Kleinman A. Patients and Healers in the Context of Culture. Berkeley, CA: University of California Press.; 1980.
(60) Byock I. Peace and Possibilities at the End of Life: Riverhead Books.; 1998.
(61) Payne R, Payne TR. The Harlem Palliative Care Network. J Palliat Med 2002;5:781-92.
(62) Whedon M. End-of-Life Care Across Settings: Hospital Care. In Ferrell B, Coyle N, editors. Oxford Textbook of Nursing. Oxford: Oxford University Press; 2001. p. 548-608.
(63) Tobin D. Peaceful Dying. The Step-by-Step Guide to Preserving Your Dignity, Your Choice and Your Inner Peace at the End of Life. Cambridge, MA: Perseus Publishing; 1999.

Chapter 11

OUTCOME OF AFRICAN AMERICAN WOMEN WITH BREAST CANCER IN COOPERATIVE GROUP CLINICAL TRIALS

[1]Joseph M Unger, MS, [1]Stephanie Green, PhD and [2]Kathy S. Albain, MD
[1]Southwest Oncology Group Statistical Center, Seattle, WA, [2]Loyola University Chicago Medical Center, Cardinal Bernadin Cancer Center, Maywood, IL, USA

1. INTRODUCTION

African American women with breast cancer have markedly higher population mortality in the United States compared to either whites or all races combined (1). The U.S. age-adjusted death rate from invasive breast cancer was approximately 32% higher in African American women than in white women according to data from the Surveillance and Epidemiology and End Results program (2). The disparity in mortality rates from breast cancer has been growing since 1980, when mortality rates for both groups were similar. In contrast, the incidence of breast cancer is somewhat higher in whites than in African Americans.

Reports from single institution clinical trials and/or population based databases show mixed results regarding whether race is an independent adverse predictor of survival for African American women with breast cancer. Where there are differences, they are typically ascribed to disparities in baseline disease characteristics such as stage at presentation, socio-economic status, and access to care (3-5). Accounting for stage is routinely performed in population based studies; adjusting for socio-economic status (often as a surrogate for access to care) is typically more difficult, due to the hesitation among patients to report (and therefore researchers to request) such factors as income and education.

C. Williams et al. (eds.), Breast Cancer in Women of African Descent, 267-295.

A recent review by Cross et al. grouped the historical literature pertaining to studies of the effect of race on outcome by whether socio-economic status was included in the analysis as an adjustment variable (6). Among the 13 studies which included socio-economic status as an adjustment variable, three studies (23%) found race to be a significant predictor of survival. Among the 18 studies without socio-economic status included, six (33%) found that race was a significant predictor of survival. Interestingly, if the studies that compared African Americans to whites are split in half according to sample size, seven of the 15 studies (47%) with large sample size (N>1500) showed significant results, whereas only 2 of the 16 studies (13%) with small sample size (N<=1500) showed significant results, suggesting the limitations of underpowered analyses. Cross concluded that between race and socio-economic status, socio-economic status is the more important predictor of outcome in breast cancer. In contrast, Newman et al, in a meta-analysis of 14 studies comprising 10,001 African American patients and 42,473 white patients, found that African American race was an independent predictor of adverse survival even after accounting for socio-economic status (7). African Americans had a 22% increased risk of mortality compared to whites (HR=1.22, 95% CI: 1.13-1.30). Newman suggested that biological differences between African Americans and whites, whether directly or in association with other factors, may play a role in breast cancer outcome.

Overlooked in such a debate is a consideration of the data itself. Twenty-four of the 31 studies reviewed by Cross et al., and all of the 14 studies included by Newman et al., were retrospective studies based on hospital or population registries. While analyses based on registry data are beneficial for answering certain types of questions, they are less useful for comparing outcome between different groups of patients due to the heterogeneity between patient groups with respect to stage and type of treatment received for cancer. Retrospective analyses also suffer from issues of missing data and ascertainment bias, and multiple unknown but non-random factors will impact survival of the different cohorts in different ways.

In the controlled setting of a phase III clinical trial, issues of stage and treatment heterogeneity should not apply. Since patients are uniformly staged and treated in a clinical trial, patients of all racial groups begin treatment with similar histology and stage. In addition, patients entered on clinical trials will be similar with respect to other possible confounders such as laboratory values, prior treatment, and absence or presence of co-morbid conditions. The randomisation procedure in a phase III clinical trial will, on average, distribute the effects of unknown confounders roughly equally across different treatment groups, although not necessarily across other factors such as racial groups. For cohorts not constructed through randomisation (e.g., defined by demographics or a single arm phase II trial), multiple unknown but non-random factors will impact survival differentially by cohort just as in population registries. However, in clinical trials this impact will be less due to the strict eligibility criteria, which refine the patient samples. Finally, data from patients enrolled on clinical trials are obtained prospectively, minimizing data collection bias and the (potentially large) problem of missing data inherent in retrospective studies.

African Americans comprise approximately 10% of patients enrolled to breast cancer clinical trials, a proportion which has been shown to be representative of the proportion of women in the U.S. cancer population who are African American (8). Thus a large sample size is needed when the intent is to analyse potential survival differences by race. The small fraction of African Americans accrued to clinical trials precludes meaningful reports on African American outcomes in the phase II setting. Therefore, this chapter focuses on analyses from randomised phase III trials. In addition, while analyses using clinical trials data benefit in general from the advantages mentioned above, the cooperative clinical trial groups of the National Cancer Institute (NCI), drawing on institutions from around the country, have the distinct advantage of being able to accrue large numbers of patients to trials which are adequately powered to detect treatment differences, although generally not race by treatment interactions.

African American women are generally younger in age at diagnosis and present with later stage disease (9-14). African American women are also at higher risk for developing oestrogen receptor negative disease (15-17). Racial discrepancies in income and education are also well documented (18). On the assumption differences in outcome by race can be completely accounted for by adjusting for differences in baseline disease characteristics or socio-economic status, one would expect no differences by race in the clinical trial setting. Yet in some instances African American women had a poorer outcome even in the context of a single large, randomised clinical trial (19-21).

The reporting of the effect of race on survival in the randomised trial literature has been inconsistent. This is understandable, since the main reports for large phase III studies typically follow strict reporting guidelines per the original design specifications (22). In addition, single-phase III breast cancer clinical trials are generally designed to detect moderate to large treatment differences, and most often have too few patients to assess the effect of race on survival or the interaction of race and treatment. Therefore the largest of the cooperative groups studying breast cancer, the National Surgical Adjuvant Breast and Bowel Project (NSABP) and the Southwest Oncology Group (SWOG), along with others, have each utilized their historical databases to combine multiple phase III trials of similar histology and stage with the objective of analysing potential outcome differences between African Americans and other patients. In this chapter we will review the reported literature of the Cancer and Leukemia Group B (CALGB), the Eastern Cooperative Oncology Group (ECOG), NSABP, and SWOG single large randomised phase III trials in breast cancer for reports on analyses of outcome by race and will compare the exploratory combined database analyses by SWOG and NSABP.

2. REPORTS FROM MAJOR PHASE III COOPERATIVE GROUP CLINICAL TRIALS

A review of selected major phase III randomised clinical trials in the past three decades performed by CALGB, ECOG, NSABP, and SWOG are shown in Tables 1-4, respectively, and are discussed below.

2.1 Selected Phase III Breast Cancer Trials from Cancer and Leukemia Group B

Table 1 depicts seven historical phase III trials for breast cancer from CALGB, (21), (23-28) four of which were performed in metastatic disease (21), (24-26). The analysis of outcome by race was performed in three of the studies, twice in the main trial report (21), (25) and once in a follow-up analysis (29). These three trials are now reviewed in detail.

2.1.1 CALGB 8541

CALGB 8541 tested different dose levels and dose intensities of the combination of adjuvant Cyclophosphamide, Doxorubicin, and Fluorouracil in a 3-arm randomised trial of stage II node positive breast cancer (23). A follow-up report showed that African Americans had shorter overall survival (HR=1.35, p=.04) and slightly shorter disease-free survival (HR=1.24, p=.09) in univariate analysis. African Americans were also more likely to be oestrogen receptor negative, have larger tumours, and be younger. In a multivariate Cox regression model adjusting for these and other factors, African American race was no longer a significant predictor of either overall or disease-free survival (29).

Table 1. The Reporting of Outcome by Race in Cancer and Leukemia Group B Phase III Breast Cancer Trials.

CALGB Study No. (Accrual Period)	Disease Type	Outcome by Race Reported	No. Pts in Race Analysis (% AA)	Outcome Diff. by Race?	Notes
8541[23] (1985-1991)	Node (+)	Yes, follow-up analysis [29]	1,572 (12%)	Univariate only	AAs had worse DFS and OS in univariate analysis; no significant differences in multivariate analysis.
8642 [24] (1986-1993)	Metastatic	No	NA	NA	
8741 [25] (1987-1991)	Metastatic	Yes, main trial report	366 (11%)*	Univariate only	Non-whites had worse PFS (p=.03) in univariate analysis; no significant PFS differences in multivariate analysis. No differences by race in OS.
9140 [26] (1991-1995)	Metastatic	No	NA	NA	
9342 [21] (1994-1997)	Metastatic	Yes, main trial report	474 (22%)	Yes	AAs had worse OS (HR=1.44, 1.13-1.84) in multivariate analysis but similar TTP and response
9344 [27] (1994-1999)	Node (+)	No	NA	NA	
9741 [28] (1997-1999)	Node (+)	No	NA	NA	

AA = African American; DFS = disease-free survival; OS = overall survival; TTP = time to progression.
* Race coded white vs. other, percent indicates percent of non-white patients.

2.1.2 CALGB 8741

CALGB 8741 randomised 366 patients with metastatic breast cancer to three different doses of Megestrol Acetate (160 mg/day vs. 800 mg/day vs. 1,600 mg/day) (25). Race was coded white vs. other and 11% of patients were non-white. The non-white patients had worse progression-free survival (p=.03) in the univariate setting, but in the multivariate setting the difference in PFS between whites and non-whites was not significant. There were no differences in overall survival between whites and non-whites.

2.1.3 CALGB 9342

CALGB 9342 studied three different doses of Paclitaxel (175 mg/m2, 210 mg/m2, and 250 mg/m2 by 3 hour infusion every 3 weeks) for the treatment of metastatic breast cancer. The percentage of African Americans participating in the trial was high (22%). There were no statistically significant differences in either response or time to treatment failure between African Americans vs. all other races combined. However, there was a difference in overall survival, with African American women surviving an average of 10.1 months and non-African-American women 13.0 months (HR 1.44, 95% CI: 1.13-1.84) (21).

2.2 Selected Phase III Breast Cancer Trials from the Eastern Cooperative Oncology Group

Table 2 shows five historical phase III trials from ECOG, all in node-positive disease (30-34). An analysis of outcome by race was performed in two of the studies as follows (33-34).

2.2.1 ECOG 5181

ECOG 5181 was a study of CMFPT (Cyclophosphamide, Methotrexate, Fluorouracil, Prednisone, and Tamoxifen) versus CMFPT plus Halotestin alternated with Adriamycin, Thiotepa, Halotestin, and Tamoxifen in premenopausal, node-positive breast cancer (33). Patients were subsequently randomised to Tamoxifen vs. observation after 12 cycles of induction therapy and again after 5 years. Race was coded white vs. non-white. In the analysis of 533 patients, 10% of whom were non-white, race did not predict either time to relapse or overall survival.

Table 2. The Reporting of Outcome by Race in Eastern Cooperative Oncology Group Phase III Breast Cancer Trials.

ECOG Study No. (Accrual Period)	Disease Type	Outcome by Race Reported?	No. Pts in Race Analysis (% AA)	Outcome Diff. by Race?
2190 [30] (1991-1998)	Node (+) (>=10 nodes)	No	NA	NA
3189 [31] (1989-1993)	Node (+), ER (-)	No	NA	NA
4181 [32] (1982-?)	Postmenopausal, node (+)	No	NA	NA
5181 [33] (1982-1987)	Premenopausal, node (+)	Yes, main trial report	533 (10%*)	No
6177 [34] (1978-1981)	Postmenopausal, node (+)	Yes, main trial report	222 (7%*)	No

AA = African American; ER = estrogen (oestrogen) receptor.

* Race coded white vs. other; percent indicates percent of non-white patients.

2.2.2 ECOG 6177

ECOG 6177 compared CMFP (Cyclophosphamide, Methotrexate, Fluorouracil, and Prednisone) to CMFP plus Tamoxifen to observation in patients with postmenopausal, node-positive disease (34). Race was again coded white vs. non-white. In the analysis of 222 analysable patients, 7% of whom were non-white, race did not predict time to relapse or overall survival.

2.3 Selected Phase III Breast Cancer Trials from the National Surgical Adjuvant Breast and Bowel Project

Table 3 shows 10 historical phase III trials from NSABP, the earliest two of which were surgical studies (35-36) in patients with primary operable breast cancer, and the remainder of which studied adjuvant chemotherapy in patients with either node-positive or node-negative breast cancer (37-44). Analysis of outcome by race was performed in the following five studies (35-39).

2.3.1 NSABP B-04

NSABP trial B-04 was a randomised study comparing radical mastectomy and total mastectomy with or without radiation. (35). An analysis of 191 black and 442 white patients from this trial who had undergone radical mastectomy was performed (45). In node-positive breast cancer patients, 5-year survival for African Americans was 45 % versus 61% for whites. Although African American patients were younger at diagnosis and had larger tumours, a survival deficit remained after adjusting for the prognostic factors. In node-negative breast cancer patients, survival was similar for the two groups (45).

Table 3. The Reporting of Outcome by Race in National Surgical Adjuvant Breast and Bowel Project (NSABP) Phase III Breast Cancer Trials

NSABP Study No. (Accrual Period)	Disease Type	Outcome by Race Reported	No. Pts in Race Analysis (% AA)	Outcome Diff. by Race?	Notes
B-04 [35] (1971-1974)	Primary operable	Yes, follow-up analysis [45]	633† (30%)	Node (+): Yes Node (-): No	Node (+): 5-yr OS 45% for AA and 61% for whites Node (-): 5-yr OS 85% for both groups
B-06 [36] (1976-1978)	Primary operable	Yes, follow-up analysis [46]	950 (7%)	Yes	5-year survival 74% for AA and 89% for whites (p=.007 in multivariate model)
B-09 [37] (1977-1980)	Node (+)	Yes, follow-up analysis [49]	1,042 (14%)	Univariate Only	AAs had worse OS in univariate analysis; no significant difference in multivariate analysis.
B-13 [38] (1981-1988)	Node (-), ER (-)	Yes, follow-up analysis [50]	1,024 (11%)	No	
B-14 [39] (1982-1988)	Node (-), ER (+)	Yes, follow-up analysis [50]	3,912 (5%)	No	
B-15 [40] (1984-1988)	Node (+), "TAM non-responsive"	No	NA	NA	
B-16 [41] (1984-1989)	Node (+), "TAM responsive"	No	NA	NA	
B-19 [42] (1988-1990)	Node (-), ER (-)	No	NA	NA	
B-20 [43] (1988-1993)	Node (-), ER (+)	No	NA	NA	
B-22 [44] (1989-1991)	Node (+)	No	NA	NA	

AA = African American; ER = estrogen (oestrogen) receptor; TAM = Tamoxifen; OS = overall survival.
† Radical mastectomy arm only.

2.3.2 NSABP B-06

NSABP trial B-06 established breast-conserving lumpectomy as equivalent therapy to mastectomy for Stage I and II breast cancer patients with <=4 cm tumours (36). In contrast to the results for patients with node-negative breast cancer from B-04, for study B-06 subsequent analyses focusing on prognostic pathologic and clinical features showed that African Americans had worse survival than whites in multivariate modelling (46-48). In particular, among the 950 patients with node-negative stage I breast cancer, five year survival was 74% in African Americans and 89% in whites (p=.007 in a multivariate Cox regression model) (46).

2.3.3 NSABP B-09

NSABP trial B-09 compared Melphalan plus 5-FU (PF) to Melphalan plus 5-FU plus Tamoxifen (PFT) in women with positive nodes following a radical mastectomy (37) A follow-up analysis of 896 white patients and 146 African American patients was performed, revealing that African Americans patients had worse unadjusted 5-year survival than white patients in both arms of the study (PF: 52% African American and 68% white; PFT: 57% African American and 65% white) (49). After adjusting for prognostic factors, however, survival between the two groups was not significantly different.

2.3.4 NSABP B-13 and NSABP B-14

As a follow-up to the conflicting results in node-negative breast cancer from NSABP B-04 and B-06, protocols B-13 and B-14, both conducted in node-negative breast cancer, were subsequently analysed to assess the predictive value of race on survival.

NSABP B-13, a study in oestrogen-receptor negative breast cancer, compared surgery alone to surgery followed by Methotrexate and 5-FU, with an additional non-randomised group of patients registered to receive Methotrexate and 5-FU (38). One hundred eight of the 1024 patients analysed in this group were African American (11%) (50). Five-year disease-free survival estimates were similar between the two groups (71% for African Americans and 74% for whites). African Americans were more likely to have negative PgR receptors (PgR < 10 fmol/mg). After adjusting for these and other prognostic factors, there was no difference in disease-free survival (HR=0.98, 95% CI, 0.70-1.37). Five-year overall survival was also similar (83% for African Americans and 85% for whites; HR = 1.02, 95% CI, 0.66 – 1.56).

NSABP B-14, a study in node-negative/oestrogen-receptor positive breast cancer, compared surgery plus placebo to surgery plus 5 years of Tamoxifen, with a follow-up randomisation of placebo vs. an additional 5 years of Tamoxifen (39). Two hundred twenty-four of the 3,912 patients analysed in this group were African American (5%). As with the oestrogen-receptor negative trial B-13, five-year disease-free survival estimates were similar between African Americans (81%) and whites (80%). African Americans were more likely to be younger and to have larger tumours. After adjusting for these and other prognostic factors, there was no difference in disease-free survival (HR=0.97, 95% CI, 0.89 – 1.43). Five-year overall survival was also similar (93% for African Americans and 92% for whites; HR = 1.14, 95% CI, 0.84 – 1.54) (50).

2.4 Selected Phase III Breast Cancer Trials from the Southwest Oncology Group.

Table 4 shows 5 historical phase III trials from SWOG, most in node-positive disease (20, 51-56) and one in node-negative disease (57). One of the studies (7827) had three main reports, one for the oestrogen receptor negative breast cancer patients (52) and separate reports for the oestrogen receptor positive breast cancer patients by menopausal status (20, 53). Outcome by race is noted for all 5 studies.

2.4.1 SWOG 7436

SWOG 7436 compared continuous CMFVP (Cyclophosphamide, Methotrexate, 5-FU, Vincritine, and Prednisone) to L-PAM (L-Phenylalanine Mustard) as adjuvant therapy for women with operable breast cancer and positive nodes (51). Race (coded white vs. other) was not found to predict survival in subsequent analyses (58-59).

2.4.2 SWOG 7827

SWOG 7827 tested adjuvant combinations in patients with node-positive disease. The oestrogen receptor negative breast cancer patients of SWOG 7827 were randomised to one vs. two years of CMFVP. Race was coded white vs. non-white. In the 411 patients analysed, 15% were non-white. Neither disease-free survival nor overall survival differed by race (52). The oestrogen receptor positive patients of SWOG 7827 were analysed separately by menopausal status. Premenopausal patients were treated with adjuvant CMFVP vs. CMFVP plus ovariectomy (53) Of the 288 patients analysed, 7% were African American. White patients were found to have

Table 4. The Reporting of Outcome by Race in Southwest Oncology Group Phase III Breast Cancer Trials.

SWOG Study No. (Accrual Period)	Disease Type	Outcome by Race Reported?	No. Pts in Race Analysis (% AA)	Outcome Diff. By Race?	Notes
7436 [51] (1975-1978)	Node (+)	Yes, follow-up analysis [59]	364 (8%)	No	
7827 [52] (1979-1984)	Node (+), ER (-)	Yes, main trial report	411 (15%)*	No	
7827 [53] (1979-1989)	Premenopausal, Node (+), ER (+)	Yes, main trial report	288 (7%)	Yes	White patients had longer OS (p<.01)
7827 [20] (1979-1989)	Postmenopausal, Node (+), ER (+)	Yes, main trial report	898 (9%)	Yes	Difference in DFS by race (no direction indicated, p=.01); OS similar
8313 [54] (1984-1990)	Node (+), HR (-)	Yes, main trial report	531 (11%)	No	
8814 [55] (1989-1995)	Postmenopausal, node (+), ER (+)	Yes, follow-up analysis [56]	1,558 (9%)	Yes	AAs had worse DFS and OS
8897 [57] (1989-1993)	Node (-)	Yes	2,690 (11%)	No	

AA = African American; ER = estrogen (oestrogen) receptor; HR = hormone receptor; DFS = disease-free survival; OS = overall survival.

* Race coded white vs. other; percent indicates percent of non-white patients.

longer overall survival (p<.01) in a multivariate Cox regression model. Postmenopausal patients were treated with adjuvant CMFVP versus Tamoxifen vs. concurrent CMFVP and Tamoxifen (20). Nine percent of the 898 patients were African American. Race predicted disease-free survival (p=.01) but not overall survival in a Cox regression model. The multivariate models included age, number of positive nodes, type of primary surgery, tumour size, and oestrogen-receptor status.

2.4.3 SWOG 8313

In SWOG 8313, a study of short course FAC-M vs. CMFVP (1 year) in node positive, hormone receptor negative breast cancer, the important baseline disease characteristics were tested by Cox regression modelling. Race did not predict survival in this case (54).

2.4.4 SWOG 8814

In SWOG 8814, postmenopausal patients with node-positive, oestrogen receptor positive breast cancer were randomised to Tamoxifen alone vs. CAF (Cyclophosphamide, Adriamycin, Fluorouracil) vs. CAF plus Tamoxifen (55). Nine percent of the 1,558 patients were African American. African Americans had both worse disease-free survival and worse overall survival in multivariate analysis (56).

2.4.5 SWOG 8897

SWOG 8897 randomised patients with node-negative breast cancer to CMF or CAF +/- Tamoxifen for 5 years. Of the 2,690 patients analysed, 11% were African American. There were no survival differences by race (57).

3. PERSPECTIVES FROM SINGLE PHASE III TRIALS

In summary, the effect of race on survival was reported for 15 out of the 27 phase III studies listed in Tables 1-4 (with the 3 reports for SWOG 7827 counted as a single study), either in the main trial report or in a subsequent ancillary analysis. Among these 15 studies that reported an analysis of race on outcome, 5 indicated a significant effect of race on survival after adjusting for possible confounders in a multivariate model, with white patients generally having better outcome than African Americans or non-white patients. The effect of race on outcome was significant in the univariate setting in three other studies, but disappeared with covariate adjustment for possible confounders. Thus even in analyses of single large cooperative group clinical trials, conclusions regarding outcome by race are disputed. Despite the advantages of the cooperative group clinical trial setting, the roughly 10% of African American patients enrolled to breast cancer clinical trials, while representative of the U.S. population, are insufficient for a comparison to whites or all other patients with respect to outcome. The statistical power to detect outcome differences by race is not adequate in these single-phase III trials, especially in the case where only a subgroup of the patients is analysed (60). Phase III studies are most often powered to test the single protocol-specified primary endpoint. Any additional tests, such as comparing outcome within a subset of interest such as race, will be underpowered and thus tend towards higher false negative rates (that is, unable to detect differences which do exist). Furthermore, analyses within subsets, such as treatment comparisons within a race category or a race comparison within a treatment category, are particularly problematic. These types of analyses suffer the undesirable fate of both a high false negative rate due to low power and a high false positive rate (that is, falsely identifying differences which do not truly exist) due to multiple comparisons (unless a multiple comparisons adjustment is made to the alpha level) (60).

These factors explain in large part why the results from the single trials (Tables 1-4) are mixed, with about half of the 15 studies showing either a univariate (n=3) or multivariate (n=5) effect of race on outcome, and the remainder (n=7) showing no effect. Combined datasets of historical phase III trials may be used to explore this issue further.

4. REPORTS FROM DATABASE ANALYSES COMBINING HISTORICAL PHASE III TRIALS

The approach whereby multiple similar phase III trials within a given cooperative group are combined alleviates problems of insufficient sample size within a single randomised clinical trial while maintaining the design and data advantages of the clinical trial setting. Although not necessarily homogeneous by treatment, adjustments for treatment and other important factors can be incorporated in the modelling procedure. With regard to the issue of outcome by race in breast cancer, combined database analyses have only rarely been undertaken, due to the requirement of a large historical database of well-designed phase III trials of similar histology and stage.

Recently, however, two such analyses, one by the NSABP and one by SWOG, have been performed.

4.1 NSABP

The NSABP combined historic phase III trials of similar histology and stage, grouping patients by lymph node status (61). Comparisons were between African Americans and whites/Caucasians in Cox regression models adjusting for age, tumour size, and treatment. Results for patients coded as unknown or other race were not presented. Baseline characteristics in both the lymph node negative and lymph node positive breast cancer were generally similar, with African Americans slightly younger at diagnosis (by 1 year in the ER-negative subset and 3 years in the ER-positive subset) and with somewhat larger tumours (about 0.5 cm on average).

4.1.1 Lymph Node Negative Disease

The analysis of patients with operable breast cancer and lymph node negative disease included four studies, NSABP B-13, B-14, B-19, and B-20.(38-39, 42-43) Two of the studies, NSABP B-13 and B-14, are described above. NSABP B-19 compared sequential Methotrexate and 5-FU to Cyclophosphamide, Methotrexate, and 5-FU in patients with ER-negative tumours (42). NSABP B-20 compared these same two regimens plus Tamoxifen with Tamoxifen alone in patients with ER-positive tumours (43).

A total of 8,581 patients were analysed, with 543 African Americans (6.3%), 7582 whites (88.4%), and 456 patients of unknown other race (5.3%). The main outcome measures are shown in Table 5. Patients with lymph node negative breast cancer were analysed separately by oestrogen receptor status, as well as all together. Disease-free survival did not differ by race in either the ER-negative or ER-positive patients. African Americans showed a marginally significant increased risk of death in the ER-negative subset (HR=1.30, 95% CI: 0.98 – 1.73), but not in the ER-positive subset (HR=1.17, 95% CI: 0.92 – 1.49). African Americans had a significantly worse risk of death than whites of 21% (HR=1.21; 95% CI, 1.01 – 1.46) when all patients were combined.

Table 5. NSABP and SWOG Multivariate Analyses of Outcome by Race in Combined Phase III Trials.

Analysis	Eligibility (Accrual Period)	N	No. AAs (%)	Adjustment variables	Group	Hazard Ratio (95% Confidence Interval)# Disease-free Survival	Overall Survival	Cause-Specific Survival
NSABP [61]	Node(-)* (1981-1993)	8,125	543 (6.7%)	treatment, age, tumour size	ER -	1.07 (0.84-1.35)	1.30 (0.98-1.73)	
					ER +	1.05 (0.87-1.27)	1.17 (0.92-1.49)	
					All		1.21 (1.01-1.46)	1.08 (0.88-1.33)
	Node(+)** (1984-1991)	5,534	548 (9.9%)	treatment, age, tumour size	ER -	1.02 (0.88-1.18)	1.06 (0.90-1.25)	
					ER +	1.09 (0.90-1.32)	1.46 (1.18-1.80)	
					All		1.18 (1.03-1.34)	1.09 (0.96-1.25)
SWOG [62-63]	Premenopausal† (1975-1995)	2,360	249 (10.6%)	study, age, tumour size, # LN+, receptor status	All	1.39 (1.12-1.73) p=.003	1.41 (1.10-1.82) p=.007	1.42 (1.09-1.85) p=.01
	Postmenopausal‡ (1975-1995)	4,316	413 (9.6%)	study, age, tumour size, #LN+, receptor status	All	1.45 (1.27-1.66) p<.0001	1.49 (1.28-1.73) p<.0001	1.39 (1.17-1.66) p=.0002

NSABP=National Surgical Adjuvant Breast and Bowel Project; SWOG = Southwest Oncology Group; AA = African American.
* NSABP node negative trials were B-13, B-14, B-19, and B-20. Coded AA vs. white/Caucasian.
** NSABP node positive trials were B-15, B-16, B22. Coded AA vs. white/Caucasian.
† SWOG pre-menopausal patients were from studies S7436, S7827, S8313, and S8897. Coded AA vs. all other patients combined.
‡ SWOG post-menopausal patients were from studies S7436, S7827, S8313, S8814, and S8897.
# Hazard Ratios represent risk of failure of African Americans to Caucasians (NSABP) or African Americans to all other patients combined (SWOG).
£ For cause-specific mortality, patients are defined to have experienced an event if death followed documented cancer relapse. Deaths without prior relapse were censored. Thus patients were defined as "most likely" to have died from breast cancer.

4.1.2 Lymph Node Positive Disease

The analysis of patients with operable breast cancer and lymph node positive disease included three studies, NSABP B-15, B-16, and B-22 (40-41, 44). NSABP B-15 patients were <50 years old (or 50-59 if tumours were PgR-negative). The study compared CMF and AC for four courses with AC followed by CMF.(40) NSABP B-16 included patients 50-59 years old with PgR-positive tumours and all patients 60-70 years old. The study compared Tamoxifen alone to Tamoxifen and AC (41). NSABP B-22 included patients of any age; patients were treated with either four courses of standard AC; intensified therapy with the total Cyclophosphamide dose delivered over just two courses; or intensified and increased therapy with a double dose of Cyclophosphamide (44).

A total of 5,851 patients were analysed, with 548 African Americans (9.4%), 4986 whites (85.2%), and 317 unknown or other race (5.4%). As in the group with lymph node negative breast cancer, patients were analysed separately by oestrogen receptor status, as well as all together. Disease-free survival did not differ by race. However, race had a significant effect on overall survival in the ER-positive subset, with a 46% increased risk of death for African Americans (HR=1.46; 95% CI, 1.18 – 1.80). In the ER-negative subset, mortality risk did not differ significantly between African Americans and white patients (HR=1.06, 95% CI, 0.90 – 1.25). With all patients combined, African Americans had significantly worse overall survival (HR=1.18; 95% CI, 1.03 – 1.34).

4.2 SWOG

The Southwest Oncology Group analysed historical large phase III trials conducted in multiple malignancies (not just breast cancer) (62). Ten major diseases were analysed: after adjustment for the major clinical prognostic factors, African Americans were found to have worse survival than all other races combined in both pre- and postmenopausal breast cancer, as well as in advanced ovarian cancer and advanced prostate cancer. No survival differences were found for the 6 other diseases (limited small cell lung cancer, advanced non-small cell lung cancer, multiple myeloma, adjuvant colon cancer, advanced non-Hodgkin's lymphoma, and acute myeloid leukaemia). The authors of the SWOG analysis observed that African Americans had worse survival in the two major hormone-related cancers (breast and prostate). They hypothesized that an unknown biologic mechanism, possibly hormone-related, may negatively impact survival in African Americans.

The analyses of the breast cancer studies were expanded (63). Five studies in adjuvant breast cancer were combined, all indicated in Table 4 (20, 51-57). A prior analysis by SWOG using recursive partitioning to evaluate the impact prognostic factors on outcome included patients from the CMFVP arms of two of these studies, 7436 and 7827, found no impact of race on outcome (64). However, this prior analysis represented only a small fraction of the patients in the analysis described here (12%).

Patients were divided by menopausal status for the recent larger analysis. Primary comparisons were between African Americans and all other patients combined, with a subsequent analysis comparing African Americans to whites only. A total of 2,360 patients with premenopausal breast cancer were analysed, of which 10.6% were African American; 4,316 postmenopausal patients were analysed, of which 9.6% were African American. The Cox regression model adjusted for study, age, tumour size, number of positive nodes, treatment, and receptor status. African Americans had significantly worse overall survival compared to all other patients, with a 41% increased risk of death among premenopausal patients (HR=1.41, 95% CI, 1.10 – 1.82, p=.007) and a 49% increased risk of death among postmenopausal patients (HR=1.49, 95% CI, 1.28 – 1.73, p<.0001). African Americans also had a significantly worse disease-free survival (63). African Americans had a 39% increased risk of relapse among premenopausal patients (HR=1.39, 95% CI, 1.12 – 1.73, p=.003) and a 45% increased risk of relapse among postmenopausal patients (HR=1.45, 95% CI, 1.27 – 1.66, p<.0001).

The SWOG analysis went one step further in an attempt to understand these apparent differences by adding the socio-economic factors income and education to the Cox multivariate regression models. Investigators matched patient zip-code to U.S. census data, assigning to each patient the median income and education level from the zip-code area within which they resided at the time of trial enrolment. This method of adjusting for income and education is commonly performed, and has been validated as an effective surrogate for patient-specific data (65). Results were unchanged after adjusting for socio-economic factors. In addition, results were unchanged with baseline body mass index (BMI) included in the model (as a surrogate for nutrition status), and when race was coded as African American vs. white only.

5. ANALYSIS OF CAUSE-SPECIFIC MORTALITY

According to data from the National Center for Health Statistics, racial disparities in life expectancy, while decreasing, still remain (66). An African American female born in 1950 had a life expectancy nearly 10 years less than that of a white female (62.9 years vs. 72.2 years). Today the gap is smaller: an African American female born in 2000 can expect to live an average of 74.9 years, compared to 80.0 years for a white female. The increased baseline mortality risk for African American women can be a concern to researchers attempting to analyse survival due to specific causes. Comparison of overall survival (defined as death due to any cause) between racial groups may be confounded by differential rates in baseline mortality, especially in the adjuvant setting where, with a greater likelihood of having extended survival from cancer, patients are more likely to experience competing risks for death.

Concerns about differential baseline mortality rates in the population motivate cause-specific survival analyses, which attempt to analyse deaths due to cancer separately from other failure types. Cause-specific survival was investigated in both the NSABP and SWOG analyses. In the NSABP analyses in particular, analysis of cause-specific survival was motivated by the excess mortality for African Americans compared to whites despite similar outcomes in disease-free survival. The author hypothesized that the excess in mortality must be due to causes other than cancer. However, actual patient-specific cause of death data was not collected on the majority of these trials. Instead, investigators approximated this endpoint by coding deaths following documented recurrence of breast cancer as a cancer-specific event; other deaths were considered censored observations (and would thus count the same as if the patient had been, say, lost to follow-up at that point). This approach is therefore a best approximation of a cause-specific analysis.

In the NSABP analysis, survival including only deaths most likely attributable to cancer was no longer significantly different by race in either lymph node negative breast cancer (HR=1.08; 95% CI, 0.88 – 1.33) or lymph node positive breast cancer (HR=1.09; 95% CI, 0.96 – 1.25) (61). In the SWOG analysis, however, the survival disadvantage for African Americans remained highly significant for both premenopausal patients (HR=1.42; 95% CI, 1.09 – 1.85, p=.01) and for postmenopausal patients (HR=1.39; 95% CI, 1.17 – 1.66) (63).

Analyses of this sort must be viewed with caution (60). The premise behind such analyses relies on the assumption that one type of outcome can be considered independently of competing outcomes. With respect to breast cancer, for instance, the same factors that reduce the likelihood of death following distant relapse might influence death due to other causes. Assessment of cause of death is itself a major issue for researchers due to problems with misclassification (67-69). In addition, difficulties with missing data may require model-based inference (70). The exploratory approach used by the NSABP and SWOG is one further step removed, since cause of death was estimated for each patient.

Differential baseline mortality may play a larger role in studies based on population or hospital registries, which include patients of diverse stages and treatments. It should be less of an issue in the clinical trial setting, however, since patients have a reasonably uniform health status at study enrolment regardless of race, due to strict eligibility criteria governing the protocol, and will have uniform treatment administered. Therefore patients of any race are more likely to have similar underlying mortality risk.

6. DISCUSSION

Survival deficits for African American women with breast cancer have been observed in a multitude of settings. Analyses of outcome by race based on databases from health maintenance organizations, local or regional hospitals, and population-based registries (e.g., SEER), while sometimes very large, typically utilize retrospective data. Additionally, these databases generally pertain to populations that are heterogeneous with respect to baseline disease characteristics such as stage, socio-economic status, and other important factors including co-morbidity status. In analyses by race, issues regarding possible differential access to care and stage at presentation will play a major role, and may be difficult to account for due to retrospective data subject to ascertainment bias and with substantial missing data. Analyses based on clinical trials, on the other hand, have the crucial advantages of prospective data and uniformly staged and treated patients who have accessed good oncologic care sites offering clinical trials. The largest randomised clinical trials are performed by the cooperative groups of the NCI, which draw on patients receiving treatment from any of a network of institutions around the country. The cooperative groups are able to perform randomised trials with large sample requirements. It has been argued that, given these advantages, differences in outcome by race will be less apparent in the clinical trials setting.

Historically, the effect of race on outcome has only occasionally been reported in the main reports of randomised trials of the cooperative groups, in part because the issue was ancillary to the main treatment question, but also because race has not been considered an important prognostic variable for the disease and so was not incorporated into the protocol design. Where it has been analysed, either in the main reports for studies or in subsequent analyses, racial differences have been found in some instances and not in others. But because single, even very large, trials are not ideal for analysing this question due to the inherent instability of underpowered analyses, both the NSABP and SWOG designed databases from a combination of historical phase III trials of similar histology and stage in order to achieve a large sample with better power. Only the largest cooperative groups with a record of successfully conducted phase III trials can perform such analyses.

The NSABP study analysed patients separately by lymph node status (61). Adjustment variables included treatment type, age, and tumour size. Although no differences in disease-free survival were found, analyses of overall survival showed at least marginally significant differences by race in 2 of the 4 subsets. Of particular importance is that this study showed a survival deficit for African Americans with all patients combined. African Americans had about a 20% increased risk of death in both lymph node negative breast cancer (HR=1.21, 95% CI, 1.01-1.46) and lymph node positive breast cancer (HR=1.18, 95% CI, 1.03-1.34). The hypothesis that the excess risk of death in African Americans without an increased risk of recurrence represents the impact of death due to other causes motivated an approximate cause-specific survival analysis, in which deaths following documented recurrence were defined as deaths due to cancer. The survival disadvantage for African Americans was no longer significant. Based on this analysis, Dignam speculates that different baseline mortality rates are the cause of survival deficits in African Americans(61).

The SWOG analysis grouped patients by menopausal status (62-63). Adjustment variables included study, age, tumour size, number of positive nodes, and receptor status. In the main analyses of both the pre- and postmenopausal patients, African Americans had worse disease-free survival, overall survival, and cause-specific survival (the latter by the same method as performed by Dignam). The SWOG analysis also adjusted for socio-economic status through patient zip-code match to census data, a method that has been validated in other studies (5,71). Even after adjusting for socio-economic status, African Americans had worse survival compared to all other races combined, and compared to whites only. The authors suggest that biologic differences, possibly hormone-related, may play a role in explaining the survival deficit for African Americans with breast cancer.

Can the results of these two large analyses be reconciled and a conclusion drawn? Broadly speaking, the SWOG and NSABP analyses have many similarities. Both groups are large cooperative clinical trial consortiums of the NCI drawing on patients from institutions around the country. Both analyses included patients from multiple adjuvant treatment trials of breast cancer, including treatments such as Tamoxifen and different combinations of chemotherapy. Patients from the NSABP studies were registered from the 1980's and 1990's, while the SWOG analysis included studies beginning in 1975. Both also adjusted for similar variables, including treatment, age, tumour size, lymph node status, and receptor status (either as an adjustment variable or through stratification). Some differences are also notable. The NSABP analysis divided patients by lymph node status and, secondarily, by oestrogen receptor status, while the SWOG analysis divided patients by menopausal status. In addition, while there is some overlap, the NSABP consortium enrolls many of its patients from a different set of institutions than the SWOG consortium; whether this might differentially impact analyses of race is difficult to guess. Another major difference is that SWOG adjusted for differences in SES status through patient zip-code match to census data.

A brief discussion of the relative validity of different endpoints in clinical trials is appropriate here. Relapse-free survival is the primary endpoint in most adjuvant breast cancer trials due to the necessity of evaluating drug efficacy in a timely fashion with adequate power based on number of events. However, proper ascertainment of relapse may be difficult in the adjuvant setting, especially once patients are removed from protocol therapy. Overall survival is generally regarded as the most reliable outcome in cancer clinical trials, (60) and the use of historical phase III trials, with much longer follow-up, allows analyses such as those by NSABP and SWOG the opportunity to use overall survival as the endpoint. But the interpretation of overall survival can be difficult if many patients are dying from causes other than cancer. Such situations, especially in the adjuvant setting, will sometimes motivate cause-specific survival analyses (time until death due to disease). As previously discussed, deaths specific to cancer are also difficult to ascertain, and regardless, cause-specific death is difficult to interpret except in the situation where competing risks of death are known to be independent of cancer death. Even instances where cause of death seems clearly independent of cancer may be difficult to interpret – for instance, a cancer patient who dies in a motor vehicle accident may be assumed to have had a death unrelated to cancer, whereas patients made more feeble due to cancer may be more likely to suffer accidents.

The SWOG analysis indicates significant differences for each of the 3 endpoints, while the NSABP study indicates significant differences in overall survival. Given the results of both analyses, it seems reasonable to conclude that African Americans have shorter survival than other patients or whites only in the adjuvant breast cancer setting, and may have shorter relapse-free survival as well. The expectation that survival differences between African Americans and other patients might disappear with adjustments for socio-economic status is belied by the results of the SWOG analysis, in which the inclusion of U.S. census surrogates for income and education, as factors in multivariate regression did not alter the conclusions.

These analyses are not based on studies designed specifically to test differences between races with regard to outcome, so they must still be considered hypothesis generating rather than definitive. It should be noted that any analysis, even if socio-economic variables are included, will not fully account for socio-economic differences between races. The advantages of randomisation in distributing unknown factors influencing the outcome only apply to the object of randomisation (e.g., treatment, not race). Therefore access to care issues may still disproportionately impact African Americans more than other patients even in the randomised clinical trial setting. However, given the large differences of the SWOG analysis in particular, the suggestion that biologic factors may also impact outcome between African Americans and other patients must be considered a plausible hypothesis worth further investigation. In parallel with the analysis of adjuvant breast cancer trials, the SWOG analysis also observed survival deficits for African Americans in prostate cancer, suggesting that hormonal factors may also play a role (62). In adjuvant breast cancer, for instance, African American women have been shown to have significantly lower white blood cell counts than white women at diagnosis. Thus African American women require longer duration of treatment and have lower dose intensity than white women, possibly reducing efficacy of adjuvant treatment (72). Studies are underway to explore this observation in the large SWOG database. The American Association for Cancer Research recently recognized the importance of biology, race and outcome by conducting a special minisymposium on this subject at its annual meeting.

Recent editorials in the NEJM illustrate the difficulty in casting the issue of research on racial differences in medicine in strictly scientific terms. Cooper et al. argues that race is not a valid surrogate for genetic differences, and that regardless the issue of race in science has the potential to perpetuate negative social consequences (73). Burchard et al., countering Cooper, contends that although some social costs may exist, they are outweighed by the potential benefits of improving health outcomes (74). Phimister likewise acknowledges the potential social consequences but agrees the potential benefits outweigh the potential risks (75). In any event, researchers should take care to avoid the further stigmatisation of race in study or analysis reports. Kaplan and Bennett recently proposed a set of guidelines for the reporting of race and ethnicity in biomedical publication, noting that the categories of race and the reasons for using race should be described and justified, and emphasizing in particular the necessity to include adjustments for socio-economic factors where possible (76). This latter point echoes the observations of others that socio-economic status accounts for at least some of the survival disadvantages in African Americans (3-4).

Differences in outcome by race will not simply disappear if they are ignored, and science may through prospective study help alleviate these differences. Ultimately, race may prove to be an interim surrogate for molecular epidemiological differences or for a class of genomic signatures, making self-reported race an antiquated indicator of prognosis. In the meantime, research suggests that African Americans with breast cancer have worse prognosis after receiving state of the art adjuvant therapy even after accounting for the factors previously assumed to explain this deficit, including socio-economic status, and in the setting of clinical trials where the influence of other baseline health disparities is minimized. Research regarding disparities in access to care and early detection for this subgroup should continue, complemented as well by research into potential biologic and hormonal differences.

REFERENCES

(1) Ries LAG, Eisner MP, Kosary CL, Hankey BF, Miller BA, Clegg L, Mariotto A, Fay MP, Feuer EJ, Edwards BK (eds). *SEER Cancer Statistics Review, 1975-2000*, National Cancer Institute. Bethesda, MD, http://seer.cancer.gov/csr/1975_2000, 2003.

(2) Weir HK, Thun MJ, Hankey BF, Ries LA, Howe HL, Wingo PA, Jemal A, Ward E, Anderson RN, Edwards BK. Annual report to the nation on the status of cancer, 1975-2000, featuring the uses of surveillance data for cancer prevention and control. J Natl Cancer Inst. 2003 Sep 3;95(17):1276-99.

(3) Baquet CR, Horm JW, Gibbs T, Greenwald P. Socioeconomic factors and cancer incidence among blacks and whites. J Natl Cancer Inst. 1991 Apr 17;83(8):551-7.

(4) Freeman H. Race, poverty, and cancer. J Natl Cancer Inst. 1991 Apr 17;83(8):526-7.

(5) Underwood SM. Minorities, women, and clinical cancer research: the charge, promise, and challenge. Ann Epidemiol 2000 Nov;10(8 Suppl):S3-12.

(6) Cross CK, Harris J, Recht A. Race, socioeconomic status, and breast carcinoma in the U.S. Cancer. 2002 Nov 1;95(9):1988-99.

(7) Newman LA, Mason J, Cote D, Vin Y, Carolin K, Bouwman D, Colditz GA. African-American ethnicity, socioeconomic status, and breast cancer survival: a meta-analysis of 14 studies involving over 10,000 African-American and 40,000 White American patients with carcinoma of the breast. Cancer. 2002 Jun 1:94(11):2844-54.

(8) Hutchins LF, Unger JM, Crowley JJ, Coltman CA Jr, Albain KS. Underrepresentation of patients 65 years of age or older in cancer-treatment trials. N Engl J Med. 1999 Dec 30;341(27):2061-7.

(9) Bibb SC. The relationship between access and stage at diagnosis of breast cancer in African American and Caucasian women. Oncol Nurs Forum. 2001 May;28(4):711-9.

(10) Chen F, Trapido EJ, David K. Differences in stage at presentation of breast and gynecologic cancers among whites, blacks, and Hispanics. Cancer. 1994 Jun 1;73(11):2838-42.

(11) El-Tamer MB, Wait RB. Age at presentation of African-American and Caucasian breast cancer patients. J Am Coll Surg. 1999 Mar;188(3):237-40.

(12) English WP, Cleveland KE, Barber WH. There is no difference in survival between African-American and white women with breast cancer. Am Surg. 2002 Jun;68(6):594-7.

(13) Joslyn SA, West MM. Racial differences in breast carcinoma survival. Cancer. 200 Jan 1;88(1):114-23.

(14) Lyman GH, Kuderer NM, Lyman SL, Cox CE, Reintgen D, Baekey P. Importance of race on breast cancer survival. Ann Surg Oncol. 1997 Jan;4(1):80-7.

(15) Elledge RM, Clark GM, Chamness GC, Osborne CK. Tumor biologic factors and breast cancer prognosis among white, Hispanic, and black women in the United States. J Natl Cancer Inst. 1994 May 4;86(9):705-12.

(16) Joslyn SA. Hormone receptors in breast cancer: racial differences in distribution and survival. Breast Cancer Res Treat. 2002 May;73(1):45-59.

(17) Kotwall CA, Brinker CC, Covington DL, Hall TL, Maxwell JG. Prognostic indices in breast cancer are related to race. Am Surg. 2003 May;69(5):372-6.

(18) McKinnon J. The black population in the United States: March 2002. U.S. Census Bureau, Current Population Reports, Series P20-541. Washington, DC. 2003.

(19) Fisher ER, Anderson S, Tan-Chiu E, Fisher B, Eaton L, Wolmark N. Fifteen-year prognostic discriminants for invasive breast carcinoma: National Surgical Adjuvant Breast and Bowel Project Protocol B-06. Cancer. 2001 Apr 15;91(8 Suppl):1679-87.

(20) Rivkin SE, Green S, Metch B, Cruz AB, Abeloff MD, Jewell WR, Costanzi JJ, Farrar WB, Minton JP, Osborne CK. Adjuvant CMFVP versus tamoxifen versus concurrent CMFVP and tamoxifen for postmenopausal, node-positive, and estrogen receptor-positive breast cancer patients: a Southwest Oncology Group study. J Clin Oncol. 1994 Oct;12(10):2078-85.

(21) Winer E, Berry DA, Woolf S, Duggan D, Kornblith A, Harris LN, Michaelson RA, Kirshner JA, Fleming GF, Perry MC, Graham ML, Sharp SA, Keresztes R, Henderson IC, Hudis C, Muss H, Norton L. Failure of higher-dose paclitaxel to improve outcome in patients with metastatic breast cancer: results from CALGB 9342. In press.

(22) Green SJ, Fleming TR. Guidelines for the reporting of clinical trials. Semin Oncol. 1988 Oct;15(5):455-61.

(23) Wood WC, Budman DR, Korzun AH, Cooper MR, Younger J, Hart RD, Moore A, Ellerton JA, Norton L, Ferree CR, et al. Dose and dose intensity of adjuvant chemotherapy for stage II, node-positive breast carcinoma. N Engl J Med. 1994 May 5;330(18):1253-9.

(24) Costanza ME, Weiss RB, Henderson IC, Norton L, Berry DA, Cirrincione C, Winer E, Wood WC, Frei E 3rd, McIntyre OR, Schilsky RL. Safety and efficacy of using a single agent or a phase II agent before instituting standard combination chemotherapy chemotherapy in previously untreated metastatic breast cancer patients: report of a randomized study – Cancer and Leukemia Group B 8642. J Clin Oncol. 1999 May;17(5):1397-406.

(25) Abrams J, Aisner J, Cirrincione C, Berry DA, Muss HB, Cooper MR, Henderson IC, Panasci L, Kirshner J, Ellerton J, Norton L. Dose-response trial of megestrol acetate in advanced breast cancer: Cancer and Leukemia Group B phase III study 8741. J Clin Oncol. 1999 Jan;17(1):64-73.

(26) Parnes HL, Cirrincione C, Aisner J, Berry DA, Allen SL, Abrams J, Chuang E, Cooper MR, Perry MC, Duggan DB, Szatrowski TP, Henderson IC, Norton L; Cancer and Leukemia Group B. Phase III study of cyclophosphamide, doxorubicin, and fluorouracil (CAF) plus leucovorin versus CAF for metastatic breast cancer: Cancer and Leukemia Group B 9140. J Clin Oncol. 2003 May 1;21(0):1819-24.

(27) Henderson IC, Berry DA, Demetri GD, Cirrincione CT, Goldstein LJ, Martino S, Ingle JN, Cooper MR, Hayes DF, Tkaczuk KH, Fleming G, Holland JF, Duggan DB, Carpenter JT, Frei E 3rd, Schilsky RL, Wood WC, Muss HB, Norton L. Improved outcomes from adding sequential Paclitaxel but not from escalating Doxorubicin dose in an adjuvant chemotherapy regimen for patients with node-positive primary breast cancer. J Clin Oncol. 2003 Mar 15;21(6):976-83.

(28) Citron ML, Berry DA, Cirrincione C, Hudis C, Winer EP, Gradishar WJ, Davidson NE, Martino S, Livingston R, Ingle JN, Perez EA, Carpenter J, Hurd D, Holland JF, Smith BL, Sartor CI, Leung EH, Abrams J, Schilsky RL, Muss HB, Norton L. Randomized trial of dose-dense versus conventionally scheduled and sequential versus concurrent combination chemotherapy as postoperative adjuvant treatment of node-positive primary breast cancer: first report of Intergroup Trial C9741/Cancer and Leukemia Group B Trial 9741. J Clin Oncol. 2003 Apr 15;21(8):1431-9. Epub 2003 Feb 13.

(29) Roach M 3rd, Cirrincione C, Budman D, Hayes D, Berry D, Younger J, Hart R, Henderson IC. Race and survival from breast cancer: based on Cancer and Leukemia Group B trial 8541. Cancer J Sci Am. 1997 Mar-Apr;3(2):107-12.

(30) Tallman MS, Gray R, Robert NJ, LeMaistre CF, Osborne CK, Vaughn WP, Gradishar WJ, Pisansky TM, Fetting J, Paietta E, Lazarus HM. Conventional adjuvant chemotherapy with or without high-dose chemotherapy and autologous stem-cell transplantation in high-risk breast cancer. N Engl J Med. 2003 Jul 3;349(1):17-26.

(31) Fetting JH, Gray R, Fairclough DL, Smith TJ, Margolin KA, Citron ML, Grove-Conrad M, Cella D, Pandyr K, Robert N, Henderson IC, Osborne CK, Abeloff MD. Sixteen-week multidrug regimen versus cyclophosphamide, doxorubicin, and fluorouracil as adjuvant therapy for node-positive, receptor-negative breast cancer: an Intergroup study. J Clin Oncol. 1998 Jul;16(7):2382-91.

(32) Falkson HC, Gray R, Wolberg WH, Gillchrist KW, Harris JE, Tormey DC, Falkson G. Adjuvant trial of 12 cycles of CMFPT followed by observation or continuous tamoxifen versus four cycles of CMFPT in postmenopausal women with breast cancer: an Eastern Cooperative Oncology Group phase III study. J Clin Oncol. 1990 Apr;8(4):599-607. Erratum in: J Clin Oncol 1990 Sep;8(9):1603.

(33) Tormey DC, Gray R, Abeloff MD, Roseman DL, Gilchrist KW, Barylak EJ, Stott P, Falkson G. Adjuvant therapy with a doxorubicin regimen and long-term tamoxifen in premenopausal breast cancer patients: and Eastern Cooperative Oncology Group trial. J Clin Oncol. 1992 Dec;10(12):1848-56.

(34) Taylor SG 4th, Kalish LA, Olson JE, Cummings F, Bennett JM, Falkson G, Tormey DC, Carbone PP. Adjuvant CMFP versus CMFP plus tamoxifen versus observation alone in postmenopausal, node-positive breast cancer patients: three-year results of an Eastern Cooperative Oncology Group study. J Clin Oncol. 1985 Feb;3(2):144-54.

(35) Fisher B, Montague E, Redmond C, et al. Comparison of radical mastectomy with alternative treatments for primary breast cancer: a first report of results from a prospective randomized clinical trial. Cancer 1977:39:2827-39.

(36) Fisher B, Bauer M, Margolese R, Poisson R, Pilch Y, Redmond C, Fisher E, Wolmark N, Deutsch M, Montague E, et al. Five-year results of a randomized clinical trial comparing total mastectomy and segmental mastectomy with or without radiation in the treatment of breast cancer. N Engl J Med. 1985 Mar 14;312(11):665-73.

(37) Fisher B, Redmond C, Brown A, et al. Treatment of primary breast cancer with chemotherapy and tamoxifen. N Engl J Med. 1981 Jul 2;305(1):1-6.

(38) Fisher B, Redmond C, Dimitrov NV, Bowman D, Legault-Poisson S, Wickerham DL, Wolmark N, Fisher ER, Margolese R, Sutherland C, et al. A randomized clinical trial evaluating sequential methotrexate and fluorouracil in the treatment of patients with node-negative breast cancer who have estrogen-repector-negative tumors. N Engl J Med. 1989 Feb 23;320(8):473-8.

(39) Fisher B, Costantino J, Redmond C, Poisson R, Bowman D, Couture J, Dimitrov NV, Wolmark N, Wickerham DL, Fisher ER, et al. A randomized clinical trial evaluating tamoxifen in the treatment of patients with node-negative breast cancer who have estrogen-receptor-positive tumors. N Engl J Med. 1989 Feb 23;320(8):479-84.

(40) Fisher B, Brown AM, Dimitrov NV, Poisson R, Redmond C, Margolese RG, Bowman D, Wolmark N, Wickerham DL, Kardinal CG, et al. Two months of doxorubicin-cyclophosphamide with and without interval reinduction therapy compared with 6 months of cyclophosphamide, methotrexate, and fluorouracil in positive-node breast cancer patients with tamoxifen-nonresponsive tumors: results from the National Surgical Adjuvant Breast and Bowel Project B-15. J Clin Oncol. 1990 Sep;8(9):1483-96.

(41) Fisher B, Redmond C, Legault-Poisson S, Dimitrov NV, Brown AM, Wickerham DL, Wolmark N, Margolese RG, Bowman D, Glass AG, et al. Postoperative chemotherapy and tamoxifen compared with tamoxifen alone in the treatment of positive-node breast cancer patients aged 50 years and older with tumors responsive to tamoxifen: results from the National Surgical Adjuvant Breast and Bowel Project B-16. J Clin Oncol. 1990 Jun;8(6):1005-18.

(42) Fisher B, Dignam J, Mamounas EP, Costantino JP, Wickerham DL, Redmond C, Wolmark N, Dimitrov NV, Bowman DM, Glass AG, Atkins JN, Abramson N, Sutherland CM, Aron BS, Margolese RG. Sequential methotrexate and fluorouracil for the treatment of node-negative breast cancer patients with estrogen receptor-negative tumors: eight-year results from National Surgical Adjuvant Breast and Bowel Project (NSABP) B-13 and first report of findings from NSABP B-19 comparing methotrexate and fluorouracil with conventional cyclophosphamide, methotrexate, and fluorouracil. J Clin Oncol. 1996 Jul;14(7):1982-92.

(43) Fisher B, Dignam J, Wolmark N, DeCillis A, Emir B, Wickerham DL, Bryant J, Dimitrov NV, Abramson N, Atkins JN, Shibata H, Deschenes L, Margolese RG. Tamoxifen and chemotherapy for lymph node-negative, estrogen receptor-positive breast cancer. J Natl Cancer Inst. 1997 Nov 19;89(22):1673-82.

(44) Fisher B, Anderson S, Wickerham DL, DeCillis A, Dimitrov N, Mamounas E, Wolmark N, Pugh R, Atkins JN, Meyers FJ, Abramson N, Wolter J, Bornstein RS, Levy L, Romond EH, Caggiano V, Grimaldi M, Jochimsen P, Deckers P. Increased intensification and total dose of cyclophosphamide in a doxorubicin-cyclophosphamide regimen for the treatment of primary breast cancer: findings from National Surgical Adjuvant Breast and Bowel Project B-22. J Clin Oncol. 1997 May;15(5):1858-69.

(45) Dignam JJ. Differences in breast cancer prognosis among African-American and Caucasian women. CA Cancer J Clin. 2000 Jan-Feb;50(1):50-64.

(46) Fisher ER, Redmond C, Fisher B. Prognostic factors in NSABP studies of women with node-negative breast cancer. National Surgical Adjuvant Breast and Bowel Project. J Natl Inst Monogr. 1992;(11):151-8.

(47) Fisher ER, Anderson S, Redmond C, Fisher B. Pathologic findings from the National Surgical Adjuvant Breast Project protocol B-06. 10-year pathologic and clinical prognostic discriminants. Cancer. 1993 Apr 15;71(8):2507-14.

(48) Fisher ER, Anderson S, Tan-Chiu E, Fisher B, Eaton L, Wolmark N. Fifteen-year prognostic discriminants for invasive breast carcinoma: National Surgical Adjuvant Breast and Bowel Project Protocol B-06. Cancer. 2001 Apr 15;91(8 Suppl):1679-87.

(49) Costantino JP, Redmond CK, Wickerham DL, et al: A comparison of survival between white and black breast cancer patients: Results from NSABP B-09. Proc Am Soc Clin Onc. 1987;6:230.

(50) Dignam JJ, Redmond CK, Fisher B, Costantino JP, Edwards BK. Prognosis among African-American women and white women with lymph node negative breast carcinoma: findings from two randomized clinical trials of the National Surgical Adjuvant Breast and Bowel Project (NSABP). Cancer. 1997 Jul 1;80(1):80-90.

(51) Glucksberg H, Rivkin SE, Rasmussen S, Tranum B, Gad-el-Mawla N, Costanzi J, Hoogstraten B, Athens J, Maloney T, McCracken J, Vaughn C. Combination chemotherapy (CMFVP) versus L-phenylalanine mustard (L-PAM) for operable breast cancer with positive axillary nodes: a Southwest Oncology Group Study. Cancer. 1982 Aug 1;50(3):423-34.

(52) Rivkin SE, Green S, Metch B, Jewell WR, Costanzi JJ, Altman SJ, Minton JP, O'Bryan RM, Osborne CK. One versus 2 years of CMFVP adjuvant chemotherapy in axillary node-positive and estrogen receptor-negative patients: a Southwest Oncology Group study. J Clin Oncol. 1993 Sep;11(9):1710-6.

(53) Rivkin SE, Green S, O'Sullivan J, Cruz AB, Abeloff MD, Jewell WR, Costanzi JJ, Farrar WB, Osborne CK. Adjuvant CMFVP versus adjuvant CMFVP plus ovariectomy for premenopausal, node-positive, and estrogen receptor-positive breast cancer patients: a Southwest Oncology Group study. J Clin Oncol. 1996 Jan;14(1):46-51.

(54) Budd GT, Green S, O'Bryan RM, Martino S, Abeloff MD, Rinehart JJ, Hahn R, Harris J, Tormey D, O'Sullivan J, et al. Short-course FAC-M versus 1 year of CMFVP in node-positive, hormone receptor-negative breast cancer: an intergroup study. J Clin Oncol. 1995 Apr;13(4):831-9.

(55) Albain K, Green S, Osborne K, Cobau C, Levine E, Ingle J, Pritchard K, Schneider D, O'Sullivan J, Hess E, Martino S. Tamoxifen (T) versus cyclophosphamide, adriamycin and 5-FU plus either concurrent or sequential T in postmenopausal, receptor(+), node(+) breast cancer: a Southwest Oncology Group phase III intergroup trial (SWOG-8814, INT-0100). Proc Am Soc Clin Oncol. 1997.

(56) Albain KS, Green SJ, Ravdin PM, Cobau CD, Levine EG, Ingle JN, Pritchard KI, Schneider DJ, Abeloff MD, Norton L, Henderson IC, Lew D, Livingston RB, Martino S, Osborne CK. Adjuvant chemohormonal therapy for primary breast cancer should be sequential instead of concurrent: initial results from intergroup trial 0100 (SWOG-8814). Proc Am Soc Clin Oncol 2002.

(57) Hutchins L, Green S, Ravdin P, Lew D, Martino S, Abeloff M, Lyss A, Henderson C, Allred C, Dakhil S, Pierce I, Goodwin W, Caton J, Rivkin S, Chapman R, Osborne K. CMF versus CAF with and without tamoxifen in high-risk node-negative breast cancer patients and a natural history follow-up study in low-risk node-negative patients: first results of intergroup trial INT 0102. J Clin Oncol. In press.

(58) Rivkin SE, Green S, Metch B, Glucksberg H, Gad-el-Mawla N, Constanzi JJ, Hoogstraten B, Athens J, Maloney T, Osborne CK, et al. Adjuvant CMFVP versus melphalan for operable breast cancer with positive axillary nodes: 10-year results of a Southwest Oncology Group Study. J Clin Oncol. 1989 Sep;7(9):1229-38.

(59) Rivkin SE, Green SJ, Lew D, Costanzi JJ, Athens JW, Osborne CK, Vaughn CB, Martino S. Adjuvant chemotherapy with cyclophosphamide, methotrexate, and 5-fluorouracil, vincristine, and prednisone compared with single-agent L-phenylalanine mustard for patients with operable breast carcinoma and positive axillary lymph nodes: 20-year results of a Southwest Oncology Group study. Cancer. 2003 Jan 1;97(1):21-9.

(60) Green SG, Benedetti J, and Crowley J, 2002, Clinical Trials in Oncology, Second Edition, Chapman and Hall/CRC, London, USA.

(61) Dignam JJ. Efficacy of systemic adjuvant therapy for breast cancer in African-American and Caucasian women. J Natl Cancer Inst Monogr. 2001;(30):36-43.

(62) Unger JM, Coltman Jr. CA, Appelbaum FR, Barlogie B, Crawford ED, Fisher RI, Gandara DR, Livingston RB, Markman M, Albain KS. African Americans have worse survival in hormone-related cancers. A Southwest Oncology Group (SWOG) study. Proc Am Soc Clin Oncol 22; page 529, 2003 (abstr 2127).

(63) Albain KS, Unger JM, Hutchins LF, Rivkin SE, Martino S, Livingston RB, Osborne CK. Outcome of African Americans on Southwest Oncology Group (SWOG) breast cancer adjuvant therapy trials. Proc San Antonio Breast Cancer Symposium. 2003.
(64) Albain KS, Green S, LeBlanc M, Rivkin S, O'Sullivan J, Osborne CK. Proportional hazards and recursive partitioning and amalgamation analyses of the Southwest Oncology Group node-positive adjuvant CMFVP breast cancer data base: a pilot study. Breast Cancer Res Treat. 1992;22(3):273-84.
(65) Krieger N. Overcoming the absence of socioeconomic data in medical records: validation and application of a census-based methodology. Am J Public Health 1991 May;82(5):703-10.
(66) National Center for Health Statistics. National Vital Statistics Reports, Vol. 51, No. 3; December 19, 2002.
(67) D'Amico M, Agozzino E, Biagino A, Simonetti A, Marinelli P. Ill-defined and multiple causes of death certificates – a study of misclassification in mortality statistics. Eur J Epidemiol. 1999 Feb;15(2):141-8.
(68) Feuer EJ, Merrill RM, Hankey BF. Cancer surveillance series: interpreting trends in prostate cancer – part II: Cause of death misclassification and the recent rise and fall in prostate cancer mortality. J Natl Cancer Inst. 1999 Jun 16;91(12):1025-32.
(69) Percy C, Stanek E 3rd, Gloeckler L. Accuracy of cancer death certificates and its effect on cancer mortality statistics. Am J Public Health. 1981 Mar;71(3):242-50.
(70) Andersen J.W., Goetghebeur E.J. and Ryan L. (1996). Analysis of survival data under competing risks with missing cause of death information : application and implications for design. In : Lifetime Data : Models in Reliability and Survival Analysis, N.P. Jewell, A.C. Kimber, M-L. Lee and G.A. Whitmore (eds.). Kluwer Academic Publishers, pp.13-19.
(71) Simon MS, Severson RK. Racial differences in survival of female breast caner in the Detroit metropolitan area. Cancer. 1996;77:308-314.
(72) Hershman D, Weinberg M, Rosner Z, Alexis K, Tiersten A, Grann VR, Troxel A, Neugut AI. Ethnic neutropenia and treatment delay in African American women undergoing chemotherapy for early-stage breast cancer. J Natl Cancer Inst. 2003 Oct 15;95(20):1545-8.
(73) Cooper RS, Kaufman JS, Ward R. Race and genomics. N Engl J Med. 2003 Mar 20;348(12):1166-70.
(74) Burchard EG, Ziv E, Coyle N, Gomez SL, Tang H, Karter AJ, Mountain JL, Perez-Stable EJ, Sheppard D, Risch N. The importance of race and ethnic background in biomedical research and clinical practice. N Engl J Med. 2003 Mar 20;348(12):1170-5.
(75) Phimister EG. Medicine and the racial divide. N Engl J Med. 2003 Mar 20;348(12):1081-2.
(76) Kaplan JB, Bennett T. Use of race and ethnicity in biomedical publication. JAMA. 2003 May 28;289(20);2709-16.

Chapter 12

TRADITIONAL MEDICINE IN BREAST CANCER MANAGEMENT WITH A FOCUS ON THE CHINESE INTEGRATIVE APPROACH

Steven K. H. Aung, MD, FAAFP
Departments of Medicine and Family Medicine, University of Alberta, Edmonton, Canada

1. INTRODUCTION

Cancer, from whatever medical perspective—north, south, east, west, African, North or South American, Asian, traditional, modern and/or synergistic—remains a poorly understood disease and illness. Disease pertains to the actual organic condition, whereas illness pertains to one's daily experience and suffering of this invasion of one's physical, mental and spiritual system *(1)*. Certainly, carcinogens are involved in the disease and illness of cancer, while other biological mechanisms contribute to its initiation and progression.. Breast cancer is a leading form of cancer in women but is much less common in men. A family history of the disorder is often obtainable, just as is the case in prostate cancer that afflicts middle-aged men.

Cancer is such a serious, life-threatening condition that it is not advisable that it be managed with unproved remedies such as alternative therapies. Cancer patients should be under the expert diagnostic and therapeutic guidance of their biomedical family physicians and oncology specialists. Increasingly, such medical experts are choosing to refer some of their cancer patients for various complementary therapies to help with symptom control intervention associated with the disease and illness of cancer as well as the various post-cancer therapies such as surgery, radiation and chemotherapy. Moreover, these measures are also useful for the complications of cancer. This is the complementary medical perspective, and it is appropriate to refer

C. Williams et al. (eds.), Breast Cancer in Women of African Descent, 297-311.

to this holistic initiative and endeavour in terms of integrative medicine *(2-6)*. It is a concept, which highlights the ongoing, proactive, and heartfelt communication process for the benefit of quality of life for all concerned.

2. TRADITIONAL CHINESE MEDICINE (TCM)

TCM has never claimed to have the answer to the cure for cancer or to have developed a perfect set of diagnostic and therapeutic modalities for this complex condition *(7)*. Generally, in the TCM theoretical system perspective, cancer is viewed in terms of severe localized disruption of Qi flow. Qi flow is perceived as vital energy circulation along nerve and blood bioenergetic pathways/meridians with specific nexuses/gates of accumulation and dispersion—acupoints *(8-9)*. Good health entails smooth and harmonious Qi, Blood and Body Fluid flow within the preventive and therapeutic context of tonification (methods to increase Qi flow) and sedation (methods to decrease Qi flow) as may be necessary for each patient according to their own unique vital energetic condition as assessed and constantly re-assessed by the TCM practitioner.

Severe blockage of Qi flow—to the extent that 'mass' (cancer cells and the resulting tumour formations) accumulates and, indeed, congeals at specific sites and then may proceed to proliferate throughout a person's physiological system (process of metastasis)—often results in systemic acute and chronic trauma and even death. In the TCM system, this severe disease and illness condition is known as Ai Zhi (Figure 1) *(10-11)*.

While phlegm in biomedicine is largely viewed as mucous accumulation in the throat or chest due to the impact of the common cold or influenza and other viruses as well as antigens/histamines reactions generated by the body in response to allergens (ordinary or common, acute or chronic allergic reactions), Phlegm in TCM is viewed as a serious condition resulting from the stagnant flow of Qi, Blood and Body Fluid—especially in the major Zang Fu organ⇔meridian systems—due to the impact of external (pathogenic/carcinogenic) and internal (emotional) factors (*12-13*). This severe disruption and imbalance may result in various disorders, including breast cancer. However, as yet, there is no comprehensive theory of cancer causation and cure in biomedicine, TCM or any other medical system.

Figure 1. The character of Ai Zhi: a. 'pathology/severe illness,' b. 'mountain pushing up' and c. 'lump/mass of stagnant energy.' (Original calligraphic artwork by Dr Steven KH Aung)

Phlegm in TCM encompasses the following synergies, resonances and cycles:

a) It is a thick fluid—somewhat like 'mud' particles or waves—causing Qi stasis in the organ⇔meridian systems.
b) This leads to the general TCM syndrome known as Qi and Blood stasis (pre-cancerous stage), which, in turn, may result in mass formation (tumour/metastasisation stages).
c) Tongue diagnosis is important, since a shiny, thick tongue indicates the accumulation of Phlegm in the body. When the Qi and Blood and Body Fluid become further blocked/stagnated, the tongue appears to manifest a white coating.
d) This may then become evident as a purplish color in the middle of the tongue, which is the holographic area of the Stomach, indicating that the condition is worsening and becoming severe Blood stasis.
e) This process embodies a pre-cancerous state of the Stomach as well as affecting the breast, since in TCM the breast region is regulated by the Stomach meridian, which anatomically flows through the middle of the breast.

f) The Stomach meridian is also the major meridian for removing the Phlegm of the whole body. Key acupoints in this regard are ST.8 (located in both temporal regions) for localized Phlegm removal such as cysts, mass, fibroids and ST.40 (located on the lateral side of the mid-legs) is indicated for generalized Phlegm removal.(Figure 2)

g) When the Stomach meridian becomes chronically disordered due to exterior or interior pathogenesis, it causes the ongoing difficulty of the removal of Phlegm, and therefore this situation may lead to ever more accumulation of Phlegm, which might cause even more Qi Stagnation, which in turn may cause Blood and Body Fluid Stagnation in chronic and severe cases and, finally, tumour mass formation and possible metastasisation in the later stages.

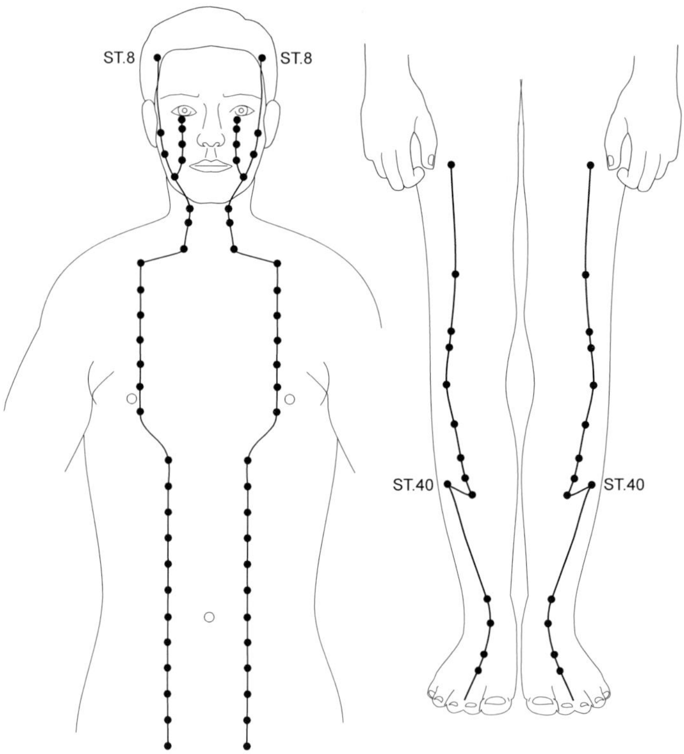

Figure 2. General Qi circulation of the Stomach (ST) meridian showing the acupoints ST.8 and ST.40

Blood Stagnation is a general TCM syndrome that can also be caused by immune disorders, severe depletion of the immune system and fluid retention, due to Spleen Qi Deficiency (the Spleen is the pertaining/linked organ of the Stomach), which in turn might generate weakness of the

Stomach organs and meridians. This pathogenesis may repeat over and over again as a cycle resulting in breast and other cancerous formations and permutations.

The following three case examples serve to illustrate the impact of the difficult, complex and sometimes manageable condition of breast cancer with respect to patients, their families, friends and colleagues as well as their physicians and healers. These patients had previously experienced mastectomies, radiation and chemotherapy according to standard biomedical treatment protocols and were referred to me for adjunctive treatment for basic symptom control, immunoenhancement and enhanced quality of life. The usual procedure is to offer a brief trial course of acupuncture, to see if the patients like it and respond positively, and then to develop a comprehensive treatment and follow-up plan customized for each and every patient according to their condition and needs. The actual tumour site is never needled and deleterious side effects are negligible when fully qualified and licensed practitioners perform acupuncture as illustrated with the following examples.

3. SELECT CASE EXAMPLES

Case 1

A 50-year-old woman with a diagnosis of invasive intraductal carcinoma was referred by her family physician for TCM consultation for post-operative mastectomy pain and for immune enhancement therapy. There was no known family history of breast cancer in her life. Clinically, she appeared healthy except that she reported severe distress and pain—and her anxiety and emotional anger was evident. Physically, this patient was found to be slightly obese with general fluid retention/oedema. Her tongue manifested a purplish colour in the middle area (Stomach) with teeth marks at the area along the front edge (Spleen). Her pulse indicated 'knotting' in the Stomach and Spleen with pathophysiological resonance in the Liver and Heart. A diagnosis was made according to TCM in terms of Blood Stasis in the Stomach with Spleen Qi Deficiency and also Liver Yang Excess Qi syndromes. She was treated with the general TCM approach for pain control and emotional stability by acupuncture, dietetics, herbal medicine and Qi Gong/Tai Chi Chuan. She received acupuncture treatment regularly twice a week for pain control, then once a week for 12 weeks and she is presently receiving regular monthly acupuncture treatments for immune enhancement, removal of Phlegm and Blood Stasis. She appears to be presently maintaining a relatively pain-free, enhanced quality of life with stable physiological parameters.

Case 2

A 70-year-old woman was admitted to a major general hospital in Edmonton, Alberta, Canada, with severe pain in her chest cage region as well as her back due to severe compression fractures of the thoracic and lumbar vertebrae. She was suffering from advanced breast carcinoma with metastasis to her bones and liver. She was given high doses of morphine and was placed in the intensive care unit several times due to her ongoing severe pain and respiratory distress. I was called to see her for TCM consultation for pain control. Clinically, she appeared drowsy due to her cancer situation and the administration of morphine. She was not interested in being treated with acupuncture for pain management and she initially refused to see me (according to her choice) over the subsequent several days and weeks. Finally—after ongoing requests within the context of ongoing consultation with her family, her family physician, oncologist, nurses and social worker—she finally chose to give me the opportunity to treat her with TCM techniques. I started to treat her with acupuncture in order to control her pain of the chest cage and the vertebrae, and her response was immediately positive. Her quality of life subsequently improved over the course of my daily, weekly and monthly treatment sessions, as indicated by her better sleep, less constipation and more mental alertness. She was discharged from the hospital after a few months, and I administered follow-up acupuncture therapy to sustain her well being. She later passed away as a result of pneumonia, unfortunately.

Case 3

A 56-year-old-woman was referred to me by her family physician and oncology specialist for severe pain in the upper chest due to herpes zoster ('shingles') infection. She was suffering from adenocarinoma with multiple small peritoneal implants due to liver and lung metastases. She was also unable to sleep, and this entire situation made her both depressed and angry. She was under the influence of strong analgesics, notably morphine, but without complete pain control. When I first saw her, she was in great pain and distress in the upper chest region. She had tears flowing from her eyes, and it was even difficult for her to wear clothing due to the hypersensivity of her skin. I treated her with acupuncture for harmonization and relaxation and at the same time for chest wall pain, stimulating several acupoints on the Triple Energizer (a TCM organ⇔meridian system encompassing the chest, stomach and bowel regions) and Kidney meridians for her chest/breast pain. Her pain currently appears to be under good control after a few months of this TCM treatment protocol, and she is presently receiving monthly acupuncture follow-up maintenance therapy for her condition. She reports an enhanced quality of life.

4. TCM MANAGEMENT STRATEGIES AND PROTOCOLS

The TCM therapeutic strategies and modalities work well together, especially medical acupuncture and ancillary techniques, dietetics and herbal medicine as well as the practice of medical Qi Gong and Tai Chi Chuan exercise techniques, and it must be noted that the TCM system has always remained open in a pragmatic sense to other beneficial techniques from different systems of medicine *(14)*. The specific acupuncture points to be stimulated vary according to the specific pathology (Figure 3). For example, as suggested above, ST.8 and ST.40 are essential 'pearls' indicated for Phlegm removal. Other relevant acupoints may be indicated for the removal of Blood Stagnation, immunoenhancement and energy balancing and harmonization. These and other relevant acupoints may be stimulated with the normal medical acupuncture protocol involving shallow needle insertion of solid stainless steel needles (individually pre-packaged, sterilised, safe and disposable medical instruments). Together with their packaging, these needles are disposed of in an environmentally safe manner after being used once only on any human beings or animals. This is done in accordance with local standards of community waste disposal and recyclement.

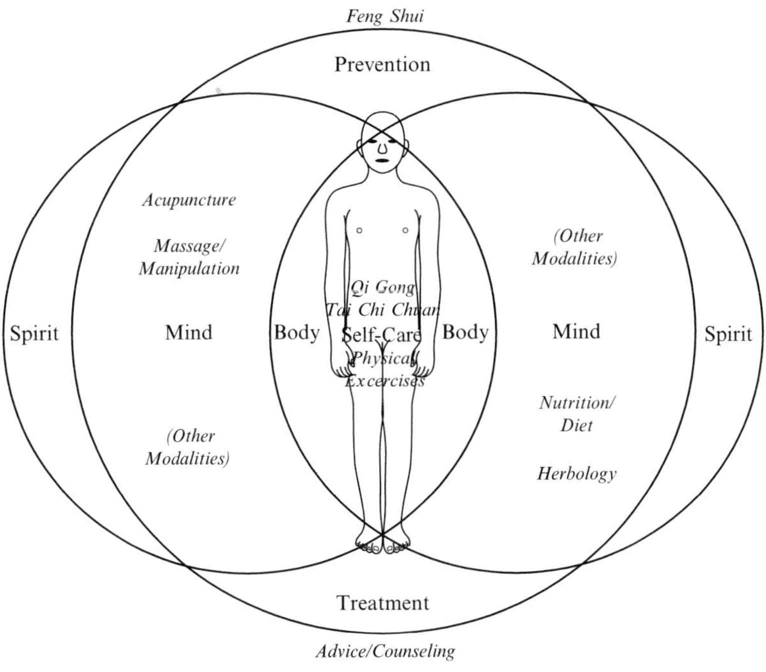

Figure 3. The general TCM integrative, holistic perspective.

A mild electrical current may be applied through the inserted needles in order to control and enhance the healing effect, a procedure known as electroacupuncture *(15).* Moreover, it is sometimes beneficial to utilise microsystems of acupuncture, which are meridian and acupoint pathways on the ear, scalp, face, hand and other areas representing holograms of the human body. The ear, for example, is a major microsystem, which is thought to represent, an 'inverted fetus' *(16)* (Figure 4). The ear is supplied by the cranial nerve and is region of rich vascularity and innervation.

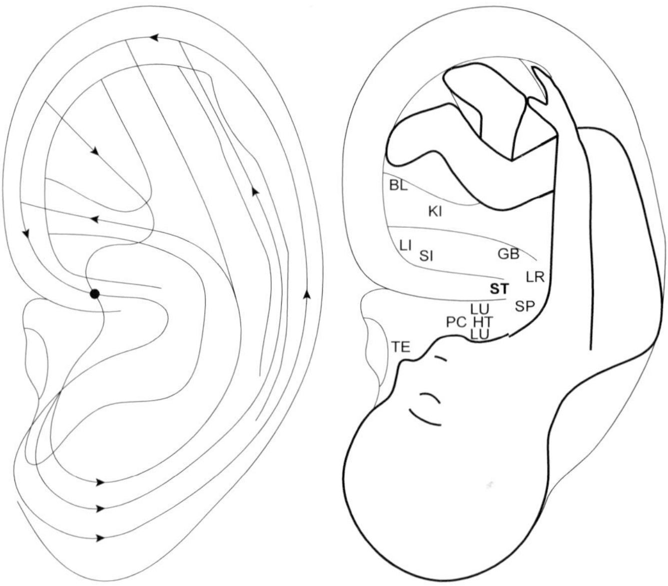

Figure 4. Auricular acupuncture Qi flow and 'inverted fetus' hologram—the Stomach (ST) meridian is near the centre

Dietetically, it is generally suggested for breast cancer patients to avoid eating excessively sweet, cold or raw foods and drinks and to eat yellow vegetables and fruits such as carrots, squash, papayas and persimmons. Yellow teas such as Chinese Five Flower Tea are also recommended. Moreover, mustard greens, Brussels sprouts, artichokes and other green vegetables appear to be beneficial. It would be best to work toward a vegetarian and, indeed, 'vegan' diet (avoiding any and all products derived from living or dead animals). The key is to eat more grains, vegetables and fruit, avoiding meat, sugar and salt (both sodium chloride and monosodium glutamate). It is appropriate to avoid chemically altered or irradiated food as well as the intake of tobacco smoke, alcohol and other drugs (both prescription and non-prescription products). Polypharmacy is also to be avoided. It is recommended to fast for a few days a few times a year to help promote restoration of the Stomach, Small/Large Intestine, Liver and Kidney organs⇔meridians. During fasting sessions, the intake of the cleanest water possible is recommended.

Herbology is the main classical and modern TCM therapeutic modality, and there are many relevant formulas for breast and other cancers *(17-19)*. This is a major ongoing TCM oncology research initiative *(4)*. Certainly, general tonics such as the various Ling Zhi fungi species (the ancient TCM elixir) *(20)* may be efficacious as well as Cordyceps (Dong Chong Xia Cao) and others. Specific herbal remedies may also be indicated, such as in the following two examples of prescriptions for breast cancer in the early stages, which are typically used daily in a decoction with 2 pieces of ginger and 2 pieces of red dates in 2 cups of water:

1. Jiawei Xiaoyao
- Radix Glycyrrhizae 3g
- Radix angelicae Sinensis (baked) 3g
- Radix Paeoniae 3g
- Poria 3g
- Rhizom Atractylodis Macrocephalae 3g
- Radix Bupleuri 1.5g
- Cortex Moutan 1.5g
- Fructus Gardeniae 1.5g

2. Yiqi Yangrong
- Radix Ginseng 3g
- Poria 3g
- Pericarpium Citri Reticulatae 3g
- Bulbus Fritillariae Cirrhosae 3g
- Rhizoma Cyperi 3g
- Radix Angelicae Sinensis 3g
- Rhizoma Chuanxiong 3g
- Radix Paeoniae Latiflorae 3g
- Radix Rehmanniae 3g
- Radix Paeoniae Alba 3g
- Radix Glycyrrhizae 1.5g
- Radix Platycodi 1.5g
- Rhizoma Atractylodis Macrocephalae 6g
- Rhizoma Zingiberis 3g
- Fructus Ziziphi Inermis 2g

An eminent TCM physician developed the following formula, known as Jiawei Guipi, for late stage breast cancer during the Ming Dynasty almost half a millennium ago:

- Rhizoma Atractylodis Macrocephalae 3g
- Radix Ginseng (baked) 3g
- Poria 3g
- Radix Bupleuri 1.5g
- Rhizoma Chuanxiong 1.5g
- Fructus Gardeniae (baked) 1.5g
- Radix Paeoniae (baked) 1.5g
- Radix Glycyrrhizae (baked) 1.5g
- Radix Rehmanniae 240g
- Radix Angelicae Sinensis 240g

There are many other TCM herbal formulas for both early and late stage breast cancer, including Bulbus Bolbosemmae, Bulbus Fritillariae Cirrhosae, Herba Pteridis Multifidae, Concha Ostrea and Radix Actinadiae Chinensis.

Herbal pastes, plasters or poultices may also be beneficial for breast cancer sufferers. These largely consist of mixtures of herbal powders and rice vinegar solutions, which are spread thickly over the breast or other, affected areas of the body and then antiseptically removed and re-applied on a daily or weekly basis.

Medical Qi Gong and related techniques comprise another major TCM therapeutic and preventive approach that is beneficial for breast cancer patients *(21-25).* It is oriented around basic breathing, concentration and posture/movement exercises, which have an impact upon the body, mind and spirit—and one's total health, well-being and quality of life. Basic breathing exercises 3 and 4 are especially useful, since they facilitate the smooth flow of Qi, especially in the Stomach, a Fu, Yang, 'hollow' organ⇔meridian system. These are apparently simple exercise that encompasses inhalation, holding the breath and exhalation. (Fig. 5) For breast cancer patients, concentration should be on the yellow/golden glowing colour and the phonation 'whoo' sound to unblock and remove the Phlegm from the Stomach meridian system. It is also indicated for breast cancer sufferers to regularly practice the Small and Big Circles (moving Qi up and down and around the front and back midlines, specifically, the GV and CV meridians) for spiritual cleansing of their physical, mental and spiritual system (Figure 6). These sets of exercises are a healing, preventive and self-care modality encompassing relaxation, revitalization and rejuvenation, which should only be applied to patients and taught to them by reputable Qi Gong teachers/ masters.

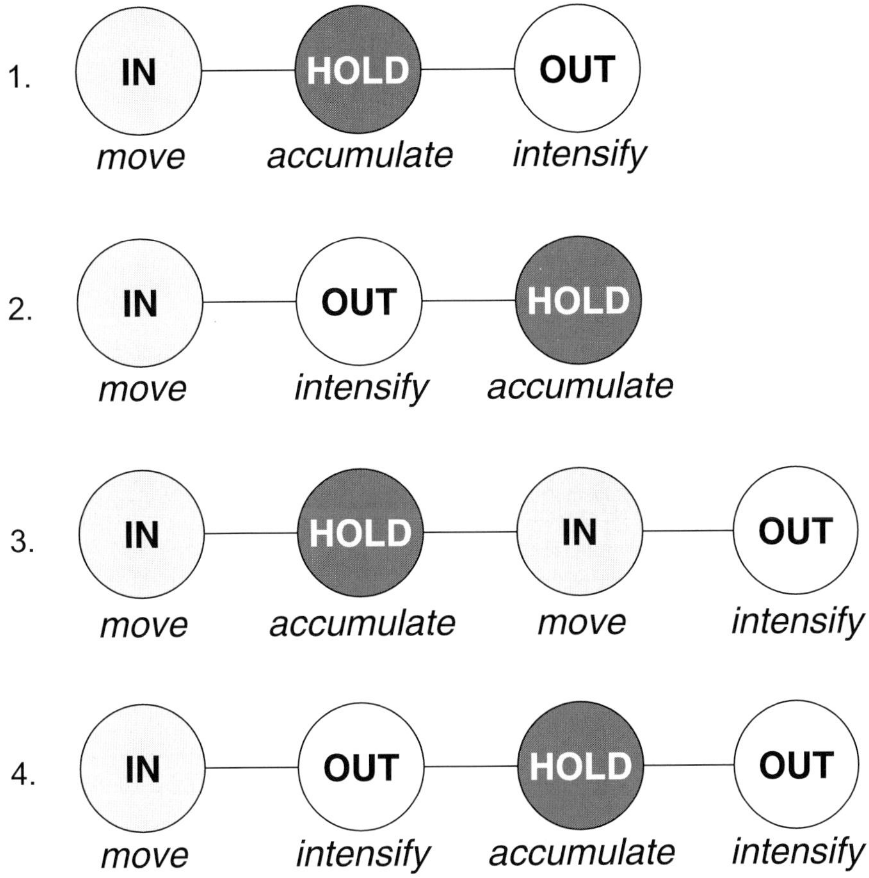

Figure 5. The four basic Aung medical Qi Gong breathing exercises

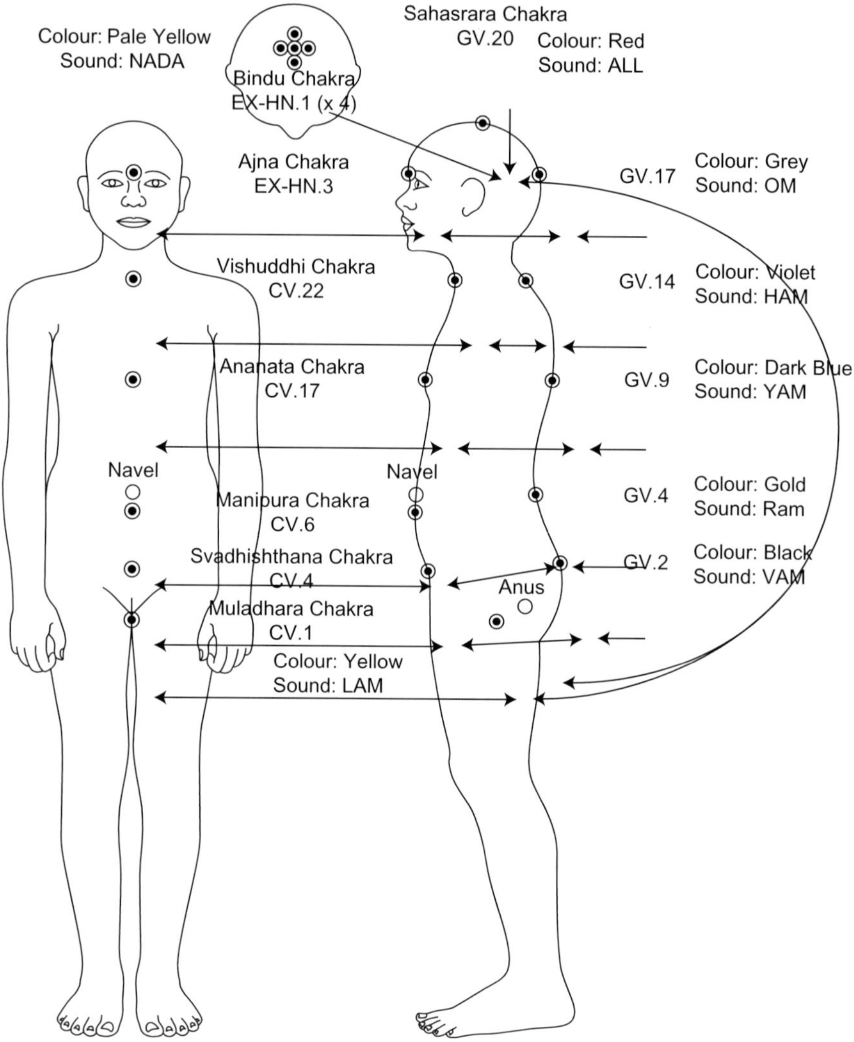

Figure 6. General Qui circulation in the Conception Vessel (CV/Ren).

Patients can be taught to perform their own meridian rubbing therapy—especially along the Stomach meridian from the face down to the toe—and also stretching and slow motion moving meditation Tai Chi Chuan exercises which will balance the body and help to keep it in a harmonious state. Acupressure, which is fingertip pressure applied by the practitioner or the patient (with proper instruction for self-care from the practitioner) on relevant acupoints, is also useful for therapeutic and preventive purposes.

5. CONCLUDING REMARKS: THE INTEGRATIVE PERSPECTIVE

Working together, hand in hand, heart to heart, no medical system can be reduced to the one or the other. They should remain open to include all other evidence-based, safe, efficacious and cost-effective therapies. The key emphasis in the TCM approach to breast cancer is not to let the Phlegm accumulate, not to let the Qi become blocked, not to let the Blood become stagnant and always try to keep one's constitution strong and the immune system constantly alert—this will help keep the body, especially in the Stomach and pertaining meridians and organs, healthy and harmonious. This is a way to help prevent breast cancer in people everywhere around the world. Overall, TCM is not only good for traditional self-care and preventive strategies and therapies, but it can also complement Western biomedical therapies such as surgery, chemotherapy and radiation, helping patients to suffer less pain and have an enhanced total quality of life *(26)*. TCM is natural, holistic and safe—and it appears to be a relatively cost-effective approach compared to the expense of pharmaceuticals, surgery and irradiation. It may take some time for a positive response to occur, but this process does eventually help improve the well being of the patients together with their practitioners, as it has for over so many centuries. No matter how rich, famous and creative a person is, or wherever a person lives or whatever cultural or so-called racial group they identify with, or however they appear or wish to appear, breast or other types of cancers may strike them. We must be constantly alert and mindful of that fact and those effects.

Finally, no matter what medical approach or technique is utilized, it must be emphasized that compassion is at the centre of the holistic, integrative medicine endeavour for the 21st century and beyond (Figure 7).

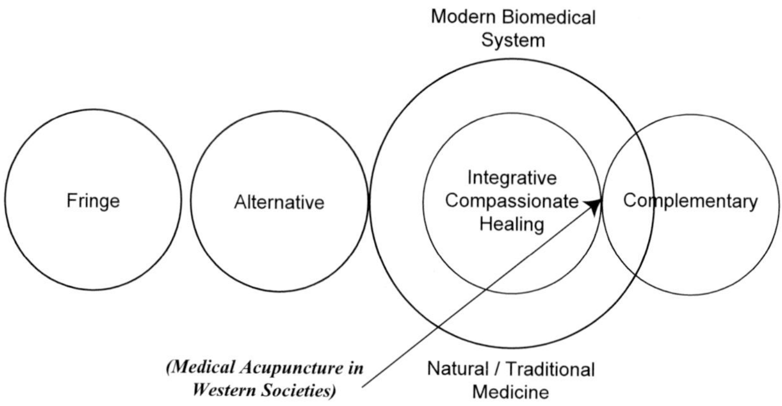

Figure 7. Overview of complementary, integrative medicine initiatives and modalities

Whether or not non-conventional approaches can be fine-tuned or customised for specific cultural or racial groups remains an interesting but problematic quest. For example, a person of Chinese, African or of any other geographical regional descent, whether male or female, young or older, living in their country of choice, may benefit from a Native North American sweat lodge spiritual healing ceremony or other indigenous humanistic practices, just as people from whatever surviving indigenous group anywhere in the world may benefit from the application of various traditional or modern complementary techniques. This is the beauty and strength of multicultural medicine and health care initiatives and practices.

REFERENCES

(1) Kleinman A. Social origins of distress and disease: depression, neurasthenia and pain in modern China. New Haven: Yale University Press; 1986.

(2) Frank BL. Medical acupuncture: a model of integrated healthcare from alternative to mainstream medicine. Colorado Med 1997;94:252-54.

(3) Ernst E (ed). The desktop guide to complementary and alternative medicine. Edinburgh: Mosby; 2001.

(4) Aung SKH. The integrative approach to cancer care: recent initiatives in China, Coherence 2000;2:20-22.

(5) Aung SKH. The clinical use of acupuncture in oncology—symptom control. Acupunct Med 1994;12:37-40.

(6) Aung SKH. Traditional Chinese medicine and cancer care: an integrated approach.Can J Clin Med 1997;4:3-7.

(7) Lerner M. Choices in healing: integrating the best of conventional and complementary approaches to cancer. Cambridge (MA): MIT Press, pp. 369-96.

(8) Dale RA. The forms of Qi (vital energy) in acupuncture. Am J Acupunct 1994;22;71-76.
(9) Aung SKH. The concept of Qi in traditional Chinese medicine: an overview and elucidation, J Austral Med Acupunct Soc 1996;14:6-13
(10) Maciocia G. The foundations of Chinese medicine: a comprehensive text for acupuncturists and herbalists. Edinburgh: Churchill Livingstone; 1989.
(11) Xibin L, Fei X, Yutan L. The treatment of fibrocystic disease Am J Acupunct 1992; 20;313-19.
(12) Helms JM. Acupuncture energetics: a clinical approach for physicians. Berkeley: Medical Acupuncture Publishers; 1995.
(13) Kaptchuk TJ. The web that has no weaver: understanding Chinese medicine. Chicago: Contemporary Books; 2000 (2nd ed).
(14) Unschuld PU. Medicine in China: a history of ideas. Berkeley: University of California Press; 1985.
(15) American Journal of Acupuncture. EAV: electroacupuncture according to Voll. Am J Acupunct 1980 (special issue).
(16) Dale RA. The principles, systems, and holograms of micro-acupuncture. In: Shores J, Oleson TD, editors. ICCAAAM '99 conference manual. Tulsa (OK): Electro Therapy Association; 1999, pp. 26-28.
(17) Bingshan H et al. Thousand Formulas and Thousand Herbs of Traditional Chinese Medicine: Volume 1 - Materia Medica and Volume 2 - Formulas. Harbin (PRC): Heilongjiang Education Press; 1993.
(18) Bingshan H et al. Syndromes of Traditional Chinese Medicine: Analysis of 338 Syndromes. Harbin (PRC): Heilongjiang Education Press; 1993.
(19) Van Benschoten MM. Acupoint Diagnostics and Chinese Herbal Support for Breast Cancer Patients. Am J Acupunct 1995;23;63-72.
(20) Aung SKH. Medicinal applications of fungi and fungal secondary metabolites with emphasis on the use of Ling Zhi in traditional Chinese medicine. Mushroom World 1992;3:17-23.
(21) Jahnke R. Qigong: awakening and mastering the profound medicine that lies within. Am J Acupunct 1989;17:139-51.
(22) Sancier KM, Bingkun H. Medical applications of Qigong and emitted Qi on humans, animals, cell cultures, and plants: review of selected research. Am J Acupunct 1991;19:367-77.
(23) Aung SKH. A brief introduction to the theory and practice of Qigong. Am J Acupunct 1994;22:335 48.
(24) Aung SKH. Qigong sounds: medical therapy through phonation,' Qi 2002;11:39-46.
(25) Motoyama H. Theories of the chakras: bridge to higher consciousness. Wheaton, (IL): Theosophical Publishing House; 1981.
(26) World Health Organization. Cancer pain relief. Geneva: WHO; 1986.

Chapter 13

IMPACT OF CULTURE, EDUCATION AND SOCIO–ECONOMIC STATUS

Rhonda S. Moore, PhD

Department of Epidemiology, Box 189, The University of Texas M. D. Anderson Cancer Center, Houston, TX, USA

> *Culture is the melting pot, which shapes living relationships between individuals, peoples and countries. It opens our minds to riches other than our own, and offers us a door to a new vision of the world. I am convinced we are enhanced by the gaze of others.*
>
> *From Dominique de Villepin, Indo French Waltz, August 1, 2002,*
> *In* http://meadev.nic.in/govt/domini_france-toi01aug.htm

Breast cancer exacts a tremendous toll on the global society. One of the commonest cancers in women, with more than half of the cases in developing nations, it remains a leading cause of cancer related death. Significant variation depending on geographic location has been observed, and an increase in incident breast cancer cases has been noted internationally over the past 40 years. Studies in geographic pathology, for instance, show that incidence rates are generally highest in North America, Western Europe, Australia and New Zealand. They are intermediate in Southern and Eastern European and South American countries, and lowest in Asia and Africa. In addition to the devastating effects on patients and their families, the economic costs of breast cancer are still enormous, both in terms of direct medical-care resources and in the loss of human life. (*1-6*).

In the year 2000, approximately 1,050,000 breast cases were reported around the world with 375,000 deaths. In developed countries the average incidence is 95 per 100,000 and in less developed countries it averaged at 20 per 100,000 population (*3-6*). Roughly 174,000 (17%) deaths in industrialized nations and 140,000 (12%) in developing countries per year are from breast cancer (3-*4*). Nonetheless, despite an increase in cancer detection and prevention efforts around the world, due to its growing

C. Williams et al. (eds.), Breast Cancer in Women of African Descent, 313-332.

prevalence in the international context, and the increasing numbers of incident cases in developing nations, breast cancer remains a global public health problem (also, see Chapter 1 in this book). It is also a growing problem in women of African descent. (*1-4), (7-12*).

Women of African descent represent a socially, economically, culturally, and ethnically diverse group and breast cancer is less common among them than in the other population groups. Nonetheless, breast cancer incidence and mortality rates are rising in these populations, with changes usually more marked in younger women (i.e. >50 years of age). Today, breast cancer is also the leading cause of cancer related death in women of African descent. In 2003 age-adjusted incidence rates in African-American women in the United States were (123.7 per 100,000). In white women, rates in South Africa of 70.2 per 100,000 are comparable to rates from other developed country populations, such as the United Kingdom (56.1 per 100,000) or the United States (140.9 per 100,000). The age-standardized rates of 11.3 per 100,000 compare well with rates from central Africa (Harare age-standardized rate, 20.4 per 100,000, Kampala, 16.4 per 100,000). Breast cancer is even rarer in the Gambia (3.4 per 100,000), and in Kenya (1.08 per 100,000) (*4), (7-12*).

While there is significant variation in patterns of incidence; there are also several important shared characteristics. These include: an increase incident cases in young women, lifestyle factors such as low SES, an aggressive histology, delay, presentation with advanced stage disease, poor diet, obesity, decreased survival time rates, lack of access to adequate medical care, and cultural beliefs and expectations that influence breast cancer decision making and prevention.

In what follows, we review the relevance of culture, and lifestyle factors such as educational and socio-economic (SES) as context effects that impact the experience of breast cancer in women of African descent. Context effects are bi-directional. They can be positive and beneficial or negative and adverse (*13-20*). In the first section, we explore the relevance of culture as a context effect that moulds individual and social beliefs, expectations, experiences and responses to breast cancer. We also address cultural variation in truth telling, and disclosure. In the second section, we further highlight the significance of context in terms of lifestyle factors such as education and socio-economic status (SES). Finally, we summarize our findings.

1. CULTURAL BELIEFS AND EXPECTATIONS: INFLUENCE ON WOMEN OF AFRICAN DESCENT

Cultural factors including beliefs and expectations about breast cancer vary dramatically by ethnic group, geographic location, and the relevance of these cultural factors is increasingly recognized. Culture influences all spheres of human life. It plays a central role in the search for meaning in life, situated in the bio-psychosocial context of health, illness and disease. The distinction between disease and illness is well documented in the literatures. Disease describes a psychopath-physiologic process. Thus, when one speaks about any disease, there is always an accompanying psychological process. Defined by the individual, illness represents an individual's unique, bio-psychosocial experience of being unwell (*21-22*). Therefore, ideas about illness in terms of what to believe, expect, who to tell, what to do, when to disclose also vary across cultural contexts. In addition, lay individuals often perceive illness quite differently from their clinicians (*22-23*).

Breast cancer survivors of African descent, similar to survivors of other chronic illnesses, also seek causes for their illness situated within their personal and socio-cultural frameworks of health, illness, and disease (*24-28*). Nevertheless, their explanatory models of illness and attributions of cancer causation may conflict with Western biomedical concepts of illness and evidence-based medicine. Thus an appreciation of the importance of cultural beliefs and expectations is important since despite advances in treatment and survival, studies still show that lay populations in both Western and non western contexts, still have a universal dread of cancer (*23-27*). The stigma associated with cancer has not (in most contexts) dramatically decreased. For these reasons, cancer across cultural contexts remains an illness heavily invested with symbolic meaning, where cancer and related symptoms threaten irreparable destruction of the patient's body/self, death, pain, and loss of previous sources of meaning in life (*23-28*). For women of African descent, in particular, cultural beliefs and expectations regarding breast cancer often contrast their explanatory models of good health. Good health is perceived in terms of integration of body and mind, consisting of a healthy body as well as a healthy social, emotional, and spiritual life. It is the consequence of harmony in the universe. Further, while health reflects an integration of the individual and the social world, breast cancer is often experienced as stress, conflict, or as disruption, particularly in social relationships (*7), (28-31*). Due to this disruption, both in and between the self and the social world, the survivor of breast cancer is often left to attribute a multitude of causes and meanings as to why this cancer happened to them. This often begins with the stream of thoughts and behaviours, segmented into events that are culturally and personally meaningful (13), (*30*). Breast cancer, for instance, could be caused by the disapproval of ancestral spirits, the embodiment and projection of others' negativity, actions of the jealous, or the non-observance of a taboo (*29-34*). Additional reasons include sorcery, bad behaviours, high-blood pressure, bad blood, worms, incestuous acts, adultery, bad diet, seeding of tumour by surgery, exposure to "wind or air", an inability for the cancer to leave (due to radiotherapy), stepping on poison, crossing an evil line, worry, lump can't be cancer if it is not painful, heartbreak, bad genes, fate, and the devil (*4), (35-42*). Such rich cultural variation is also mirrored in the meanings associated with the word cancer across the different cultural groups that exist within populations of women of African descent. Cancer, generally defined as loss of genomic stability, comes from the Latin word for crab, a clawed creature that reinforces the idea of a fierce, unstop-able force creeping relentlessly through the body. This one word is generally used to describe cancer in most Western contexts. However, among the Igbo (Nigeria), there is no common name for cancer among traditional healers (*dibias)*. Indeed, often more than one word describes cancer. Additional aetiologies are also associated with the disease. For instance, certain healers also use local terms to describe cancer as a disease of the breast (*mbubu ara)*, or as ulcers (*onya)* at various locations on the body. Among the Yoruba (Nigeria), the word for cancer (*aarun jejere*) literally means "the disease that consumes one" or a condition, which cannot be cured (C.K.O. Williams, personal communication, 2003). Healers may also believe that the causes of cancer lay outside the body, others may note that it exists inside the body, and still others claim it is both outside and inside. What this means is that the treatment of patients is highly influenced by individual subjective perceptions about the causes of the cancer. In the case of the Igbo, it also raises additional questions and concerns, as to whether or not cancer is a foreign illness not known in *Igbo* terminology or if it is a new disease that has not yet been classified by the *Igbo* (*22*). In Egypt, for instance, cancer is often called the "insidious, the wicked, cruel or very bad disease." In Zimbabwe, there is no word in the Shona vocabulary for cancer. And few people use words meaning borrowing in the ground or parasitic mistletoe (*39*). Likewise, only three of the nine ethnic black languages in South Africa (Zulu, Swazi and Xhosa) are there words for cancer. Consequently, the reactions of patients to these words may not include any concept of a disease that may spread to other sites of the body, or that requires any special treatments to effect a cure. Parallel to this is the finding that the majority (65%) of 100 African women presenting for the first time with breast cancer surveyed by the authors were found to have advanced disease at the time of referral (*39*). Other studies describe the potential power of negative thoughts given that an angry or jealous person with enough power can send breast cancer to another person. In South Africa, for example, an evil *sangoma* (sorcerer) is believed to have poisoned the breast cancer patient by placing the cancer in her food, taking it to her when she slept, or leaving it on the ground for her to walk over. Once this poison enters the body, it moves to a specific site (in this case the breast), and unless the cancer is drawn out through indigenous medicines and healing rituals (*imbizas)* it will kill the patient (*30*). Other findings in African-American women in North America also emphasise this point since many women believe that once diagnosed that only a "chosen few" actually survive breast cancer. Others claim to "know" that tumours are caused by being hit in the breast during an act of violence, that cancer spreads when the air hits it (surgery), that cancer is a disease of white women, and that it is a condition of the mind and can be cured by prayer (*40-47*). Of particular relevance is the fact that there is the belief that if the individual has enough mindfulness, faith and hope, they can alter the course of the disease in the body. Such beliefs express fundamentally American notions of personhood, autonomy and the power of thought for good or ill to transform body function. On the other hand, this can lead to demoralisation, depression and self-blame when (given advanced stage at diagnoses in these patients) the cancer does return (*30), (43-44*).

Still other investigations assess the role of stress as repeated heartbreak, or perceptions that breast cancer as a disease of white women. Research using narrative analysis, suggests that African-American women report that oncology care and treatments are designed to enhance the care of white women, but not for African-American women. Comparable studies on rural women in the US, and Xhosa women in South Africa note that there is the underlying belief that breast cancer is a fatal disease and thus one should abscond from treatment since there is really no point in having hospital treatment (*29-30*), (*41-47*). In South Africa, many patients with cancer, teachers and secondary school students still believe that a special witchcraft causes cancer, and, thus, their first priority is to first reverse the sorcery before presenting to hospital to be treated by modern medical methods. At first the patient seeks help from a traditional healer as a way of dealing with the cause of the disease, for instance, to appease an angry ancestor. If the treatment fails then they can conclude that the patient did not follow the instructions given by the healer because the *inyanga* is never wrong (*29*). In Nigeria, cancer is a condition that cannot be cured, thus many patients prefer to consult traditional healers as these healers can communicate effectively and instil hope. While such practices across do lead to delays in treatment and advanced stage at diagnosis, in the patient's cultural and biomedical understandings, this often does not imply delay in medical treatment (*29*). Likewise, in other rural communities in South Africa, indigenous healers are often perceived to be the only legitimate and successful healers of cancer given their expert knowledge of the causes and cures of cancer (*22), (46*). In other contexts, the failure of curative efforts represent a sign, reinforcing a reliance on traditional therapies, at the expense of Western biomedicine. In both Nigeria and Tanzania, for instance, the failure of Western medicine to obtain a cure for advanced stage cancer is often perceived as a signal to initiate other alternative health care services. For these reasons many patients and families request a discharge, perhaps even against medical advice to seek the care of traditional healers (*22), (46*). These cultural beliefs and associated practices, often termed c*ancer fatalism* in the oncology literatures, are enduring, working, against Western cancer detection and prevention efforts since why would anyone endure cancer screening tests that to them seem to serve only as heralds of a disease that will ultimately kill them. This often leads to doubts regarding the efficacy of Western biomedical treatments, if prevention is truly worthwhile, or if the breast cancer can actually be cured using western biomedical tools (*29), (41-47*).

In sum, despite apparent advances in medical science that have led to effective treatments for cancer, such myths influence beliefs about causation, expectations regarding the meanings and the course of illness, treatment can serve as barriers to early detection, treatment and recovery. Reinforcing these points are negative attitudes towards cancer or the environment where cancer is treated which can also perpetuate cultural barrier to communication between patients and clinicians and may influence decision making about referrals and treatments (*22*), (*29-30*), (*41-42*), (*46-48*). These beliefs and expectations are often culturally specific, and reinforce the importance of eliciting and addressing patient beliefs so that they do not cause unnecessary distress, nor reduce the chance of a clearer decision about treatment.

2. CULTURE, TRUTH TELLING AND DISCLOSURE

Cultural variation has also been observed in terms of truth telling and disclosure. Truth telling evolves in interactions over time and the synchrony of time between clinician and patient shapes the biomedical objectives and the cultural context of care (*49*).

Indeed, what the patients' believes about time has a great deal to do with underlying beliefs and expectations about health, illness and disease (*35*). There are two dimensions of truth in medicine that are equally important: the patient’s subjective perception of disease and the context, which varies according to the historical, cultural and spiritual background of the patient and clinician (*35*), (*49*).

A diagnosis of cancer can cause great suffering to patients and families. While there is significant evidence that the vast majority of cancer patients in Western and increasingly in non-western contexts may wish to be informed of their illness, nonetheless, cultural differences, in time, place and family, all play significant roles. North American studies tend to focus on patient autonomy, implying that it is unethical to not disclose a terminal cancer diagnosis, since the lack of disclosure decreases the patient's knowledge as empowerment in their interactions with clinicians (*43*), (*50*). A study on US oncologist, for example, found that patient concerns are managed over time and in the context of an evolving relationship. At each stage of an extended process of disclosure, consideration is given to the level of hope patients and families should be encouraged to maintain. As with disclosure, hope is staged, given in calibrated, achievable and realistic bits (*43*), (*50*). Hope, for these reasons, is Janus-faced (*43*) with caring conveyed through the treatment process, through the offering of new therapeutic, potentially beneficial treatment options and through the holding out hope for the development of new treatments and hopes for a cure. Then again, hope is also mediated by cultural and socio-economic factors. In the United States, for example, minority, elderly, culturally and linguistically diverse, and medically underserved patients are least likely to have adequate access to care. Often they might not even have a long-standing relationship with their primary caregiver. Others either have non-existent social networks or the networks that they do have may have limited resources or cannot respond adequately to the challenges of the illness. Thus when the clinician tells a patient with few resources the truth about their cancer, they may inadvertently demoralise the patient extinguishing hope, and this can unintentionally impede progress of patients, and in the long term affect the health decisions and behaviours of patients and other members in their communities.

In many other Western and in non-western cultures, relationships with patients are much more hierarchical, protective, and paternalistic, and the family remains the central organizing structure (*43*). Thus when time is in short supply, and the cancer becomes unbearable, an intrusion to be expunged, but only within the context of the family. The role of the family is also highlighted in medical decision-making. Familial support is a coping mechanism whereby members provide mutual economic and emotional support, with the members relying on social ties created and maintained in such groups (*7*), (*30*), (*34*), (*43*), (*46*), (*50-55*). In South Africa, for instance, patients with breast cancer (particularly in rural areas) are not necessarily the key decision-makers with regard to the different therapeutic choices available. Care and help seeking is collaborative involving family members and sometimes elders of the community. Given the role of these individuals in decision-making, patients are often encouraged to abscond from further Western treatment and to visit the traditional healers. African patients living in the milieu of an urban community, with exposure to Western medical standards of care and where there are fewer tribal ties, however, may have the necessary freedom of action and choice to obtain available medical attention. Nevertheless, in general, patients tend to conform to cultural and familial norms since in most cases, it is more important to please the family, given that patients fear rejection and a lonely death (*22*), (*30*). In Cambodia, Africa Japan, and China, for example, there is an also an additional emphasis on family reputation, lineage, and privacy, where the shame associated with cancer as a polluting force is not only inflicted on the individual by their diagnosis; but a status of stigma is also conferred onto the family. The studies that evaluate the matrilineal inheritance have highlighted the role of women (mothers) in the dissemination of information about genetic risk to daughters and sons. Recent studies however, note that in patriarchal societies, the male line of descent from a common ancestor defines the kindred. Thus a disease "running in the family" may be construed as being derived directly from a common male ancestor and this belief may impact on help seeking behaviours and patterns of disclosure (*51-52*).

Perhaps for these reasons, in disclosing a cancer diagnosis in Tanzania, clinicians often choose to invoke *therapeutic privilege.* They factor in issues such as treatment availability, age, the stigma associated with cancer that would adversely affect the family, prognosis about time left to live, and patient poverty, as moral justifications for withholding diagnostic and prognostic information from patients with cancer. Direct disclosure is seen in these instances to harm the patient, to strip them of hope, and clinicians even encourage the patient to see alternative healers to seek physical and spiritual comfort for their disease (*22), (46*). Similarly, Ethiopian immigrants' preference for nondisclosure of terminal illness also arises from cultural beliefs regarding appropriateness of space, time, and familial support. Direct disclosure of cancer status is considered "cruel," inconsiderate" and even damaging since it is deemed as a failure to properly care for the patient, give them hope, or protect them from harm (*53*). Indeed, direct action against the patient's beliefs may distress patients, leading to a deterioration of their health (*53-57*). For these reasons, if the disclosure is made, clinicians may not tell the patient, whilst they may choose, given the prevailing cultural norms, to inform the family that the person has cancer (*46), (53-54*).

It is clear that while hope is important (*43*), it is also culturally informed. Despite this, hope is associated with heightened activation of the autonomic nervous system, personal vigour and social connectedness (*43*). In contrast, demoralisation is associated with opposite effects and the negative messages from the health care environment and from the patient's social milieu reinforce the context effect. This is similar to the what Engel describes as the *giving up/giving up complex,* where following a stressful situation the person feels unable to cope and has no expectation that any change in the environment will possibly help (*14-19*), (*56*), (*58-61*).

3. INFLUENCE OF KNOWLEDGE ON HEALTH

As we have noted, survival after breast cancer remains significantly worse for women of African descent. Poorer survival in this population has been partially attributed to both competing cultural and cognitive models that adversely affect cancer detection and prevention outcomes as well as, actual knowledge gaps about risk, screening and prevention. These alternative models translate into both a lack of awareness about cancer detection (i.e. screening, knowledge about risk) and a failure of cancer prevention efforts.

Health literacy, in general and breast-cancer specific literacy in particular have been shown to be inadequate in women of African descent (please, see Chapter 14 in this book). These factors contribute to the late presentation of the cases in women of African descent. For example, African American women tend to have lower levels of knowledge about the benefits of genetic testing, even though, informed consent for BRCA1 mutation testing often requires adequate knowledge of patterns of inheritance of cancer and the benefits, limitations, and risks of DNA testing. In addition, African American women are also more likely to avoid breast cancer-related thoughts and feelings; while at the same time, they are also significantly less likely than white women to report heightened perceptions of personal risk for breast cancer after their relative was diagnosed with this disease. This may be due to the fact that until recently there was a paucity of educational and screening programs in the US (*3), (11*). In addition, culturally tailored written materials about breast cancer screening and risk for women of African descent with low literacy skills are often not available (*62-65*). This lack of knowledge not only occurs in the Western context as there is also an absence of health educational programs on cancer as well as lack of screening facilities that focus on the unique needs of African women in nearly all countries in sub Saharan Africa. (*62-63*). Thus disparities in rates of later-stage disease and death may be related to lower screening rates due to behavioural (e.g., knowledge gap) and structural (e.g., access) barriers in low SES women (*63*). This lack of transmission of information regarding risk, early signs and symptoms of breast cancer and knowledge of self-examination adversely influences breast cancer survival rates in women of African descent. In contrast, early stage at diagnosis is associated with higher education, residence in urban areas, and socio-economic levels and a positive family history of breast cancer in these populations (*63-65*).

This lack of knowledge is apparent not only in-patients, but also in cancer clinicians (*66-69*). Nzarubara in a study on breast cancer knowledge in Uganda noted that participants did not know the risk factors and could not associate any risk factor with breast cancer. Moreover, knowledge of the art of self-breast examination was practically non-existent (*67*). Other studies on female schoolteachers in Lagos, Nigeria observed than many symptoms were less well known, and that awareness about breast cancer was quite low as only one-quarter of participants were categorised as possessing a satisfactory knowledge of breast cancer (*68*). Njah in a study in Tunisia also established that that there is an absence of preventive practice by health professionals and that risk factors and opportunities for screening were rarely identified by the women who participated in this study (*69*).

These disparities exist across all socio-economic and geographic backgrounds in women of African descent (*70-79*). However; they are especially apparent in rural areas. In the US, where there is heightened awareness of breast cancer, many women, particularly minority women, and those who live in rural areas or in a low socio-economic class, are not being screened according to recommended guidelines. These rural women are also less likely than their urban counterparts to have ever received a mammogram. This difference is attributable to disparities in education level, household income, and health insurance coverage. Rural women were also less likely than urban women to have a clinical breast examination (CBE) (*80*). Pillay in a study on cancer awareness in rural and urban South Africa found that almost one-fifth of the women had not heard of breast cancers, and almost half were unaware of the breast self-examination techniques. Generally lower awareness levels have been found in older and rural women in the African context who were also significantly more inclined to consult traditional healers (than doctors) about lumps in their breast or abnormal cervical bleeding (*71*). Reinforcing these points and often in conflict with evidence based Western biomedicine, in certain rural communities in South Africa, indigenous healers are still perceived to be the only legitimate and successful healers of cancer because of their expert knowledge of the cause and cures of cancer. In contrast, modern medical approaches, including methods of screening such as mammography may be viewed with some suspicion (*22), (30), (46*).

4. IMPACT OF SOCIO-ECONOMIC STATUS ON HEALTH

The relevance of environmental factors such as socio-economic status (SES) on health in general and on breast cancer outcomes in particular has been well characterized in the oncology literatures. SES is a broad term describing economic and social circumstances. SES is difficult to measure directly and studies have previously used a variety of proxy measures to indicate different socio-economic groups (*70-79*). SES may also be determined by background and by early and late life experiences. Equally, SES itself is likely to affect behaviours throughout life.

Lifestyle factors, such as SES strongly correlate with an increased risk of breast cancer and with other negative health outcomes (*75-77), (81-85*). Epidemiological studies have described the socio-economic correlates of breast cancer risk in a variety of settings. In the US, for instance African-American women tend to be poorer than their non-Hispanic white counterparts, and are also 40% more likely to be given initial breast cancer treatments that were below national standards than were non-Hispanic White women (*77*). Moreover, the effectiveness of culturally appropriate interventions is likely to be reduced if women's ability to respond is limited by inadequate insurance coverage in urban and rural populations (*80), (84), (86*). In many other countries including the US, urban areas also carry a greater burden than rural areas, given changes in environmental exposures, which elevate the risk of breast cancers in migrant populations. However as we have noted the opposite is also true (*22), (30), (46), (74-76), (86*).

Disparities in health occur because privilege and power remain unequal in racially and ethnically stratified societies. Indeed, the globalisation of complex chronic diseases seems to confirm the view that all populations are susceptible and that variation in rates can often be understood as the result of differential exposure to environmental causes (*82-85*). Affluence for instance, has been shown to increase the risk for breast cancer (*79*). Richer nutrition, later first pregnancy, delayed menopause, greater use of hormone replacement therapy (HRT), increased tobacco and alcohol consumption in the Western and increasingly in non Western cultural contexts are thought to modulate the increased risk conferred by lifestyle factors (*78-79*). The opposite is also true, given that the actual prevalence of most adverse health risk behaviours remains higher among those with lower levels of education and income (*83-85), (87-93*). Poverty leads to a higher incidence, morbidity and mortality for a number of diseases, including cancers (*92*) probably because environments differentially expose the individual to a variety of risk factors, including chronic stress, which in turn affects neural-immune function. Put more succinctly, systematic variations in health and mortality across the range of income, ethnicity and education, collectively referred to as SES, contributes to the consequences of cancer (*83-93*).

Thus, in both sociological and epidemiological terms, breast cancer is a disease of both poverty and affluence. However, while affluence elevates the risks associated with cancer both in the individual and in nations, low SES is also associated with a variety of adverse health outcomes. Minority, poor, elderly, rural, and other underserved breast cancer patients of African descent continue to disproportionately suffer from decreased survival rates from cancer, often due to delays in time to diagnoses, in referrals for treatment, and in actual treatments offered (*83-92*). In Nigeria, for example, the most common reasons for delay was found to be fears of mastectomy, preference for prayer houses and traditional healers, and economic reasons (*9*). Additionally, due to the lack of universal medical care, few facilities, cancer specialists and centres devoted to oncology care, African patients also often present with advanced stage at diagnosis. For instance, in South Africa, and Kenya, the majority of black women also present with advanced stage disease. Another important environmental factor is a lack of adequate medical care. Individuals from minority, culturally, elderly, linguistically diverse and low SES communities are also least likely to have adequate medical insurance or receive appropriate medical care. Medically underserved patients from the United States, for example, due to an inability to afford adequate medical coverage, often use emergency room services to obtain care for chronic conditions including cancer. As a consequence given the delay in time to treatment, many are diagnosed with advanced stage tumours. There is also evidence that even among those who actually do receive treatment in US medical facilities that the medical therapies, particularly invasive medical techniques are under-utilised in the treatment of African-American, elderly, female, and poor patients with chronic conditions including coronary artery disease and cancer. These cases highlight the need for sensitivity to and awareness of the particular difficulties faced by those with limited resources (*93-111*).

5. WHAT WE DO NOT KNOW ABOUT CANCER IN WOMEN OF AFRICAN DESCENT: UNDERLYING MECHANISMS

The oncology literatures have emphasized the social, cultural and economic factors associated with adverse breast cancer outcomes in women of African-descent. However, we are aware of no studies that have explored how adverse and health-resiliency promoting factors related SES contribute to patterns of the body's principal adaptive responses to psycho -

physiological stress (e.g. poverty, discrimination) stimuli in terms of breast cancer risk in women of African descent. Nonetheless, the body's principal adaptive responses to psycho-physiological stress (e.g. poverty, discrimination) stimuli are mediated by an intricate stress system involving the hypothalamic-pituitary-adrenocortical (HPA) axis and the sympathoadrenal system (SNS). Evidence from animal models also indicate that dysregulation of the system, caused by the cumulative burden of repetitive or chronic environmental stress challenges contributes to the development of a variety of illnesses including hypertension, atherosclerosis, the metabolic syndrome, as well as certain disorders of immune function, including cancer. Furthermore, while the underlying mechanisms of these effects remain uncertain, it is clear that both adverse and health-resiliency promoting factors related to SES and health, influence the developmental process, and start early in life (*93-113*).

Beginning early in life, various interactions between social and environmental factors and genetic susceptibility lead to large individual differences in susceptibility to stress and disease. The sympathoadrenal system (SNS) plays a crucially important role since alterations in sympathoadrenal function because of exposures in early life aid in the development of a phenotype that is adapted to the challenges of the local environment. Sympathetic innervation of peripheral tissues, the responsiveness of sympathetic nerves and adrenal medulla to standard stimuli are susceptible to modification by exposures in early life, such as environmental temperature, nutrition and stress. Social and environmental exposures at crucial points in development have been shown to permanently alter sympathoadrenal function in mammals. For example, very low birth-weight at birth that is often found in low SES and minority women has been shown to elevate the relative risk of developing diabetes, asthma, respiratory problems, heart disease and cancer as the child grows older (*109-121*). Under these circumstances, adaptations in early life may actually prove to be maladaptive in adulthood and, as a consequence, might provide a basis for developmental origins of paediatric and adult diseases or in the permanent impairments in neural regulatory pathways. In addition, even though the SNS is capable of responding to stressors during foetal life, the long-term effects of such exposures on the adult organism are unknown. What we do know is that the hypothalamic-pituitary-adrenal (HPA) axis is fundamental for long-term survival and protection from the ravages of autoimmune disease. However, in response to long-term chronic stress over the life course, physiological and psychological systems fluctuate, in the attempt to meet the taxing demands of an ever-changing environment. Exposure to many types of physical and psychological stressors *over time (*i.e. poverty, discrimination, and other diseases) can initiate a cascade of neuroendocrine events in the HPA axis. This results in the sequential release of corticotrophin releasing hormone (CRH) from the hypothalamus, adrenocorticotropic hormone (ACTH) from the anterior pituitary and, ultimately, glucocorticoids (cortisol in primates and corticosterone in rodents) from the adrenal cortex. Indeed, the stress-induced persistent genetic instability in the context of carcinogenesis may be a general response of tumour cells to a wide range of stress conditions (*113*). This state of fluctuation, as severe variation from homeostasis, has been termed allostasis, and over time, allostatic load builds up. Allostatic load defined as the physiological and psychological costs of chronic exposure to fluctuating or heightened neuroimmune or neuroendocrine response results from repeated or chronic environmental challenges that an individual reacts to as stressful. Poverty or affluence are merely two of the more important environmental factors, which can adversely impact allostatic load, including the individual and social experience of suffering from chronic stress which then enhances susceptibility to disease, or decreased risk for survival once diagnosed with cancer (*114-121*).

6. SUMMARY

Breast cancer, remains the most common malignancy in women of African descent, and is a genetically heterogeneous disease that is difficult to treat. Breast cancer incidence and survival in these populations is modulated through a combination of cultural, lifestyle and biological factors. These factors influence: either the individual's perceived or actual risk for this disease, if they engage behaviours that enhance the ability to detect or prevent this disease in the early stages, or could influence the severity of the phenotype, once the individual has been diagnosed and or treated for breast cancer. For these reasons, the importance of cultural, and lifestyle factors such as SES and education that influence the beliefs, expectations and practices in terms of breast cancer in women of African descent cannot be determined. They also remain central to our understandings of this disease. Nevertheless, women of African descent, particularly young women, who are less educated, uninsured/underinsured, and women who reported not having family history are less likely to have adequate information regarding detection and prevention (*122*). Hence, an emphasis on achievable cancer prevention strategies with a concentration on modifiable lifestyle factors is imperative.

ACKNOWLEDGEMENTS

We are grateful to CKO Williams, M.D. for his insight and comments on this paper. This research was supported by NCI CURE Supplement # 3R25 CA5770-8S2.

REFERENCES

(1) Etzioni R, Urban N, Ramsey S, McIntosh M, Schwartz S, Reid B, et al: The case for early detection. Nat Rev Cancer 2003;3:243-52.

(2) Kelsey JL, Horn-Ross PL: Breast cancer: magnitude of the problem and descriptive epidemiology. Epidemiol Rev 1993;15:7-16.

(3) Sandelin K, Apffelstaedt JP, Abdullah H, Murray EM, Ajuluchuku EU: Breast Surgery International--breast cancer in developing countries. Scand J Surg 2002;91:222-6.

(4) American Cancer Society: Facts and Figures. Atlanta: Georgia; 2003.

(5) Parkin DM, Bray F, Ferlay J, Pisani P: Estimating the world cancer burden: Globocan 2000. Int J Cancer 2001;94:153-6.

(6) Parkin DM, Bray FI, Devesa SS: Cancer burden in the year 2000. The global picture. Eur J Cancer 2001;37 Suppl 8:S4-66.

(7) Bjerregaard B, Kung'u A: Breast cancer in Kenya: a histopathologic and epidemiologic study. East Afr Med J 1992;69:22-6.

(8) Vorobiof DA, Sitas F, Vorobiof G: Breast cancer incidence in South Africa. J Clin Oncol 2001;19:125S-27S.

(9) Ajekigbe AT: Fear of mastectomy: the most common factor responsible for late presentation of carcinoma of the breast in Nigeria. Clin Oncol (R Coll Radiol) 1991;3:78-80.

(10) De Jesus MA, Fujita M, Kim KS, Goldson AL: Retrospective analysis of breast cancer among young African American females. Breast Cancer Res Treat 2003;78:81-7.

(11) Shavers VL, Harlan LC, Stevens JL: Racial/ethnic variation in clinical presentation, treatment, and survival among breast cancer patients under age 35. Cancer 2003;97:134-47.

(12) Mbonde MP, Amir H, Schwartz-Albiez R, Akslen LA, Kitinya JN: Expression of estrogen and progesterone receptors in carcinomas of the female breast in Tanzania. Oncol Rep 2000; 7:277-83.

(13) Di Blasi Z, Harkness E, Ernst E, Georgiou A, Kleijnen J: Influence of context effects on health outcomes: a systematic review. Lancet 2001;357:757-62.

(14) Brody H: The placebo response. Recent research and implications for family medicine. J Fam Pract 2000;49:649-54.

(15) Balint M: The Doctor, His Patient and the Illness. London: Pitman; 1957.

(16) Benson H, Friedman R: Harnessing the power of the placebo effect and renaming it "remembered wellness". Annu Rev Med 1996;47:193-9.

(17) Benedetti F: How the doctor's words affect the patient's brain. Eval Health Prof 2002;25:369-86.

(18) Barsky AJ, Saintfort R, Rogers MP, Borus JF: Nonspecific medication side effects and the nocebo phenomenon. Jama 2002;287:622-7.

(19) Hahn RA: The nocebo phenomenon: concept, evidence, and implications for public health. Prev Med 1997;26:607-11.

(20) Spiegel H: Nocebo: the power of suggestibility. Prev Med 1997;26:616-21.
(21) Engel GL: The need for a new medical model: a challenge for biomedicine. Science 1977;196:129-36.
(22) Nwoga IA: Traditional healers and perceptions of the causes and treatment of cancer. Cancer Nurs 1994;17:470-8.
(23) Moore RJ, Spiegel D.: Cancer Culture and Communications: Kluwer Academic Press; 2003.
(24) Sontag S: Illness As Metaphor: Picador; 2001.
(25) Frank A: The Wounded Storyteller. Chicago: University of Chicago Press; 1985.
(26) Mayer M: Examining Myself: One Woman's Story of Breast Cancer Treatment and Recovery: Faber & Faber; 1994.
(27) Fallowfield L: Truth sometimes hurts but deceit hurts more. Ann N Y Acad Sci 1997;809:525-36.
(28) Bury M: Chronic illness as biographical disruption. Sociol Health Illn 1982;4:167-82.
(29) Hacking A: Breast Cancer in Xhosa Women. SA Journal of CME 1998;6:57-62.
(30) Wright SV: An investigation into the causes of absconding among black African breast cancer patients. S Afr Med J 1997;87:1540-3.
(31) Moore RJ, Spiegel, D.: Uses of guided imagery for pain control by African-American and white women with metastic breast cancer. Integr. Med 2000;2:115-26.
(32) Dunnavant S: African American Women Speak Out About Breast Cancer. Celebrating Life: Usfi, 1995.
(33) Lorde A: Cancer Journals. San Francisco: Aunt Lute Press; 1980.
(34) Spiegel D: Living Beyond Limits: Fawcett Books; 1994.
(35) Lee RV: Doctoring to the music of time. Ann Intern Med 2000;132:11-7.
(36) Lannin DR, Mathews HF, Mitchell J, Swanson MS, Swanson FH, Edwards MS: Influence of socioeconomic and cultural factors on racial differences in late-stage presentation of breast cancer. Jama 1998;279:1801-7.
(37) Ohaeri JU, Campbell OB, Ilesanmil AO, Ohaeri BM: Psychosocial concerns of Nigerian women with breast and cervical cancer. Psychooncology 1998;7:494-501.
(38) Eisenbruch M, Handelman L: Cultural consultation for cancer: astrocytoma in a Cambodian adolescent. Soc Sci Med 1990;31:1295-9.
(39) Colvin HL, J.: Transcultural and language problems in communicataing with cancer patients in Southern Africa. In Zwitter Sa, editor. Communication with the Cancer Patient. NY: The NY Academy of Sciences, 1997, p. 211-22.
(40) Kinney AY, Emery G, Dudley WN, Croyle RT: Screening behaviors among African American women at high risk for breast cancer: do beliefs about god matter? Oncol Nurs Forum 2002;29:835-43.
(41) Phillips JM: Breast cancer and African American women: moving beyond fear, fatalism, and silence. Oncol Nurs Forum 1999;26:1001-7.
(42) Moore RJ: African American women and breast cancer: notes from a study of narrative. Cancer Nurs 2001;24:35-42; quiz 43.
(43) Delvecchio-Good MJ, Good BJ, Schaffer C, Lind SE: American oncology and the discourse on hope. Cult Med Psychiatry 1990;14:59-79.
(44) Spiegel D: Effects of psychotherapy on cancer survival. Nat Rev Cancer 2002;2:383-9.
(45) Mayne L, Earp J: Initial and repeat mammography screening: different behaviors/different predictors. J Rural Health 2003;19:63-71.
(46) Harris JJ, Shao J, Sugarman J: Disclosure of cancer diagnosis and prognosis in Northern Tanzania. Soc Sci Med 2003;56:905-13.

(47) Gregg J, Curry RH: Explanatory models for cancer among African-American women at two Atlanta neighborhood health centers: the implications for a cancer screening program. Soc Sci Med 1994;39:519-26.
(48) Kearney N, Miller M, Paul J, Smith K, Rice AM: Oncology health care professionals' attitudes to cancer: a professional concern. Ann Oncol 2003;14:57-61.
(49) Surbone A: Communication with the Cancer Patient. NY Academy of Sciences 2000.
(50) Blackhall LJ, Murphy ST, Frank G, Michel V, Azen S: Ethnicity and attitudes toward patient autonomy. Jama 1995;274:820-5.
(51) Meiser B, Eisenbruch M, Barlow-Stewart K, Tucker K, Steel Z, Goldstein D: Cultural aspects of cancer genetics: setting a research agenda. J Med Genet 2001;38:425-9.
(52) Koehly LM, Peterson SK, Watts BG, Kempf KK, Vernon SW, Gritz ER: A Social Network Analysis of Communication about Hereditary Nonpolyposis Colorectal Cancer Genetic Testing and Family Functioning. Cancer Epidemiol Biomarkers Prev 2003;12:304-13.
(53) Beyene Y: Medical disclosure and refugees. Telling bad news to Ethiopian patients. West J Med 1992;157:328-32.
(54) Hira K, Fukui T, Endoh A, Rahman M, Maekawa M: Influence of superstition on the date of hospital discharge and medical cost in Japan: retrospective and descriptive study. Bmj 1998;317:1680-3.
(55) Bennett E: Soft Truth. Anthropology & Medicine 1999;6:395-404.
(56) Engel GL: A life setting conducive to illness. The giving-up--given-up complex. Bull Menninger Clin 1968;32:355-65.
(57) Steele CM: A threat in the air. How stereotypes shape intellectual identity and performance. Am Psychol 1997;52:613-29.
(58) Classen C, Sephton SE, Diamond S,: Studies of Life Extending Interventions. Psycho-oncology. New York: OUP, 1998.
(59) Sephton SE, Sapolsky RM, Kraemer HC, Spiegel D: Diurnal cortisol rhythm as a predictor of breast cancer survival. J Natl Cancer Inst 2000;92:994-1000.
(60) McEwen BS: Protective and damaging effects of stress mediators. N Engl J Med 1998;338:171-9.
(61) Bovbjerg D: Psychoneuroimmunology. In Holland J, editor. Psycho-oncology. New York: OUP, 1998.
(62) Anim JT: Breast cancer in sub-Saharan African women. Afr J Med Med Sci 1993;22:5-10.
(63) Young RF, Waller JB, Jr., Smitherman H: A breast cancer education and on-site screening intervention for unscreened African American women. J Cancer Educ 2002;17:231-6.
(64) Coleman EA, Coon S, Mohrmann C, Hardin S, Stewart B, Gibson RS, et al: Developing and testing lay literature about breast cancer screening for African American women. Clin J Oncol Nurs 2003;7:66-71.
(65) Campbell JB: Breast cancer-race, ethnicity, and survival: a literature review. Breast Cancer Res Treat 2002;74:187-92.
(66) Hoffman M, de Pinho H, Cooper D, Sayed R, Dent DM, Gudgeon A, et al: Breast cancer incidence and determinants of cancer stage in the Western Cape. S Afr Med J 2000;90:1212-6.
(67) Nzarubara RG: Control of breast cancer using health education. East Afr Med J 1999;76:661-3.
(68) Odusanya OO: Breast cancer: knowledge, attitudes, and practices of female schoolteachers in Lagos, Nigeria. Breast J 2001;7:171-5.

(69) Njah M, Hergli R, Gloulou J, Bent Ahmed S, Marzouki M: [Knowledge, attitude and behavior of Tunisian women apropos of gynecological cancers]. Soz Praventivmed 1994;39:280-6.

(70) Santora LM, Mahoney MC, Lawvere S, Englert JJ, Symons AB, Mirand AL: Breast cancer screening beliefs by practice location. BMC Public Health 2003;3:9.

(71) Pillay AL: Rural and urban South African women's awareness of cancers of the breast and cervix. Ethn Health 2002;7:103-14.

(72) Dallabetta GA, Miotti PG, Chiphangwi JD, Saah AJ, Liomba G, Odaka N, et al: High socioeconomic status is a risk factor for human immunodeficiency virus type 1 (HIV-1) infection but not for sexually transmitted diseases in women in Malawi: implications for HIV-1 control. J Infect Dis 1993;167:36-42.

(73) Killewo JZ, Sandstrom A, Bredberg Raden U, Mhalu FS, Biberfeld G, Wall S: Incidence of HIV-1 infection among adults in the Kagera region of Tanzania. Int J Epidemiol 1993;22:528-36.

(74) Kengeya-Kayondo JF, Seeley JA, Kajura-Bajenja E, Kabunga E, Mubiru E, Sembajja F, et al: Recognition, treatment seeking behaviour and perception of cause of malaria among rural women in Uganda. Acta Trop 1994;58:267-73.

(75) Mayaud P, Grosskurth H, Changalucha J, Todd J, West B, Gabone R, et al: Risk assessment and other screening options for gonorrhoea and chlamydial infections in women attending rural Tanzanian antenatal clinics. Bull World Health Organ 1995;73:621-30.

(76) Fylkesnes K, Musonda RM, Kasumba K, Ndhlovu Z, Mluanda F, Kaetano L, et al: The HIV epidemic in Zambia: socio-demographic prevalence patterns and indications of trends among childbearing women. Aids 1997;11:339-45.

(77) Forshee RA, Storey ML, Ritenbaugh C: Breast cancer risk and lifestyle differences among premenopausal and postmenopausal African-American women and white women. Cancer 2003;97:280-8.

(78) Rock CL, Demark-Wahnefried W: Can lifestyle modification increase survival in women diagnosed with breast cancer? J Nutr 2002;132:3504S-07S.

(79) Stoll BA: Affluence, Obesity, and Breast Cancer. Breast J 2000;6:146-49.

(80) Sung JF, Alema-Mensah E, Blumenthal DS: Inner-city African American women who failed to receive cancer screening following a culturally-appropriate intervention: the role of health insurance. Cancer Detect Prev 2002;26:28-32.

(81) Keogh P, Allen S, Almedal C, Temahagili B: The social impact of HIV infection on women in Kigali, Rwanda: a prospective study. Soc Sci Med 1994;38:1047-53.

(82) Cooper RS, Kaufman JS, Ward R: Race and genomics. N Engl J Med 2003;348:1166-70.

(83) Catalano RA, Satariano WA, Ciemins EL: Unemployment and the detection of early stage breast tumors among African Americans and non-Hispanic whites. Ann Epidemiol 2003;13:8-15.

(84) Berkman LF, Glass T, Brissette I, Seeman TE: From social integration to health: Durkheim in the new millennium. Soc Sci Med 2000;51:843-57.

(85) Williams DR, Collins C: Racial residential segregation: a fundamental cause of racial disparities in health. Public Health Rep 2001;116:404-16.

(86) Cummings DM, Whetstone LM, Earp JA, Mayne L: Disparities in mammography screening in rural areas: analysis of county differences in North Carolina. J Rural Health 2002;18:77-83.

(87) Jemal A, Thomas A, Murray T, Thun M: Cancer statistics, 2002. CA Cancer J Clin 2002;52:23-47.

(88) Anderson WF, Chatterjee N, Ershler WB, Brawley OW: Estrogen receptor breast cancer phenotypes in the Surveillance, Epidemiology, and End Results database. Breast Cancer Res Treat 2002;76:27-36.
(89) Joslyn SA, West MM: Racial differences in breast carcinoma survival. Cancer 2000;88:114-23.
(90) Govindarajan R, Shah RV, Erkman LG, Hutchins LF: Racial differences in the outcome of patients with colorectal carcinoma. Cancer 2003;97:493-8.
(91) Lochner KA, Kawachi I, Brennan RT, Buka SL: Social capital and neighborhood mortality rates in Chicago. Soc Sci Med 2003;56:1797-805.
(92) Claussen B, Davey Smith G, Thelle D: Impact of childhood and adulthood socioeconomic position on cause specific mortality: the Oslo Mortality Study. J Epidemiol Community Health 2003;57:40-5.
(93) Hall SA, Rockhill B: Race, poverty, affluence, and breast cancer. Am J Public Health 2002;92:1559; author reply 60.
(94) McEwen BS, Wingfield JC: The concept of allostasis in biology and biomedicine. Horm Behav 2003;43:2-15.
(95) Bresnitz EA, Frumkin H, Goldstein L, Neumark D, Hodgson M, Needleman C: Occupational impairment and disability among applicants for Social Security disability benefits in Pennsylvania. Am J Public Health 1994;84:1786-90.
(96) Hedegaard HB, Davidson AJ, Wright RA: Factors associated with screening mammography in low-income women. Am J Prev Med 1996;12:51-6.
(97) Wile AG, Hourani L, Schell MJ: The contributions of patient factors, physician delay, and tumor biology to the outcome of gastric cancer. Am Surg 1993;59:850-4.
(98) Adler NE, Boyce WT, Chesney MA, Folkman S, Syme SL: Socioeconomic inequalities in health. No easy solution. Jama 1993;269:3140-5.
(99) Bobak M, Marmot M: East-West mortality divide and its potential explanations: proposed research agenda. Bmj 1996;312:421-5.
(100) Cannon W: The Wisdom of the Body. Physiol. Rev. 1929;9:399-431.
(101) Devesa SS, Diamond EL: Association of breast cancer and cervical cancer incidence with income and education among whites and blacks. J Natl Cancer Inst 1980;65:515-28.
(102) Felitti VJ, Anda RF, Nordenberg D, Williamson DF, Spitz AM, Edwards V, et al: Relationship of childhood abuse and household dysfunction to many of the leading causes of death in adults. The Adverse Childhood Experiences (ACE) Study. Am J Prev Med 1998;14:245-58.
(103) Marmot MG, Smith GD, Stansfeld S, Patel C, North F, Head J, et al: Health inequalities among British civil servants: the Whitehall II study. Lancet 1991;337:1387-93.
(104) McEwen BS, Stellar E: Stress and the individual. Mechanisms leading to disease. Arch Intern Med 1993;153:2093-101.
(105) Einbinder LC, Schulman KA: The effect of race on the referral process for invasive cardiac procedures. Med Care Res Rev 2000;57:162-80.
(106) Rathore SS, Berger AK, Weinfurt KP, Feinleib M, Oetgen WJ, Gersh BJ, et al: Race, sex, poverty, and the medical treatment of acute myocardial infarction in the elderly. Circulation 2000;102:642-8.
(107) Schulman KA, Berlin JA, Harless W, Kerner JF, Sistrunk S, Gersh BJ, et al: The effect of race and sex on physicians' recommendations for cardiac catheterization. N Engl J Med 1999;340:618-26.
(108) Bradley CJ, Given CW, Roberts C: Disparities in cancer diagnosis and survival. Cancer 2001;91:178-88.

(109) Freeman HP: Cancer in the socioeconomically disadvantaged. CA Cancer J Clin 1989;39:266-88.
(110) Eskandari F, Sternberg EM: Neural-immune interactions in health and disease. Ann N Y Acad Sci 2002;966:20-7.
(111) Webster JI, Tonelli L, Sternberg EM: Neuroendocrine regulation of immunity. Annu Rev Immunol 2002;20:125-63.
(112) Lupien SJ, Lepage M: Stress, memory, and the hippocampus: can't live with it, can't live without it. Behav Brain Res 2001;127:137-58.
(113) Sandhu MS, Luben R, Day NE, Khaw KT: Self-reported birth weight and subsequent risk of colorectal cancer. Cancer Epidemiol Biomarkers Prev 2002;11:935-8.
(114) Bernstein L: Epidemiology of endocrine-related risk factors for breast cancer. J Mammary Gland Biol Neoplasia 2002;7:3-15.
(115) Jaing TH, Hung IJ, Lin JN, Lien RI, Hsueh C, Lu CS: Hepatoblastoma in a child of extremely low birth weight. Am J Perinatol 2002;19:149-53.
(116) Harbuz M: Neuroendocrinology of autoimmunity. Int Rev Neurobiol 2002;52:133-61.
(117) Young JB: Programming of sympathoadrenal function. Trends Endocrinol Metab 2002;13:381-5.
(118) Seals DR, Esler MD: Human ageing and the sympathoadrenal system. J Physiol 2000;528:407-17.
(119) Kee F, Wilson R, Currie S, Sloan J, Houston R, Rowlands B, et al: Socioeconomic circumstances and the risk of bowel cancer in Northern Ireland. J Epidemiol Community Health 1996;50:640-4.
(120) Balsham M, editor: Class and Medical Authority; 1993.
(121) Taylor SE, Repetti RL, Seeman T: Health psychology: what is an unhealthy environment and how does it get under the skin? Annu Rev Psychol 1997;48:411-47.
(122) Strzelczyk JJ, Dignan MB: Disparities in adherence to recommended followup on screening mammography: interaction of sociodemographic factors. Ethn Dis 2002;12:77-86.

Chapter 14

BARRIERS TO SUCCESSFUL MANAGEMENT OF BREAST CANCER

Christopher K. O. Williams, MD, FRCPC
B.C. Cancer Agency, Vancouver Island Centre, Nanaimo, BC and University of British Columbia, Vancouver, Canada

1. HISTORY OF THE WAR AGAINST CANCER: UNDERESTIMATING THE ENEMY

A remarkable degree of advancement has been made over the last three decades in our understanding of the nature of cancer and the successful management of at least some of its variants. Much of the progress is attributable to worldwide efforts following upon the signing to law the United States of America National Cancer Act on December 23, 1971 by President Richard M. Nixon. The Act was based on a blueprint compiled by a panel of experts, the Consultants on the Conquest of Cancer, which had been convened by the Senate of the United States. The struggle against cancer, which led to this extraordinary action of the US Senate, was spearheaded by the American people, under the leadership of Mary Woodard Lasker of the "Citizens Committee for the Conquest of Cancer". Its tactics to get the attention of the US Government included provocative campaign-type full-page advertisements in national newspapers with slogans such as "Mr. Nixon, you can cure cancer" (*1*). The views of the American public and the Congress about "the war against cancer" were that with little extra effort and money, cancer would be "eliminated". It was therefore not surprising that the signing of the National Cancer Act in 1971 was welcome with euphoria and a call "for the end of cancer by 1976" as "an appropriate commemoration of the two-hundredth anniversary of the independence of our country" (*2*). The euphoria of the 1970's was to be followed twenty

C. Williams et al. (eds.), Breast Cancer in Women of African Descent, 333-363.

years later by the gloom and pessimism of the 1990's about the "lost war against cancer."

2. ACHIEVEMENTS OF THE FIRST THREE DECADES OF THE "WAR AGAINST CANCER"

The panel of the Consultants on the Conquest of Cancer was charged with the responsibility of carrying out a study of the state of cancer research in the country as well as advising on the way forward. The result of the process was enhancement and acceleration of biomedical research efforts that was already on the way at the National Cancer Institute of the National Institutes of Health in Bethesda, MD, USA (3). One of the outcomes of the activities unleashed by the implementation of the National Cancer Act is the demystification of cancer, from a conundrum to our present day knowledge of it. The achievement of the first two decades following the US National Cancer Act is best appreciated by contrasting it with what had been achieved in the preceding 100 years. Peyton Rous, who was honoured in 1968 with the Nobel Prize for his work on retroviruses, dating back to 1911 when he discovered the first retrovirus, is said to have stated in his acceptance speech that the mechanism of carcinogenesis was unknown at the time (4). This statement contrasts starkly against what is now know about the process 30 years thereafter. Cancer is now characterised as a disease of genetic instability (5) involving a complexity of biochemical interplay in what is now commonly referred to as the signal transduction pathways (6). Many of the discoveries of biomedical research are leading to interventions that are already impacting on treatment outcomes. Some examples include our current understanding of the molecular basis of chronic myeloid leukaemia (7-9), and the role of HER2/neu in breast cancer pathogenesis (10), observations, which have become models for targeted therapeutic intervention in cancer control (11-13).

Other achievements of the last three decades of biomedical research include major improvements in disease taxonomy through the use of specific agents such as monoclonal antibodies. Cancers can now be more accurately diagnosed and categorised by the definition of their differentiation antigens, otherwise known as tumour markers. The new technology of DNA microarray analysis is providing valuable insights into differences in an individual's tumour. This in turn is providing individual tissue-specific disease signatures that provide diagnosis based on hundreds of informative genes (5). Thus, it is no longer enough to characterise a case of diffuse large B-cell lymphoma as being, for instance, CD4 or CD10 positive. The use of the powerful diagnostic potentials of the DNA microarray analysis has now enabled the recognition of three new subtypes of this disease with two very different prognoses (14). The tool, therefore, is potentially useful in enabling individualised approach to management, by tailoring and design (15-16).

Some have argued that the struggle to understand the mysteries of cancer should not have been characterised as a "war" in the first place. It is universally agreed that the National Cancer Act has led to unprecedented achievements. These include:

a) Strengthening of all types of biomedical research,
b) Growth of the new industry of biotechnology,
c) Unlocking of many secrets of the working of mammalian cell in general, and the cancer cell in particular,
d) Lifesaving advances and positive impact on the lives of millions of cancer survivors in North America and other parts of the economically developed world,
e) Improved quality of live for cancer patients on treatment regimens,
f) More effective pain control, and most importantly
g) Emergence of the new discipline of translational research (3).

Commenting on the achievement in the field of translational research in the period of 1971-1991, Margaret L. Kripke, 1993-1994 President of the American Association for Cancer Research said: "There is absolutely universal agreement that the greatest opportunity in cancer research today -- and perhaps the greatest opportunity of all time in this field – is the area that has been termed translational research; that is, making use of our findings in molecular genetics and basic biological research for the benefit of patients with cancer and other diseases. Translational research truly represents the chance to reap the benefit of our first 20 years of investment in the National Cancer Program. The next 10 to 20 years can be, and undoubtedly should be, a period of enormous advances in cancer diagnosis, prevention, and treatment" (Kripke, 1994 - unpublished). The resources directed to the National Cancer Act have had a direct impact on some 8 million cancer survivors. Many cancers, such as leukaemia, lymphoma, Hodgkin's disease, cervical cancer, and childhood malignancies, which were almost always fatal, are now most often curable. Adjuvant chemotherapy, a direct result of the innovative drug research supported under the National Cancer Act, has decreased the recurrence rate and the mortality associated with many cancers, such as breast cancer (3) (also, see Chapter 8 in this book). It has however been estimated that chemotherapy saves only 2% of the lives of those who receive it over an observation period of 10 years.

3. THE PERSISTENT CHALLENGE

Unfortunately, in spite of the unprecedented advances of cancer in understanding of the biology of cancer and its management, the ancient fearsome connotation of the word "cancer" persists among much of the world societies. This is so not only in cultures in much of the developing world, but also in areas of developed countries (*17*). In fact, it is the disease that Americans fear most (*3*). The statistics of cancer mortality remains extremely grim. It has been likened by Donald Coffey, former President of the American Association for Cancer Research, to the lives lost in five 747 jumbo jets crashing every day for one year (*3*). The persistence of cancer as a public health problem of the most affluent societies in the world has been attributed to the existence of a gap between knowledge and its application at the community level where millions of lives are being lost needlessly. This has reawakened the United States of America to find ways of controlling cancer as a major health problem. The result of this new commitment has led to the birth in the country of the National Dialogue on Cancer in 1999 (*3*).

The National Dialogue on Cancer (NDC) hopes to develop a lasting plan to eradicate cancer as a major public health problem by identifying barriers, gaps and opportunities related to eradicating cancer. Its underlying principles are: a commitment to:

a) Research as the underlying engine that drives the increased understanding of the disease;
b) Effective collaboration;
c) Moral obligation to reach all people, with special attention to those at greatest risk and, in particular, ethnic minorities and the medically underserved who disproportionately carry the greatest burden of cancer (Figure. 1);
d) Staunch dedication to primary cancer prevention and early detection as well as access to quality cancer care and quality of life for those who develop the disease (*3*).

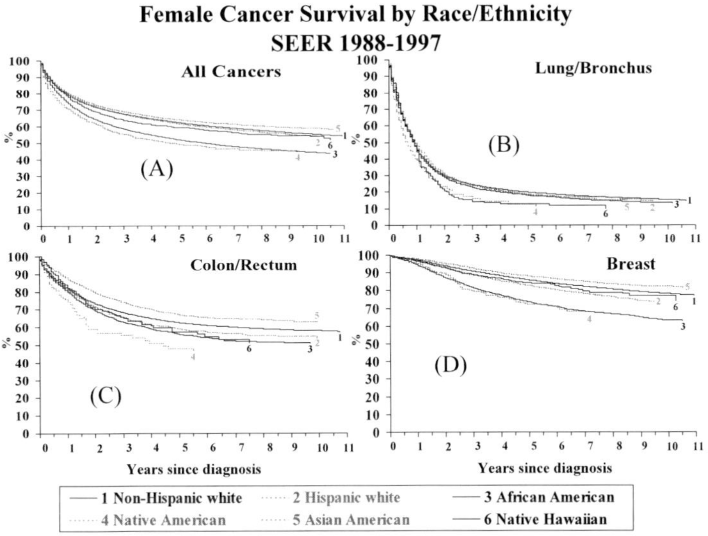

Figure 1. Female survival by race or ethnicity for all cancers combined (A) and for lung cancer (B), colorectal (C), and breast (D) cancers. Surveillance, Epidemiology, and End Results (1988-1997). Adapted from Clegg, LX et al.: Arch Intern Med 2002; 162:1985-93.

The National Dialogue on Cancer has set a target to eradication of cancer in this millennium. Priority Teams have been created to promote research in private, public and not-for-profit organisations, patient empowerment through education, access to health (cancer) care, quality care and surveillance. The overall objective is to revitalise the National Cancer Act. In a way, there seems to be recognition that the present unsatisfactory state of cancer control in the community is the result of a state of ineffective cancer communication, thereby leading to creation of barriers of access to the fruit of achievements that have emanated from of the National Cancer Act.

Cancer care includes risk assessment, primary prevention, screening, detection diagnosis, treatment, recurrence surveillance, and end-of-life care (*18*). Each step in this framework of continuum of care has been the subject of intensive and extensive research over the last three decades. The conceptual framework of quality in the continuum of cancer care (QCCC) represents a systematic approach for assessing factors that influence types of cancer care and the transition between them (Figure 2). Focusing on the steps and transitions in care where failures can occur can facilitate cancer care and ensure measures of quality that promote improved outcomes [Zapka, 2003 #277].

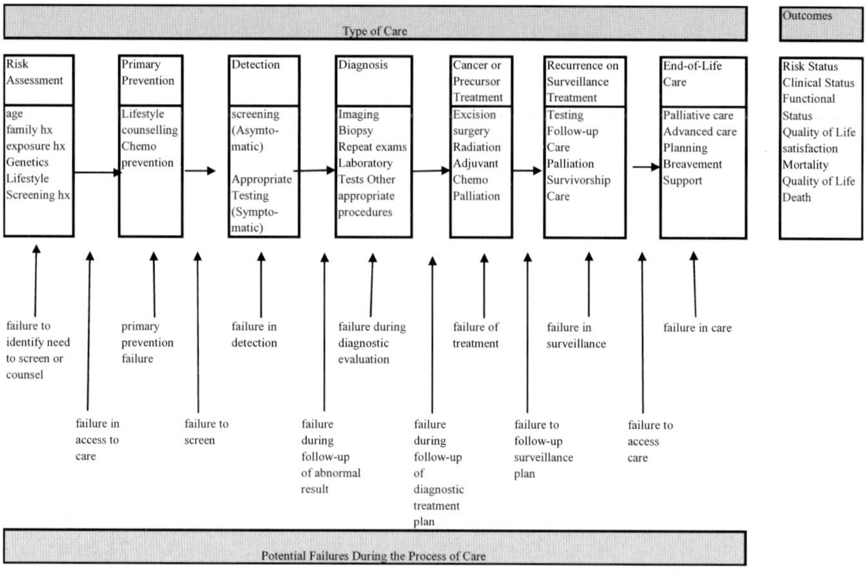

Figure 2. Concept of continuum of cancer care and the potentials of its disruption leading to failed outcomes. Adapted with permission from Zapka et al. Cancer Epidemiol Biomarkers Prev.2003; 12: 4-13.

4. BIOLOGICAL CONCEPTS AND TREATMENT FAILURE

The management of breast cancer over the centuries has changed along with the prevailing biological understanding of the disease process (*19*). Until relatively recently, cancer was view as an anatomical problem that was amenable to rectification by surgical intervention alone. With the advent of effective chemotherapeutic agents after the Second World War, and the demonstration that the natural history of hitherto uniformly fatal diseases like childhood leukaemia could be positively modified (*20*), the stage was sent for curative breast cancer management. The concept of anatomic pattern of direct and lymphatic extension of breast cancer, which was the basis of the Halstedian radical surgery as curative management of the disease, gradually yielded grounds to the modern concept of its early systemic dissemination. It is now well established that the management of early-stage breast cancer by surgery alone is inferior to surgery plus any form of systemic therapy (*21-24*). The Overviews have confirmed a modest survival advantage in patients treated in trial studies. Olivotto and his colleagues (*25*) have validated the results of the clinical trial findings with those of population based tumour registry data. It has, however, not been possible to improve upon the marginal benefit in spite of hundreds of clinical trials examining various potential opportunities ranging from trial of ever increasing list of newer agents, treatment schedules and treatment dose intensities. One major problem appears to be the limitation of our knowledge of the growth pattern of breast cancer cells in micro-metastatic state. The Gompertzian growth pattern as proposed in the Norton-Simon model *(26), (27)* explains why treatment with chemotherapy alone (following adequate local-regional therapy), even when given at above-normal doses, has failed to improve overall survival in breast cancer to any significant extent. The challenge for the clinical investigator is to identify the biological barriers that limit progress in this respect and how to overcome them.

5. GENETIC EPIDEMIOLOGICAL BARRIERS

Several studies indicate that breast cancer carries a more unfavourable prognosis among African American women as compared to Caucasian American women (*28-42*). A number of studies have also shown that when patients of different racial backgrounds receive equal treatments for identical disease stages of breast cancer, equal results are obtained (*43-44*). It has also been suggested that based on results of clinical and genetic epidemiological studies, the prospect of explaining differences in treatment response strictly on the basis of race, once known prognostic disease features are accounted, is diminishing (*45*). A President's Cancer Panel meeting in 1997 concluded that biology does not dictate race. However, the social consequences of having an identified race can have biological effects. This is because race is a surrogate for socio-economic status (*46*). In other words, genetic make up of racial groups do not constitute barriers to successful treatment in such groups. The social implications of these differences are probably responsible for the differences in treatment outcomes.

6. AGE AS BARRIER SUCCESSFUL CARE

Age is an important factor in breast cancer incidence, mortality and treatment outcomes. Two population-based studies indicate that the risk of death from breast cancer is highest among the youngest and the oldest cohorts when compared with patients of intermediate age, even when the analysis allows for differences in initial tumour stage (*47-48*). Breast cancer is rare below the age of 35 years with only 2% of patients with the disease being in this age category in the US Surveillance, Epidemiology, and End

Results (SEER) Program (*49*). However, when it does occur in this age group, the prognosis of the disease is significantly poorer than in older women (*47-48), (50-51*).

The incidence and mortality of the disease increases significantly with age and the majority of breast cancer deaths occur in women 60 years old or older in the United States and other industrialised countries (Figure 3) (*34*). Thus, breast cancer is likely to become more prevalent in the future in view of the ageing trend in the population of these countries. The biology of the breast cancer variant in the elderly is associated with favourable prognosis, including a higher tendency for node negative, oestrogen and receptor positive status as well as relatively high prevalence of low S phase component and low prevalence of HER2 positivity (*52*). Tamoxifen in the elderly is beneficial in terms of improvement of disease free and overall survival (*53*). However, Tamoxifen related symptoms tend to increase with age with the oldest age group having the greatest risk of serious adverse effect (*54*).

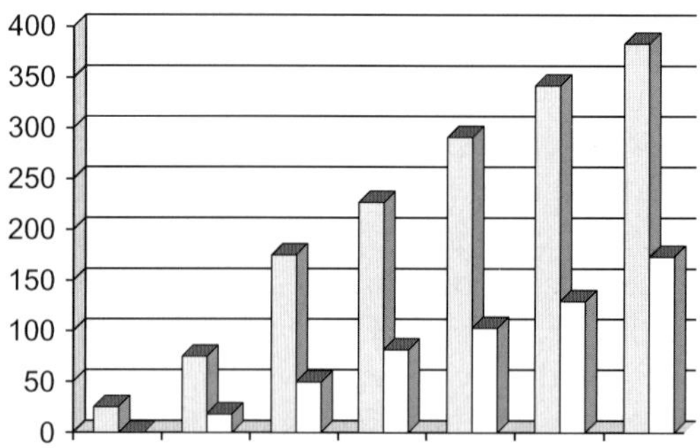

Figure 3: Incidence of and mortality from breast cancer and age [modified from Yancik et al (13)]. The **x-axis** shows incidence (per 100,000); the **y-axis shows** age range in years. Adapted from Muss; HB: JNCI Monographs 2001:30; 52-55.

□ Incidence □ Mortality

Rising age is also associated with rising prevalence of co-morbid conditions and mortality from such conditions, which are not related to breast cancer. The efficacy of chemotherapy diminishes with advancing age as co-morbidities increase in prevalence. The co-existence of these two factors is probably one of the reasons that elderly women tend not to be offered chemotherapy as part of their treatment modalities of early breast cancer as their presence significantly predict mortality from non cancer causes. This would seem to be at least one of the reasons why physicians/clinical investigators fail to solicit elderly patients for clinical trials (*34*). Thus, although breast cancer is highly prevalent in the elderly, management in this age group may be suboptimal. Clinical trials are required to address the question of optimisation of treatment cancer in the elderly.

7. VERY YOUNG AGE AS A BARRIER TO SUCCESSFUL TREATMENT

In order to investigate whether young age at diagnosis is a negative prognostic factor in breast cancer and how stage of disease at diagnosis and treatment influence such an association, the Danish Breast Cancer Cooperative Group carried out a retrospective cohort study based on a population-based database of patients enrolled in their clinical trials studies between 1977 and 1996 (*55*). The study included 10,356 women who were less than 50 years old at diagnosis. These consisted of 867 (8.4%) who were less than 35 years of age, and 9489 that were aged between 35 and 49 years. They observed that young women who were judged to have low risk disease, and therefore did not receive adjuvant chemotherapy had a significantly increased risk of dying. The risk increased with decreasing age at diagnosis as follows (adjusted relative risk - RR - with the 45-49 year old cohort as the reference group having RR of 1.00) 40-44: 1.12 (95% confidence interval - CI- of 0.89-1.40); 35-39 years: 1.40 (CI of 1.10-1.78); <35 years: 2.18 (CI of 1.64-2.89). However, no similar trend was observed in patients who received chemotherapy. The increased risk in younger women who did not receive adjuvant treatment compared with those who did remained when women were grouped according to presence of node negative disease and by tumour size. The authors concluded that young women, on the basis of age alone should be regarded as high-risk patients and be given adjuvant cytotoxic treatment (*55*).

The International Breast Cancer Study Group (IBCSG) reported similar findings based on 3700 (including 314 or 8.5% women under the age of 35 years) premenopausal and perimenopausal women from a series of their clinical trials in which CMF based regimen of up to 12 monthly cycles were administered (*56*). They observed that relapse and death occurred earlier and more often in younger (<35 years old) than older (≥35 years old) women. The 10-year disease-free survival (DFS) for younger patients was 35% (SE 3) versus 47% (1) for older patients (hazard ratio: 1.41 [95% CI 1.22-1.62], P<0.001), and the 10-year overall survival (OS) was 49% (3) versus 62% (1) (1.50 [1.28-1.77] p<0.001) respectively. Younger patients with ER-positive tumours had a significantly worse prognosis than did younger patients with ER-negative tumours, their 10-year DFS and OS being 25% and 47% (p = 0.014), respectively. In contrast, however, the prognosis among older patients was similar for ER-positive and ER-negative tumours (10-year DFS:

45% versus 46% - p = 0.27). These data have been interpreted to mean that young premenopausal breast cancer patients treated with adjuvant CMF chemotherapy had a higher risk of relapse and death than older premenopausal patients did, especially if they expressed oestrogen receptors. The retrospective analysis of the IBCSG data would tend to suggest that the endocrine effects of chemotherapy alone were insufficient for the younger patients with endocrine responsive breast cancer. Such patients should strongly consider additional endocrine therapies (Tamoxifen or ovarian ablation) if their tumours express oestrogen receptor (*56*).

Goldhirsch and colleagues (*49*) have provided extensive analysis of data from the clinical trials of the various collaborative groups on the possible interaction of age and ER status vis-à-vis treatment outcomes. These are reproduced in tables 1and 2. As shown in table 1, the relative risk of relapse is substantially higher for young patients with ER-positive tumours. In contrast, the difference in outcome with respect to age group is much smaller for patients with ER-negative tumours. The interaction between age and ER status was statistically significant for the cohorts in the IBCSG, NSABP and SWOG trials. A similar observation is reflected in the lower half of the table, which shows markedly poorer 5-year DFS among the younger ER-positive women compared to the observation among ER-negative women of similar age range. Table 2 shows that for both young and older premenopausal women with ER-positive tumours, failure to achieve the state of amenorrhoea is associated with higher risk of relapse, although the result in younger women is statistically uncertain because of the small sample size. The difference in the proportion of patients who achieve amenorrhoea in the young age group might contribute to the poor outcome of these women with endocrine-responsive tumours treated with chemotherapy alone. In contrast, the association between failure to achieve chemotherapy-induced amenorrhoea and the risk of relapse is not statistically significant for patients with ER-negative primaries. The authors have concluded that there is a need for tailored treatment investigations, especially in the population of younger women with early-stage breast cancer. Such investigations should explore the efficacy (and risks) of a combination of chemotherapy and "optimal" endocrine therapy because of evidence that the "current approaches are suboptimal".

Table 1. Relative Risk of relapse* and corresponding 5-year disease-free survival* for Premenopausal women in chemotherapy-only groups in trials conducted by IBCSG, NSABP, ECOG and SWOG†

Group	Total Patients	ER positive		ER negative		Interaction P
		< 35 y	≥35y‡	<35 y	≥35y‡	
		Relative risk of relapse (No. events/No. of patients)				
IBCSG	2233	1.84(72/96)	1.00 (referent) (737/1353)	1.13(50/88)	1.02(37/696)	.009
NSABP	5849	1.72(254/402)	1.00 (referent) (1210/2716)	1.27(214/441)	1.12(1045/2290)	.0001
ECOG	1112	1.54(42/71)	1.00 (referent) (274/602)	1.40(40/73)	1.26(195/366)	.17
SWOG	670	2.67(11/29)	1.00 (referent) (48/293)	0.81(7/55)	1.13(52/293)	.012
		5-y disease-free survival, %*				
		39	60	56	59	
IBCSG		49	66	59	63	
NSABP		53	66	55	58	
ECOG		73	89	91	88	
SWOG						

*Includes breast cancer relapses, second primary breast tumors, and deaths without relapse for IBCSG (also includes non-breast second primaries), ECOG, and SWOG; includes only breast cancer relapses (other events are censored) for NSABP.

† Cohorts defined by age and estrogen-receptor (ER) status are compared with the reference population of older women with ER-positive tumors (number of events/number of patients are shown in parentheses).

IBCSG = International Breast Cancer Study Group; NSABP = National Surgical Adjuvant Breast and Bowel Project; ECOG = Eastern Cooperative Oncology Group; and SWOG = Southwest Oncology Group.

‡Premenopausal women ≥35 years old for IBCSG, ECOG, and SWOG; 35-49 years old for NSABP.

Adapted from: Goldhirsch et al. JNCI Monographs 2001; 30: 44-51.

Table 2. 9-month landmark analysis of premenopausal patients with lymph node-positive disease treated with at least 3 months of classical CMPF* on International Breast Cancer Study Group Trials: relative risk of relapse comparing patients with no amenorrhea at.9 months versus patients with chemotherapy-induced amenorrhea at 9 months.

	No. of Patients				
Age, y, and receptor status	**No amenorrhea**	**Amenorrhea**	**Relative Risk**	**95% confidence interval**	**P**
< 35, Estrogen receptor (ER) positive	61	21	1.53	0.80 to 2.94	.20
≥ 35, (ER) positive	320	773	1.34	1.12 to 1.61	.0014
< 35, ER negative	54	23	1.14	0.42 to 3.15	.79
≥ 35, ER negative	152	387	1.11	0.85 to 1.45	.45

CMF = combination of cyclophoshamide, methotrexate, and 5-fluorouracil.
Adapted from Goldhirsch et al.
JNCI Monographs 2001; 30: 44-51.

8. MAN-MADE BARRIERS TO SUCCESSFUL MANAGEMENT OF BREAST CANCER

The differences in breast cancer survival among racial/ethnic groups that have been well documented (*57-58*) have been attributed to factors such as disease characteristics at diagnosis, economic resource inequalities and other social factors, and disparity in treatment access and quality care. The existence of cancer outcome disparities between race groups despite improvements in diagnosis and treatment over recent decades is considered by most to be largely a consequence of personal and institutional resource limitations in specific racial/ethnic communities rather than of any intrinsic aspect of race itself, although cultural and social factors associated with ethnicity may also play a role (*45*), (*59-62*) (also, see Chapter 13 in this book). There is evidence that where resource disparities are absent, cancer screening participation, disease stage at diagnosis, treatment, and subsequent outcomes are similar, regardless of race. This indicates that economic inequality is a major underlying factor in outcome disparities.

9. CULTURAL BARRIERS TO SUCCESSFUL MANAGEMENT

The fear of cancer is a universal phenomenon. Recent advances in understanding the nature of the disease and its treatment has contributed to demystifying the disease in certain communities. In much of the world communities, however, it is probably safe to say that fear and misunderstanding of what we know about the disease is the norm rather than the exception. Good education and acceptable societal values of a community do not necessarily exclude age-old cultural believes and attitudes about cancer as a study among American Samoans by Ishida and colleagues (*17*) showed. In spite of the cohort's education level, which ranged from grade 9 to some college, and priorities that included health, family and education, all expressed negative perception of cancer. Reported major barriers to mammography were fear, opinion that "it is not a priority", and pain. Over half of the women reported concern over their breasts being touched either by themselves or by others. The study participants had strong belief that cancer meant death and, therefore, no cure. Among the authors' recommendations of cultural sensitivity were the need for health providers to emphasise to Samoans that examination of the breast is not sexual, admonition about gentle handling of the breasts during examinations as important considerations for the prevention of breast cancer in Samoan women. In their studies among low-income African American women, Gregg and Curry (*63*) came to the realisation that patients and clinicians must each understand how the other perceives cancer, its prevention and its treatment, and that the cancer models held by the patient population differs significantly from those held by physicians.

10. HEALTH LITERACY AND CANCER COMMUNICATION AS BARRIERS TO SUCCESSFUL MANAGEMENT

Health literacy is increasingly recognised as a critical factor affecting communication across the continuum of cancer care (*64*). According to the United States National Adult Literacy Survey (NALS), considered to be the most accurate portrait of literacy in the American society, about one in five American adults may lack the necessary literacy skills to function adequately in the society. An additional 27% with marginal literacy may also struggle with cancer communication (*65*). As patients, such individuals are at a disadvantage in their ability to obtain, process, and understand cancer information and services needed to make appropriate health care decisions.

Patients with poor health literacy have a complex array of difficulties with written and oral communication that may limit their understanding of informed consent for routine procedures and clinical trials (*66*). These patients report worse health status, and have less understanding about their medical conditions and treatment (*67*).

An individual's health literacy may be worse than his or her general literacy (*67*). Functional literacy is context specific; it is therefore likely that many individuals at all literacy levels lack a clear understanding of cancer control guidelines and screening recommendations. Individuals who have been screened for cancer may lack basic understanding of test results. Likewise, cancer patients may lack adequate knowledge of treatment recommendations and clinical trial options (*66*). Given the increasing options and complexity of cancer care, which require more involvement of patients in decision-making, the magnitude of the disparity between those with adequate cancer related literacy and those without it is probably growing. The problem is further exacerbated by the greater prevalence of low health literacy among the elderly, who not only bear the greater burden of cancer, but are growing in number as well (*66*). People with inadequate literacy skills come from a variety of backgrounds; they are native-born and immigrants, come form all races and classes, and have no visible signs of disability. However, functionally illiterate adults are more likely to have more health problems, to live in poverty, to have fewer years of education, and to be older. Lack of adequate literacy is twice as common for Americans over 65 years of age and among inner city minorities, the primary users of Medicare and Medicaid (*65-66*).

Adults with lower socio-economic and older age are less likely to be screened for cancer, more likely to have advanced cancer at presentation, and suffer disproportionately high levels of cancer mortality (*18), (68-72*). A number of studies also have associated late stage at diagnosis of breast cancer with low socio-economic status (*62), (73-75*). As new treatments emerge, risk communication becomes increasingly important in public health, clinical medicine and clinical research (*66*). However, it is often difficult for physicians to quantify risks in language that is easily understood by patients (*76*), as patients with low health literacy may lack the numeracy skills needed to understand and apply cancer risk communication. According to Siminoff and colleagues (*76*), 80% of breast cancer patients made final decisions about adjuvant therapy after one visit with a medical oncologist, during which little specific risk communication occurred. The same authors found that 60% of patients overestimated their chance of cure by 20% or more compared with their physicians' estimations.

Low literacy complicates provider/patient communication as it often results in a mismatch in logic and experience between the patient and the physician. The physician's scientific and clinical background is based on facts, probability and their previous clinical experience. The detailed information that may emanate from this background is often not relevant or useful to the patient with low literacy, who are more likely to be responsive to information based on a health belief model priority and to patient action, motivation, and self-employment (*64*), (*66*).

The challenge that the average patient faces with standard consent forms has been the subject of numerous investigations. Davis and colleagues tested the hypothesis that a simplified, illustrated consent document formatted using low literacy recommendations would be less intimidating and more easily understood by individuals with marginal-to-low reading skills (*77*). Participants were tested for reading ability and then asked to first read either the standard Southwestern Oncology Group (SWOG) consent form (college senior reading level) or a simplified form (seventh-grade reading level). Patients stated that the simpler form was easier to read, less frightening, and significantly less likely to discourage them from participating in the clinical trial. They preferred it almost two to one to the SWOG consent form (*66*), (*78*). Patient comprehension of the forms, however, was essentially the same for the simplified form (58%) and the SWOG form (56%). Patient's comprehension score was related to their reading levels. Patients on an eight-grade reading level or below were able to comprehend less than half of the essential information, while those on a ninth-grade reading level or higher were able to understand about three-fourth of the consent document (*77*). The findings raise serious questions regarding the adequacy of written consent documents for clinical trials for the substantial proportion of Americans with marginal-to-low literacy skills.

Until recently, most health materials, including advisory materials published by major cancer control organisations such as the National Cancer Institute (NCI), the American Cancer Society (ACS) and others, had an average reading level between tenth and eleventh grade. Although the average American adult has achieved at least a twelfth grade education, the average reading level for American adults is estimated to be at the eighth or ninth grade (*66*), (*79*). The NCI of the United States and The National Work Group on Cancer and Literacy (NWG) (*80*) are collaborating to focus attention on the need for more effective communication with individuals who have limited literacy skills (*66*). The NWG, in its recommendations to reduce disparities in cancer screening and treatment, acknowledged that it is not yet known if using specially developed low-literacy educational materials would improve the health outcomes of patients with low-health literacy. The group recommended that when written communication with patients is essential, materials should be written at the fifth-grade reading level or lower. Specifically, the NWG noted that materials intended for low-level readers should change unfamiliar words to more commonly-used words, use large fonts, and have substantial amounts of blank (white) space to make the text look easier to read (*66*).

11. BARRIERS TO SUCCESSFUL MANAGEMENT OF BREAST CANCER IN WOMEN OF AFRICAN DESCENT

11.1 Biological Determinants of Treatment Failure

Although Caucasian American women have a higher incidence of breast cancer than their African American counterparts, the mortality rates of the latter meet or exceed those of the former, indicating a poorer survival experience of African American women (*81-84*). Differences in breast cancer survival among racial/ethnic groups have been noted in studies as well as national cancer statistic summaries (*58), (85*). Numerous factors have been implicated as sources of these differences, including disease characteristics at diagnosis, economic resource inequalities and other social factors, and disparities in treatment access and (possibly) efficacy (*45), (86*). When compared to the observations in Caucasian American women, the prognosis of breast cancer in Asian-American women (*31), (87*) is more favourable, while it is less favourable in women of Hispanic origin (*87-88*), American Indians and African Americans (Figure 4) (*45), (88-89*). The most extensively studied disparity in breast cancer prognosis is that between African-Americans and Caucasians. Dignam (*86*) documents this in considerable details. The disease characteristics that contribute to poorer outcomes in African-American women with breast cancer have been discussed elsewhere in this book (please, see Chapter 8 in this book).

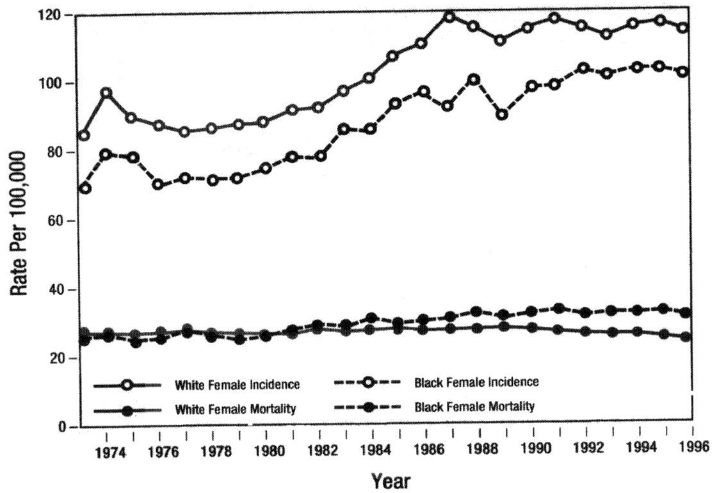

Figure 4. Breast Cancer Incidence and Mortality among African American and Caucasian Women, 1973 – 1996. Adapted from Ries et al, 1999 Modified from Dignam 2000.

African American women with breast cancer are less likely than Caucasian American women to be diagnosed while their disease is still at a localised stage. Jones and colleagues (*90*) sought to determine the extent that observed racial differences in stage at diagnosis of breast cancer could be explained by racial differences in obesity, specifically severe obesity. They observed that obesity was associated with both race and stage at diagnosis: African American women were significantly more likely than Caucasian American women to be severely obese were (26% vs 7%, respectively), and severe obesity was significantly associated with diagnosis at TNM stage II or greater. They concluded that the higher prevalence of severe obesity among black women might play an important role in explaining their relative disadvantage in stage at diagnosis of breast cancer. In a population-based study in North Carolina comprising 791 breast cancer cases (302 in African American and 489 in Caucasian American women), African American women were more likely to have later stage breast cancer, and much more likely to be severely obese as well as being in the highest tertile of waist:hip ratio. The observations suggest that obesity and body fat distribution, in addition to socio-economic and medical care factors, contribute to racial differences in stage at breast cancer diagnosis (*91*). Similar results were reported in a study in an urban Nigerian population, in which Adebamowo and his colleagues showed a significant association between the highest tertile of waist:hip ratio and the risk of breast cancer (odds ratio = 2.67, 95%

confidence interval = 1.05-6.80) among postmenopausal but not in premenopausal Nigerian women (*92*).

In a related study, the relationship of selected demographic, lifestyle, antecedent medical experience, and health care access factors to breast cancer stage at diagnosis in the two-racial/ethic groups was evaluated by Hunter and colleagues (*93*). Factors associated with cancer staging were differentially expressed in African American and Caucasian Americans. These factors, including an increased body mass ($p<0.02$), contributed to stage differences in African Americans, whereas income was marginally associated ($p = 0.06$) with stage for Caucasian women only. Their findings suggested that no single factor or group of factors could explain more than half of the race-stage differences noted in their study with respect to African American and Caucasian American breast cancer patients.

The higher incidence of breast cancer before the age of 40 years and other features of poor prognosis of the disease among African American compared to Caucasian American women probably contributes to the poorer outcomes of breast cancer management in African American women. More than 10% of African American women with breast cancer were diagnosed before age 40 years compared to 5% of Caucasian patients. The incidence of breast cancer in the 30-39-age bracket for African Americans and Caucasian American women was 48.9% and 40.2%, while the proportion of African American and Caucasian women reported by the Census Bureau was not too dissimilar, 15.8% and 14.6%. African American women in 30-39 year age group have twice the age-specific distribution, have a higher incidence compared to their Caucasian counterparts, and exhibit more ominous prognostic signs (*94*). Similar results have been reported on women breast cancer from an urban population of Nigeria. Their disease features have been compared with those of Finnish women with early stage breast cancer. Earlier onset of disease, predominance of premenopausal cases, higher prevalence of poor prognostic features and poorer disease outcomes featured prominently among the Nigerian breast cancer cases (*95*).

11.2 Barriers to Care Quality

Access to the continuum of care is an essential feature of breast cancer control (*18*). A study of system delay, the time between the initial consultation and the establishment of a diagnosis, in breast cancer patients revealed that almost 49% of women reported delays of at least 4 weeks. In a study of 367 female breast cancer patents from the National Cancer Institute's Black/White Cancer Survival Study, which included hospital outpatients and emergency room, private clinic, public clinic, private doctor, and health maintenance organisation, in 25% of cases, the delay was attributed by the woman to herself. The most common reason she gave was not that she felt the problem was not important. In about 45% of the cases, the provider and the health care system were said to be responsible for the delay through difficulties in scheduling or physician inaction. In another 17% both the patient and the system were responsible (*96*).

Appropriate treatments of operable breast cancer include surgery, with or without radiation therapy, generally followed by some type of systemic treatment. This may include hormonal therapy, chemotherapy, or a combination of both (*97*). Some studies have investigated whether suitable care was provided uniformly to black and white patients, including adequate diagnostic procedures and therapy according to national guidelines, and whether established therapies differ by racial groups (race interactions) (*86*).

McWhorter and Mayer (*98*) investigated the relationship between race, treatment received, and survival using 36,905 cases from nine registries of the SEER program. African Americans were significantly less likely to have received surgical treatment, probably because of the greater prevalence of advanced and inoperable cases. However, differences in treatment persisted after adjustment for age, stage, and histology, probably accounting for the survival disadvantages in African American women as noted in the SEER survey (*86*). These findings are in keeping with the observation that, omission of important diagnostic tests and radiation therapy was more prevalent in the urban hospitals in Illinois, which are the hospitals where inner city and underinsured dwellers, such as African Americans receive medical care, suggesting a substandard care in such facilities (*60*). It would seem that stage II breast cancer patients receiving care at a government agency, such as the National Cancer Institute of the United States are not spared from this differential type of care. Breen and colleagues (*62*) reported that certain aspects of care differed according to factors such as age and race, with older and black patients receiving care less frequently than is recommended in guidelines. Studies in which comparable treatments were given, such as in the clinical trials of CALGB (*44*) and NSABP (*43*), as well as in single institution studies (*99-101*) tend to show similar outcome between African American and Caucasian American women with early-stage breast cancer.

Ikpatt and his colleagues (*95*) compared outcomes in cohorts of Nigerian women treated in Nigeria and Finnish women treated in Finland. They observed 2-year survival figures of 72.8% and 96.4% respectively in the Nigerian and Finnish women, and 80.6% and 96.4% respectively for stages III to I combined. Stage-associated analysis did not suggest big differences between stages 1 and 2, but there was a clear difference in survival among patients with stage 3 tumours, in which 2-year survival of 64.5% and 94.7%

were observed, probably reflecting differences in access to proper care for this stage of the disease.

11.3 Disparities Resulting From Barriers to Breast Cancer Screening

An investigation by Pillay (*102*), which sought to establish the awareness of breast and cervical cancers among women of African descent, in both rural and urban areas, came up with findings of considerable concern. In a study that was carried out in South Africa, two groups of randomly selected women in a rural (n = 70) and urban (n = 70) area were interviewed using a structured questionnaire assessing knowledge and attitudes regarding breast and cervical cancer and screening options. The age range of the sample was 21-59 with a mean of 35.23 years. Almost one-fifth of the women had not heard of these cancers, and almost half were unaware of the breast self-examination technique. Over one-third did not know about tests for breast cancer and more than half were unaware of tests for cervical cancer. Generally, lower awareness levels were found in older and rural women who were also significantly more inclined to consult traditional healers (rather than doctors) about lumps in their breast or abnormal cervical bleedings. The author attributed the findings to the effects of oppression and deprivation experienced by South Africans of African descent and the persistence of its effect in the post-apartheid South Africa. The study highlights the role of socio-economic hardships in this (and most certainly in much of other regions) of Africa. Bassett (*103*) has suggested that "the United States share with Zimbabwe and South Africa a history of racial subjugation". She opines that comparisons of illness and death rates across race groups have shown "the health consequences of White privilege and Black disadvantage". The problem of poor screening and deficient early diagnosis of breast cancer is at least equally unsatisfactory in a homogeneously indigenous African country like Nigeria. A study comparing the features of breast cancer in Finnish and Nigerian women reported the mean tumour size in Nigerian women to be 4.8 (sd 2.4) cm compared to 2.0 (sd 1.9) cm in Finnish women (*95*). They suggested that the difference is likely due to differences in the availability of screening mammography in Finland (*104*), and its virtual total absence in the part of Nigeria in which the studied women lived.

The potential role of breast cancer screening in eliminating survival disparity of breast cancer was illustrated by the outcome of a study of the Health Insurance Plan of New York. Participants were randomised to either a screening group where periodic examinations were provided, or to a control group where patients received their usual care with no particular emphasis on breast cancer. Five-year relative survival rates indicated that mortality due to breast cancer was lower in the screened group by nearly 40%, with both Caucasian American and African American women having roughly equal survival rates. In the control group, however, mortality rates differed substantially by race (*86*), (*105*). Thus, barriers to breast cancer screening have the potential of impacting negatively on the outcomes of the continuum of care of the disease. However, delay in care caused by greater time from consultation to diagnosis and treatment recommendation have not been shown to be major contributing factors in stage differences between African and Caucasian Americans (*106-107*). The role of cultural believes and personal attributes in impeding access to screening and early breast cancer diagnosis has been the subject of numerous publications (*17*), (*63*), (*108-110*).

11.4 Socio-economic Factors as Barrier to Successful Management

In their study comparing the features of breast cancer in Nigerian and Finnish women, Ikpatt and his colleagues attributed their observation of larger tumour sizes and higher frequency of positive axillary lymph nodes to late diagnosis and delayed treatments resulting from socio-cultural beliefs, poor health-care access and patronage of herbalists and spiritualists before conventional care (*95*).

Earlier in this chapter, the point was made that race is largely a social rather than biological construct. Dignam (*86*) quoting Harold Freeman, the Chairman of the President's Cancer Panel, has also pointed out that race can serve as a weak discriminant of outcome at best once its confounders are accounted for (*111-112*). It is therefore not surprising that treatment outcomes in breast cancer are virtually similar between African and Caucasian Americans if given similar treatments for the disease. This is consistent with the results of genetic epidemiological studies that suggest that race has no biological relevance (*113-115*). Race is a surrogate for several social disadvantages and deprivations deriving from geopolitical history, not only in the Western world, but elsewhere, such as Africa (*103*), (*114*), (*116*).

Strong associations between race, low economic status and poor survival have been demonstrated in studies (*116-118*). An association has also been established between income and stage of disease at presentation, whereby the disparity in stage between African American and Caucasian women is minimal among women from higher income areas of New York State (*119*). The observation has subsequently been confirmed in another study of breast cancer in African American women of higher income community, who had only a slight (about 1.15 times) higher risk than Caucasian American women of dying of breast cancer (*120*). Other studies in keeping with the association of higher socio-economic status and minimisation of the disparities of breast cancer survival include those of Wells and Horm (*75*), and Farley and Flannery (*121*).

Low socio-economic status invariably translates into factors that limit the access of the individual to the continuum of care. These include poor education, low literacy, inadequate health literacy, and poor communication ability. Other disadvantages include under-insurance and reliance on social assistance and subsidised health care (Medicare and Medicaid), and lack of access to cutting-edge and state-of-the-art interventions that are available in investigational environment. Breen and colleagues have presented evidence suggestive of the influence of socio-economic factors in the access to minimum of expected therapy among female breast cancer patients in the United States National Cancer Institute. They reported that older women and women with no usual source of care were significantly less likely to receive minimum expected therapy. Overall, 21% of black women did not receive the minimum expected therapy compared to 15% of white women (*62*). Tejeda and colleagues (*122*) have, however, provided evidence that accrual of American cancer patients to NCI-sponsored treatment trials generally parallels the incident burden of disease among African Americans, Hispanic Americans and Caucasian Americans, and that there is equal access to NCI clinical trials. While the data provided information on accrual of patients with cancer to cancer treatment trials, it did not speak to accrual of healthy minority subjects to cancer prevention trials. Minority accrual to several large-scale NCI cancer prevention trials has been lower than desired (*123-124*). Thus, the same institutions that have provided near racial/ethnic proportionality in cancer treatment trials have not been able to provide similar enrolment to cancer prevention clinical trials (*122*). This suggests that the dynamic of recruitment and accrual to cancer treatment trials differs from that of recruitment and accrual to cancer prevention trials (*125-126*).

11.5 Onco-political Issues as Barriers to Treatment Success

Successful treatment of breast requires efficient interactions with a multidisciplinary team. The management of locally advanced breast cancer probably best illustrates the need for a team approach in breast cancer care. Each of the disciplines involved, surgery, radiation, medical treatment, radiology, treatment facility administration etc has its own benchmark of excellence. Unless carefully integrated and balanced, unsatisfactory treatment outcomes may be the end product (*127-128*).

11.6 Gaps in Research of the Barriers to Successful Treatment of Breast Cancer in Women of African Descent

Attempt has been made to highlight several factors that militate against the successful management of breast cancer in general, and in women of African descent in particular. Some of these factors are among those that were identified at a recent 2-day "Summit Meeting on Breast Cancer Among African American Women (Washington, DC, September 8-10, 2000) (*129*). The basis or backgrounds of several of these factors are poorly understood, and further elucidation of the nature of these problems should promote the improvement of breast cancer in women of African descent in general. The following observations and questions need to be addressed:

a) Several factors probably contribute to the barriers of successful treatment of breast cancer in African American women. However, it would be useful to be able to prioritise them for interventional policy development.
b) Clinical research studies in breast cancer have been designed and implemented with research subjects that are predominantly young women of the developed countries of North America and Western Europe. The relevance of the results of these trials to the large population of elderly women of all ethnic backgrounds, who share the bulk of the burden of breast cancer, and to the large population of women of low socio-economic status in the developed countries (including the majority of African American women) and developing countries is largely unknown.
c) Given the association between race/ethnicity-related body fat distribution and the risk of breast cancer, is it conceivable that there might be associated differences in response to hormonal agents, both for treatment and prevention of breast cancer?
d) Are variants of breast cancer associated with socio-economic factors? If so, what is the biological basis of the association?
e) What are the genetic determinants of the nature of breast cancer in women of African descent?
f) What is the biological mechanism underlying race/ethnicity associated differences in the biological markers of breast cancer (e.g. increased incidence of oestrogen receptor negative breast cancer in African American women)? Is this a common feature of breast cancer in other groups of women of African descent?
g) What biological factors influence or promote early onset premenopausal breast cancer in women of African descent?
h) Is there a relationship between outcomes and the cultural backgrounds of patients vis-à-vis primary care providers?

i) Can treatment outcomes be improved by promoting culturally sensitive and balanced care atmosphere?
j) What non-cancer related mortality factors are responsible for the reduced survival among African American women treated for breast cancer?
k) What role does nutrition play in the abnormal ethnicity-related fat distribution (hip:waist) ratio and its association with poor prognostic features of breast cancer?
l) What is the best treatment for the high-risk early-onset premenopausal breast cancer of women of African descent?
m) Should African descent /ethnicity be a predictive marker for treatment?
n) Is the breast cancer research information in women of African descent more relevant to the African American women, than data gathered on their Caucasian compatriots and other Caucasian populations?
o) How can chemotherapy be made safer and more applicable in the elderly breast cancer patient?
p) Should systemic management of early-onset breast cancer involve ovarian ablation ± hormonal agents?

REFERENCES

(1) Rettig RA: The story of the national cancer act of 1971. Princeton; 1977.

(2) Langone J: Cautious optimism. Discover 1986:47.

(3) Ultmann JE, Baxter MD, Lierman T: The Government and Cancer Medicine. In Bast RC, Kufe DW, Pollock RE, Weichselbaum RR, Holland JF, Frei III E, editors. Cancer Medicine. Hamilton and London: B.C. Decker Inc., 2000, p. 1024-34.

(4) Klausner RD: Keynote Address: Rethinking our National Therapeutic Discovery Program: From Biology to Clinical Trials. AACR-NCI-EORTC International Conference. Washington, DC: CD-Rom produced by the American Association for Cancer Research (Selected Sessions), 1999.

(5) Mohr S, Leikauf GD, Keith G, Rihn BH: Microarrays as cancer keys: an array of possibilities. J Clin Oncol 2002;20:3165-75.

(6) Pawson T, Saxton TM: Signaling networks--do all roads lead to the same genes? Cell 1999;97:675-8.

(7) Heisterkamp N, Henster G, Ten Hoeve J, al. e: Acute leukemia in bcr-abl transgenic mice. Nature 1990;344:251.

(8) Bernards A, Rubin CM, Westbrook CA, Paskind M, Baltimore D: The first intron in the human c-abl gene is at least 200 kilobases long and is a target for translocations in chronic myelogenous leukemia. Mol Cell Biol 1987;7:3231-6.

(9) Faderl S, Talpaz M, Estrov Z, O'Brien S, Kurzrock R, Kantarjian HM: The biology of chronic myeloid leukemia. N Engl J Med 1999;341:164-72.

(10) Slamon DJ, Clark GM, Wong SG, Levin WJ, Ullrich A, McGuire WL: Human breast cancer: correlation of relapse and survival with amplification of the HER-2/neu oncogene. Science. 1987;235:177-82.

(11) Thor A: Are patterns of HER-2/neu amplification and expression among primary tumors and regional metastases indicative of those in distant metastases and

predictive of Herceptin response? [letter; comment.]. Journal of the National Cancer Institute. 2001;93:1120-1.

(12) Druker BJ, Talpaz M, Resta DJ, Peng B, Buchdunger E, Ford JM, et al: Efficacy and safety of a specific inhibitor of the BCR-ABL tyrosine kinase in chronic myeloid leukemia. N Engl J Med 2001;344:1031-7.

(13) Tan-Chiu E, Piccart M: Moving forward: Herceptin in the adjuvant setting. Oncology. 2002;63:57-63.

(14) Rosenwald A, Wright G, Chan WC, Connors JM, Campo E, Fisher RI, et al: The use of molecular profiling to predict survival after chemotherapy for diffuse large-B-cell lymphoma. N Engl J Med 2002;346:1937-47.

(15) DeVita VT, Jr., Bleickardt EW: National Oncology Forum: perspectives for the year 2000. Cancer J 2001;7 Suppl 1:S2-13.

(16) Tracey L, Villuendas R, Ortiz P, Dopazo A, Spiteri I, Lombardia L, et al: Identification of genes involved in resistance to interferon-alpha in cutaneous T-cell lymphoma. Am J Pathol 2002;161:1825-37.

(17) Ishida DN, Toomata-Mayer TF, Braginsky NS: Beliefs and attitudes of Samoan women toward early detection of breast cancer and mammography utilization. Cancer 2001;91:262-6.

(18) Zapka JG, Taplin SH, Solberg LI, Manos MM. A framework for improving the quality of cancer care: The case of breast and cervical cancer screening. Cancer Epidemiol Biomarkers Prev 2003;12:4-13.

(19) Margolese RG: 1994 Roussel Lecture: a biologic basis for changes in breast cancer management. Canadian Journal of Surgery. 1995;38:402-8.

(20) Holland J: Clinical studies of unmaintained remissions in acute lymphocytic leukemia. 21st Annual Symposium on Fundamental Cancer Research 1967. Baltimore, MD: Williams and Wilkins, 1968, p. 453-62.

(21) Group EBCTC: Effects of adjuvant tamoxifen and of cytotoxic therapy on mortality in early breast cancer. An overview of 61 randomized trials among 28,896 women. New England Journal of Medicine. 1988;319:1681-92.

(22) Anonymous: Systemic treatment of early breast cancer by hormonal, cytotoxic, or immune therapy. 133 randomised trials involving 31,000 recurrences and 24,000 deaths among 75,000 women. Early Breast Cancer Trialists' Collaborative Group. Lancet. 1992;339:1-15.

(23) Anonymous: Systemic treatment of early breast cancer by hormonal, cytotoxic, or immune therapy. 133 randomised trials involving 31,000 recurrences and 24,000 deaths among 75,000 women. Early Breast Cancer Trialists' Collaborative Group. Lancet. 1992;339:71-85.

(24) Anonymous: Polychemotherapy for early breast cancer: an overview of the randomised trials. Early Breast Cancer Trialists' Collaborative Group. Lancet. 1998;352:930-42.

(25) Olivotto IA, Bajdik CD, Plenderleith IH, Coppin CM, Gelmon KA, Jackson SM, et al: Adjuvant systemic therapy and survival after breast cancer. N Engl J Med 1994;330:805-10.

(26) Norton L: Evolving concepts in the systemic drug therapy of breast cancer. Seminars in Oncology. 1997;24:S10-3-S10-10.

(27) Gilewski T, Dang C, Surbone A, Norton L: Cytokinetics. In Bast R, Kufe D, Pollock R, Weichselbaum R, Holland J, Frei E, editors. Cancer Medicine. Hamilton and London: B.C. Decker, 2000, p. 509-38.

(28) Winchester DP, Osteen RT, Menck HR: The National Cancer Data Base report on breast carcinoma characteristics and outcome in relation to age. Cancer. 1996;78:1838-43.

(29) Simon MS, Severson RK: Racial differences in breast cancer survival: the interaction of socioeconomic status and tumor biology. American Journal of Obstetrics & Gynecology. 1997;176:S233-9.

(30) Wojcik BE, Spinks MK, Optenberg SA: Breast carcinoma survival analysis for African American and white women in an equal-access health care system. Cancer. 1998;82:1310-8.

(31) Boyer-Chammard A, Taylor TH, Anton-Culver H: Survival differences in breast cancer among racial/ethnic groups: a population-based study. Cancer Detection & Prevention. 1999;23:463-73.

(32) El-Tamer MB, Homel P, Wait RB: Is race a poor prognostic factor in breast cancer? Journal of the American College of Surgeons. 1999;189:41-5.

(33) Joslyn SA, West MM: Racial differences in breast carcinoma survival. Cancer. 2000;88:114-23.

(34) Muss HB: Factors used to select adjuvant therapy of breast cancer in the United States: an overview of age, race, and socioeconomic status. Journal of the National Cancer Institute. Monographs. 2001:52-5.

(35) Bradley CJ, Given CW, Roberts C: Race, socioeconomic status, and breast cancer treatment and survival.[comment]. Journal of the National Cancer Institute. 2002;94:490-6.

(36) English WP, Cleveland KE, Barber WH: There is no difference in survival between African-American and white women with breast cancer. American Surgeon. 2002;68:594-7.

(37) Joslyn SA: Hormone receptors in breast cancer: racial differences in distribution and survival. Breast Cancer Research & Treatment. 2002;73:45-59.

(38) Joslyn SA: Racial differences in treatment and survival from early-stage breast carcinoma. Cancer. 2002;95:1759-66.

(39) Newman LA, Bunner S, Carolin K, Bouwman D, Kosir MA, White M, et al: Ethnicity related differences in the survival of young breast carcinoma patients. Cancer. 2002;95:21-7.

(40) Newman LA, Mason J, Cote D, Vin Y, Carolin K, Bouwman D, et al: African-American ethnicity, socioeconomic status, and breast cancer survival: a meta-analysis of 14 studies involving over 10,000 African-American and 40,000 White American patients with carcinoma of the breast. Cancer. 2002;94:2844-54.

(41) Newman LA, Theriault R, Clendinnin N, Jones D, Pierce I: Treatment choices and response rates in African-American women with breast carcinoma. Cancer 2003;97:246-52.

(42) Shavers VL, Harlan LC, Stevens JL: Racial/ethnic variation in clinical presentation, treatment, and survival among breast cancer patients under age 35. Cancer 2003;97:134-47.

(43) Dignam JJ, Redmond CK, Fisher B, Costantino JP, Edwards BK: Prognosis among African-American women and white women with lymph node negative breast carcinoma: findings from two randomized clinical trials of the National Surgical Adjuvant Breast and Bowel Project (NSABP). Cancer. 1997;80:80-90.

(44) Roach M, 3rd, Cirrincione C, Budman D, Hayes D, Berry D, Younger J, et al: Race and survival from breast cancer: based on Cancer and Leukemia Group B trial 8541.[comment]. Cancer Journal From Scientific American. 1997;3:107-12.

(45) Dignam JJ: Efficacy of systemic adjuvant therapy for breast cancer in African-American and Caucasian women. Journal of the National Cancer Institute. Monographs. 2001:36-43.

(46) Kaufman JS, Cooper RS, McGee DL: Socioeconomic status and health in blacks and whites: the problem of residual confounding and the resiliency of race. Epidemiology 1997;8:621-8.

(47) Adami HO, Malker B, Holmberg L, Persson I, Stone B: The relation between survival and age at diagnosis in breast cancer. N Engl J Med 1986;315:559-63.

(48) Kollias J, Elston CW, Ellis IO, Robertson JF, Blamey RW: Early-onset breast cancer--histopathological and prognostic considerations. Br J Cancer 1997;75:1318-23.

(49) Goldhirsch A, Gelber RD, Yothers G, Gray RJ, Green S, Bryant J, et al: Adjuvant therapy for very young women with breast cancer: need for tailored treatments. J Natl Cancer Inst Monogr 2001;30:44-51.

(50) Walker RA, Lees E, Webb MB, Dearing SJ: Breast carcinomas occurring in young women (< 35 years) are different. Br J Cancer 1996;74:1796-800.

(51) Chung M, Chang HR, Bland KI, Wanebo HJ: Younger women with breast carcinoma have a poorer prognosis than older women. Cancer 1996;77:97-103.

(52) Diab SG, Elledge RM, Clark GM: Tumor characteristics and clinical outcome of elderly women with breast cancer. J Natl Cancer Inst 2000;92:550-6.

(53) Anonymous: Tamoxifen for early breast cancer: an overview of the randomised trials. Early Breast Cancer Trialists' Collaborative Group.[comment]. Lancet. 1998;351:1451-67.

(54) Ganz PA: Impact of tamoxifen adjuvant therapy on symptoms, functioning, and quality of life. Journal of the National Cancer Institute. Monographs. 2001:130-4.

(55) Kroman N, Jensen MB, Wohlfahrt J, Mouridsen HT, Andersen PK, Melbye M: Factors influencing the effect of age on prognosis in breast cancer: population based study. Bmj 2000;320:474-8.

(56) Aebi S, Gelber S, Castiglione-Gertsch M, Gelber RD, Collins J, Thurlimann B, et al: Is chemotherapy alone adequate for young women with oestrogen-receptor- positive breast cancer? Lancet 2000;355:1869-74.

(57) Greenlee RT, Hill-Harmon MB, Murray T, Thun M: Cancer statistics, 2001.[erratum appears in CA Cancer J Clin 2001 Mar-Apr;51(2):144]. Ca: a Cancer Journal for Clinicians. 2001;51:15-36.

(58) Ries LA, Wingo PA, Miller DS, Howe HL, Weir HK, Rosenberg HM, et al: The annual report to the nation on the status of cancer, 1973-1997, with a special section on colorectal cancer. Cancer 2000;88:2398-424.

(59) Diehr P, Yergan J, Chu J, Feigl P, Glaefke G, Moe R, et al: Treatment modality and quality differences for black and white breast- cancer patients treated in community hospitals. Med Care 1989;27:942-58.

(60) Hand R, Sener S, Imperato J, Chmiel JS, Sylvester JA, Fremgen A: Hospital variables associated with quality of care for breast cancer patients. Jama 1991;266:3429-32.

(61) Lannin DR, Mathews HF, Mitchell J, Swanson MS, Swanson FH, Edwards MS: Influence of socioeconomic and cultural factors on racial differences in late-stage presentation of breast cancer. Jama 1998;279:1801-7.

(62) Breen N, Wesley MN, Merrill RM, Johnson K: The relationship of socio-economic status and access to minimum expected therapy among female breast cancer patients in the National Cancer Institute Black-White Cancer Survival Study. Ethn Dis 1999;9:111-25.

(63) Gregg J, Curry RH: Explanatory models for cancer among African-American women at two Atlanta neighborhood health centers: the implications for a cancer screening program. Soc Sci Med 1994;39:519-26.
(64) Doak CC, Doak LG, Friedell GH, Meade CD: Improving comprehension for cancer patients with low literacy skills: strategies for clinicians. CA Cancer J Clin 1998;48:151-62.
(65) Kirsch I, Jungeblut A, Jenkins L, al. e: Adult literacy in America: A first look at the results of th National Adult Literacy Survey. Washington, DC: National Center for Education Statistics, US Department of Education, 1993.
(66) Davis TC, Williams MV, Marin E, Parker RM, Glass J: Health literacy and cancer communication. CA Cancer J Clin 2002;52:134-49.
(67) Anonymous: Health literacy: report of the Council on Scientific Affairs. Ad Hoc Committee on Health Literacy for the Council on Scientific Affairs, American Medical Association. Jama 1999;281:552-7.
(68) Anonymous: A summary of the American Cancer Society Report to the Nation: cancer in the poor. CA Cancer J Clin 1989;39:263-5.
(69) Baquet CR, Horm JW, Gibbs T, Greenwald P: Socioeconomic factors and cancer incidence among blacks and whites. J Natl Cancer Inst 1991;83:551-7.
(70) Freeman H: Race, poverty, and cancer. J Natl Cancer Inst 1991;83:526-7.
(71) Li A, Burton G, Glass J: Breast cancer: a socioeconomic and racial comparison in northwest Louisiana. J La State Med Soc 2001;153:420-5.
(72) Bennett CL, Ferreira MR, Davis TC, Kaplan J, Weinberger M, Kuzel T, et al: Relation between literacy, race, and stage of presentation among low- income patients with prostate cancer. J Clin Oncol 1998;16:3101-4.
(73) Ann Gilligan M, Kneusel RT, Hoffmann RG, Greer AL, Nattinger AB: Persistent differences in sociodemographic determinants of breast conserving treatment despite overall increased adoption. Med Care 2002;40:181-9.
(74) Li BD, Brown WA, Ampil FL, Burton GV, Yu H, McDonald JC: Patient compliance is critical for equivalent clinical outcomes for breast cancer treated by breast-conservation therapy. Ann Surg 2000;231:883-9.
(75) Wells BL, Horm JW: Stage at diagnosis in breast cancer: race and socioeconomic factors. Am J Public Health 1992;82:1383-5.
(76) Siminoff LA, Fetting JH, Abeloff MD: Doctor-patient communication about breast cancer adjuvant therapy. J Clin Oncol 1989;7:1192-200.
(77) Davis TC, Holcombe RF, Berkel HJ, Pramanik S, Divers SG: Informed consent for clinical trials: a comparative study of standard versus simplified forms. J Natl Cancer Inst 1998;90:668-74.
(78) Dawson LA, Chow E, Goss PE: Evolving perspectives in contralateral breast cancer. European Journal of Cancer. 1998;34:2000-9.
(79) Wilson FL: Measuring patients' ability to read and comprehend: a first step in patient education. Nursingconnections 1995;8:17-25.
(80) Anonymous: Communicating with patients who have limited literacy skills. Report of the National Work Group on Literacy and Health. Journal of Family Practice. 1998;46:168-76.
(81) Ries LAG, Kosary CL, Hankey BF, Miller BA, Clegg LX, Edwards BK, editors: Cancer prevalence and survivorship issues: analyses of the 1992 National Health Interview Survey. (NIH Publication no. 99-2789). Bethesda, MD,; 1996.
(82) Axtell LM, Myers MH: Contrasts in survival of black and white cancer patients, 1960-73. J Natl Cancer Inst 1978;60:1209-15.

(83) Young JL, Jr., Ries LG, Pollack ES: Cancer patient survival among ethnic groups in the United States. J Natl Cancer Inst 1984;73:341-52.

(84) Chu KC, Tarone RE, Kessler LG, Ries LA, Hankey BF, Miller BA, et al: Recent trends in U.S. breast cancer incidence, survival, and mortality rates.[comment]. Journal of the National Cancer Institute. 1996;88:1571-9.

(85) Greenlee RT, Murray T, Bolden S, Wingo PA: Cancer statistics, 2000. CA Cancer J Clin 2000;50:7-33.

(86) Dignam JJ: Differences in breast cancer prognosis among African-American and Caucasian women. CA Cancer J Clin 2000;50:50-64.

(87) Natarajan N, Nemoto D, Nemoto T, Mettlin C: Breast cancer survival among Orientals and whites living in the United States. J Surg Oncol 1988;39:206-9.

(88) Gilliland FD, Hunt WC, Key CR: Trends in the survival of American Indian, Hispanic, and Non-Hispanic white cancer patients in New Mexico and Arizona, 1969-1994. Cancer 1998;82:1769-83.

(89) Clegg LX, Li FP, Hankey BF, Chu K, Edwards BK: Cancer survival among US whites and minorities: a SEER (Surveillance, Epidemiology, and End Results) Program population-based study. Arch Intern Med 2002;162:1985-93.

(90) Jones BA, Kasi SV, Curnen MG, Owens PH, Dubrow R: Severe obesity as an explanatory factor for the black/white difference in stage at diagnosis of breast cancer. Am J Epidemiol 1997;146:394-404.

(91) Moorman PG, Jones BA, Millikan RC, Hall IJ, Newman B: Race, anthropometric factors, and stage at diagnosis of breast cancer. Am J Epidemiol 2001;153:284-91.

(92) Adebamowo CA, Ogundiran TO, Adenipekun AA, Oyesegun RA, Campbell OB, Akang EE, et al: Waist-hip ratio and breast cancer risk in urbanized Nigerian women. Breast Cancer Research 2002;5:18-24.

(93) Hunter CP, Redmond CK, Chen VW, Austin DF, Greenberg RS, Correa P, et al: Breast cancer: factors associated with stage at diagnosis in black and white women. Black/White Cancer Survival Study Group. J Natl Cancer Inst 1993;85:1129-37.

(94) Johnson ET: Breast cancer racial differences before age 40--implications for screening. Journal of the National Medical Association. 2002;94:149-56.

(95) Ikpatt OF, Kuopio T, Ndoma-Egba R, Collan Y: Breast cancer in Nigeria and Finland: epidemiological, clinical and histological comparison. Anticancer Res 2002;22:3005-12.

(96) Caplan LS, Helzlsouer KJ, Shapiro S, Wesley MN, Edwards BK: Reasons for delay in breast cancer diagnosis. Prev Med 1996;25:218-24.

(97) Goldhirsch A, Glick JH, Gelber RD, Coates AS, Senn HJ: Meeting highlights: International Consensus Panel on the Treatment of Primary Breast Cancer. Seventh International Conference on Adjuvant Therapy of Primary Breast Cancer.[comment]. Journal of Clinical Oncology. 2001;19:3817-27.

(98) McWhorter WP, Mayer WJ: Black/white differences in type of initial breast cancer treatment and implications for survival. Am J Public Health 1987;77:1515-7.

(99) Heimann R, Ferguson D, Powers C, Suri D, Weichselbaum RR, Hellman S: Race and clinical outcome in breast cancer in a series with long-term follow-up evaluation. J Clin Oncol 1997;15:2329-37.

(100) Briele HA, Jr., Walker MJ, Wild L, Wood DK, Greager JA, Schneebaum S, et al: Results of treatment of stage I-III breast cancer in black Americans. The Cook County Hospital experience, 1973-1987. Cancer 1990;65:1062-71.

(101) Yood MU, Johnson CC, Blount A, Abrams J, Wolman E, McCarthy BD, et al: Race and differences in breast cancer survival in a managed care population. J Natl Cancer Inst 1999;91:1487-91.
(102) Pillay AL: Rural and urban South African women's awareness of cancers of the breast and cervix. Ethn Health 2002;7:103-14.
(103) Bassett MT: The pursuit of equity in health: reflections on race and public health data in Southern Africa. Am J Public Health 2000;90:1690-3.
(104) Dean PB, Pamilo M: Screening mammography in Finland--1.5 million examinations with 97 percent specificity. Mammography Working Group, Radiological Society of Finland. Acta Oncol 1999;38:47-54.
(105) Shapiro S, Venet W, Strax P, Venet L, Roeser R: Prospects for eliminating racial differences in breast cancer survival rates. Am J Public Health 1982;72:1142-5.
(106) Coates RJ, Bransfield DD, Wesley M, Hankey B, Eley JW, Greenberg RS, et al: Differences between black and white women with breast cancer in time from symptom recognition to medical consultation. Black/White Cancer Survival Study Group. J Natl Cancer Inst 1992;84:938-50.
(107) Caplan LS, Helzlsouer KJ, Shapiro S, Freedman LS, Coates RJ, Edwards BK: System delay in breast cancer in whites and blacks. Am J Epidemiol 1995;142:804-12.
(108) Womeodu RJ, Bailey JE: Barriers to cancer screening. Med Clin North Am 1996;80:115-33.
(109) Guidry JJ, Greisinger A, Aday LA, Winn RJ, Vernon S, Throckmorton TA: Barriers to cancer treatment: a review of published research. Oncol Nurs Forum 1996;23:1393-8.
(110) Guidry JJ, Matthews-Juarez P, Copeland VA: Barriers to breast cancer control for African-American women: the interdependence of culture and psychosocial issues. Cancer 2003;97:318-23.
(111) Freeman HP: Cancer in the socioeconomically disadvantaged. CA Cancer J Clin 1989;39:266-88.
(112) Freeman HP: The meaning of race in cancer of the breast. Cancer J Sci Am 1997;3:76-7.
(113) Brawley OW, Freeman HP: Race and outcomes: is this the end of the beginning for minority health research? J Natl Cancer Inst 1999;91:1908-9.
(114) Freeman HP: The meaning of race in science--considerations for cancer research: concerns of special populations in the National Cancer Program. Cancer 1998;82:219-25.
(115) Owens K, King MC: Genomic views of human history. Science 1999;286:451-3.
(116) Bassett MT, Krieger N: Social class and black-white differences in breast cancer survival. Am J Public Health 1986;76:1400-3.
(117) Vernon SW, Tilley BC, Neale AV, Steinfeldt L: Ethnicity, survival, and delay in seeking treatment for symptoms of breast cancer. Cancer. 1985;55:1563-71.
(118) Dayal HH, Power RN, Chiu C: Race and socio-economic status in survival from breast cancer. Journal of Chronic Diseases. 1982;35:675-83.
(119) Polednak AP: Breast cancer in black and white women in New York State. Case distribution and incidence rates by clinical stage at diagnosis. Cancer 1986;58:807-15.
(120) Polednak AP: Cancer mortality in a higher-income black population in New York State. Comparison with rates in the United States as a whole. Cancer. 1990;66:1654-60.

(121) Farley TA, Flannery JT: Late-stage diagnosis of breast cancer in women of lower socioeconomic status: public health implications. Am J Public Health 1989;79:1508-12.
(122) Tejeda HA, Green SB, Trimble EL, Ford L, High JL, Ungerleider RS, et al: Representation of African-Americans, Hispanics, and whites in National Cancer Institute cancer treatment trials. J Natl Cancer Inst 1996;88:812-6.
(123) Thompson IM, Coltman CA, Jr., Crowley J: Chemoprevention of prostate cancer: the Prostate Cancer Prevention Trial. Prostate 1997;33:217-21.
(124) Chlebowski RT, Butler J, Nelson A, Lillington L: Breast cancer chemoprevention. Tamoxifen: current issues and future prospective. Cancer 1993;72:1032-7.
(125) Brawley OW, Tejeda H: Minority inclusion in clinical trials issues and potential strategies. J Natl Cancer Inst Monogr 1995;17:55-7.
(126) Swanson GM, Ward AJ: Recruiting minorities into clinical trials: toward a participant-friendly system. J Natl Cancer Inst 1995;87:1747-59.
(127) Tripathy D: Multidisciplinary care for breast cancer: barriers and solutions. Breast J 2003;9:60-3.
(128) Lee CZ: Oncopolitical issues: obstacles and options for success in a comprehensive breast center. Surgical Oncology Clinics of North America. 2000;9:279-94.
(129) Jones LA, Brawley O, Johnson-Thompson M, Lythcott N, Newman L: Overview of the summit meeting evaluating research in African-American women. Cancer 2003;97:207-10.

INDEX